*Multiscale Biomechanical
Modeling of the Brain*

Multiscale Biomechanical Modeling of the Brain

Edited by

Raj K. Prabhu
USRA, NASA HRP CCMP, NASA Glenn Research Center,
Cleveland, OH, United States

Mark F. Horstemeyer
School of Engineering, Liberty University,
Lynchburg, VA, United States

Academic Press is an imprint of Elsevier
125 London Wall, London EC2Y 5AS, United Kingdom
525 B Street, Suite 1650, San Diego, CA 92101, United States
50 Hampshire Street, 5th Floor, Cambridge, MA 02139, United States
The Boulevard, Langford Lane, Kidlington, Oxford OX5 1GB, United Kingdom

Copyright © 2022 Elsevier Inc. All rights reserved.

No part of this publication may be reproduced or transmitted in any form or by any means, electronic or mechanical, including photocopying, recording, or any information storage and retrieval system, without permission in writing from the publisher. Details on how to seek permission, further information about the Publisher's permissions policies and our arrangements with organizations such as the Copyright Clearance Center and the Copyright Licensing Agency, can be found at our website: www.elsevier.com/permissions.

This book and the individual contributions contained in it are protected under copyright by the Publisher (other than as may be noted herein).

Notices
Knowledge and best practice in this field are constantly changing. As new research and experience broaden our understanding, changes in research methods, professional practices, or medical treatment may become necessary.

Practitioners and researchers must always rely on their own experience and knowledge in evaluating and using any information, methods, compounds, or experiments described herein. In using such information or methods they should be mindful of their own safety and the safety of others, including parties for whom they have a professional responsibility.

To the fullest extent of the law, neither the Publisher nor the authors, contributors, or editors, assume any liability for any injury and/or damage to persons or property as a matter of products liability, negligence or otherwise, or from any use or operation of any methods, products, instructions, or ideas contained in the material herein.

British Library Cataloguing-in-Publication Data
A catalogue record for this book is available from the British Library

Library of Congress Cataloging-in-Publication Data
A catalog record for this book is available from the Library of Congress

ISBN: 978-0-12-818144-7

For Information on all Academic Press publications visit our website at
https://www.elsevier.com/books-and-journals

Publisher: Mara Conner
Acquisitions Editor: Carrie Bolger
Editorial Project Manager: Leticia M. Lima
Production Project Manager: Sojan P. Pazhayattil
Cover Designer: Matthew Limbert

Typeset by Aptara, New Delhi, India

Contents

Contributors .... *xi*

Preface .... *xiii*

Chapter 1: The multiscale nature of the brain and traumatic brain injury 1
M.A. Murphy, A. Vo

1.1 Introduction 1
1.2 The brain's multiscale structure 2
    1.2.1 Gross anatomy 3
    1.2.2 Microanatomy 8
1.3 The multiscale nature of TBI 13
    1.3.1 Multiscale injury mechanisms 14
    1.3.2 Types of injury 15
    1.3.3 Examples of injuries 16
    1.3.4 Neurobehavioral sequelae 17
    1.3.5 TBI research methods 18
1.4 Summary 21
    References 21

Chapter 2: Introduction to multiscale modeling of the human brain 27
Raj K. Prabhu, Mark F. Horstemeyer

2.1 Introduction 27
2.2 Constitutive modeling of the brain 27
2.3 Brain tissue experiments used for constitutive modeling calibration 32
2.4 Modeling summary of upcoming chapters in the book 33
2.5 Summary 34
    References 34

Chapter 3: Density functional theory and bridging to classical interatomic force fields 39
D. Dickel, S. Mun, M. Baskes, S. Gwaltney, Raj K. Prabhu, Mark F. Horstemeyer

3.1 Introduction 39
    3.1.1 Why quantum mechanics? 39
    3.1.2 Physical chemistry of biomechanical systems 41
3.2 Density functional theory 42

v

Contents

3.3		Downscaling requirements of classical force field atomistic models	43
	3.3.1	Upscaling properties	44
3.4		Sample atomistic force fields formalism and development of an interatomic potential for hydrocarbons	46
	3.4.1	MEAMBO	46
	3.4.2	Calibration of the MEAMBO potential	47
	3.4.3	Parameterization of the interatomic potential	48
	3.4.4	Validation of the interatomic force fields	49
3.5		Summary	50
		References	51

Chapter 4: Modeling nanoscale cellular structures using molecular dynamics ... 53

M.A. Murphy, Mark F. Horstemeyer, Raj K. Prabhu

4.1		Introduction	53
4.2		Methods	57
	4.2.1	Molecular dynamics simulation method	57
	4.2.2	Atomic force fields	57
	4.2.3	Simulation ensembles of atoms	59
	4.2.4	Boundary conditions	60
	4.2.5	Current simulation details	63
	4.2.6	Molecular dynamics analysis methods for the phospholipid bilayer (neuron membrane)	64
4.3		Results and discussion for the phospholipid bilayer (neuron membrane)	69
	4.3.1	Stress–strain and damage response	70
	4.3.2	Membrane failure limit diagram	70
4.4		Summary	72
		Acknowledgments	73
		References	73

Chapter 5: Microscale mechanical modeling of brain neuron(s) and axon(s) ... 77

Mark F. Horstemeyer, A. Bakhtiarydavijani, Raj K. Prabhu

5.1		Introduction	77
5.2		Modeling microscale neurons	78
	5.2.1	Modeling neurons	79
	5.2.2	Modeling mechanical behavior of axons	81
5.3		Summary and future	81
		References	82

Chapter 6: Mesoscale finite element modeling of brain structural heterogeneities and geometrical complexities ... 85

A. Bakhtiarydavijani, R. Miralami, A. Dobbins, Mark F. Horstemeyer, Raj K. Prabhu

6.1		Introduction	85
	6.1.1	Modeling length scale	86
6.2		Methods	87
	6.2.1	Computational methods for properties	87
	6.2.2	Model validation and boundary conditions	92

Contents

6.3	Results and discussion	93
	6.3.1 Geometrical complexities	94
6.4	Summary	100
	References	101

Chapter 7: Modeling mesoscale anatomical structures in macroscale brain finite element models ... 103

T. Wu, J.S. Giudice, A. Alshareef, M.B. Panzer

7.1	Introduction	103
7.2	Macroscale brain finite element model	103
7.3	Mesoscale anatomical structures and imaging techniques	105
7.4	The importance of structural anisotropy in macroscale models of TBI	107
7.5	Material-based method	108
7.6	Structure-based method	109
7.7	Summary and future perspectives	110
	References	113

Chapter 8: A macroscale mechano-physiological internal state variable (MPISV) model for neuronal membrane damage with subscale microstructural effects 119

A. Bakhtiarydavijani, M.A. Murphy, Raj K. Prabhu, T.R. Fonville, Mark F. Horstemeyer

8.1	Introduction	119
	8.1.1 Definitions	120
8.2	Membrane disruption	121
8.3	Development of damage evolution equation	122
	8.3.1 Pore number density rate	123
	8.3.2 Pore growth rate	125
	8.3.3 Pore resealing	126
8.4	Garnering data from molecular dynamics simulations	126
8.5	Calibration of the mechano-physiological internal state variable damage rate equations	127
8.6	Sensitivity analysis of damage model at this length scale	128
8.7	Comparison of model with cell culture studies	129
8.8	Discussion	133
8.9	Summary	135
	References	135

Chapter 9: MRE-based modeling of head trauma ... 139

Amit Madhukar, Martin Ostoja-Starzewski

9.1	Introduction	139
9.2	Model formulation	140
	9.2.1 MRE acquisition and inversion	140
	9.2.2 Finite element mesh generation	141
	9.2.3 Material properties	143
	9.2.4 Experimental verification	144

vii

Contents

9.3   Results and discussion 144
9.4   Conclusion 150
      References 150

Chapter 10: Robust concept exploration of driver's side vehicular impacts for human-centric crashworthiness 153

A.B. Nellippallil, P.R. Berthelson, L. Peterson, Raj K. Prabhu

10.1   Frame of reference 153
10.2   Problem definition 155
10.3   Adapted CEF for robust concept exploration 156
10.4   Head and neck injury criteria-based robust design of vehicular impacts 158
      10.4.1 Clarification of design task—Step A 158
      10.4.2 Design of experiments—Step B 159
      10.4.3 Finite element car crash simulations for predicting injury response—Step C 160
      10.4.4 Building surrogate models—Step D 162
      10.4.5 Formulation of robust design cDSP—Step E 164
      10.4.6 Formulating the design scenarios, exercising the cDSP and exploration of solution space—Step E 168
10.5   Future: correlate human brain injury to vehicular damage 171
10.6   Summary 172
      References 172

Chapter 11: Development of a coupled physical–computational methodology for the investigation of infant head injury 177

M.D. Jones, G.A. Khalid, Raj K. Prabhu

11.1   Introduction 177
11.2   Methods 181
      11.2.1 Pediatric head development 181
      11.2.2 Material properties 182
      11.2.3 Mesh convergence 184
      11.2.4 Boundary and loading conditions 184
      11.2.5 Global validation of the FE-head against PMHS 185
      11.2.6 Global, regional, and local validation of the FE-head against the physical model 185
      11.2.7 Statistical analysis 185
11.3   Results and discussion 186
      11.3.1 Global validation of the FE-head versus the postmortem human surrogate 186
      11.3.2 Global validation of the FE-head versus the physical model 187
      11.3.3 FE-head regional and local validation versus the physical model 188
      11.3.4 Head deformation 190
11.4   Summary 191
      References 191

Contents

Chapter 12: Experimental data for validating the structural response of computational brain models **193**

A. Alshareef, J.S. Giudice, D. Shedd, K. Reynier, T. Wu, M.B. Panzer

12.1   Introduction 193
12.2   Methods 195
      12.2.1 Experimental brain pressure measurements 195
      12.2.2 Experimental brain deformation measurements 196
12.3   Challenges and limitations 203
12.4   Summary and future perspectives 205
      References 206

Chapter 13: A review of fluid flow in and around the brain, modeling, and abnormalities **209**

R. Prichard, M. Gibson, C. Joseph, W. Strasser

13.1   Introduction 209
13.2   Flow anatomy 209
      13.2.1 Ventricular system 209
      13.2.2 Ventricles and subarachnoid space 210
13.3   Characteristic numbers 211
      13.3.1 Reynolds number 211
      13.3.2 Womersley number 212
      13.3.3 Péclet number 212
13.4   Common brain flow abnormalities 213
      13.4.1 Misfolded proteins 214
      13.4.2 Injury 215
      13.4.3 Reduced arterial pulsatility 215
      13.4.4 Hydrocephalus 215
      13.4.5 Chiari malformation 216
      13.4.6 Syringomyelia and syringobulbia 216
13.5   Boundary conditions for models 216
      13.5.1 General comments 216
      13.5.2 Cardiac flow 217
      13.5.3 Respiratory flow 217
      13.5.4 Circulatory flow 217
      13.5.5 Intracranial pressure 219
13.6   Brain measurement and imaging 219
      13.6.1 Magnetic resonance imaging 219
      13.6.2 Spin/field/gradient echo MRI 219
      13.6.3 Phase contrast MRI 220
      13.6.4 MRI limitations 220
      13.6.5 Pressure monitoring 221
      13.6.6 MRI segmentation 221

Contents

13.7 Flow modeling...222
    13.7.1 CFD simplifications: rigid walls 231
    13.7.2 CFD simplifications: microstructures...232
13.8 Literature gap 233
    References 233

Chapter 14: Resonant frequencies of a human brain, skull, and head 239

T.R. Fonville, S.J. Scarola, Y. Hammi, Raj K. Prabhu, Mark F. Horstemeyer

14.1 Introduction 239
14.2 Problem set-up for the finite element simulations.....................................241
14.3 Results 243
    14.3.1 Whole head: fundamental frequency and mode shapes 244
    14.3.2 Brain: fundamental frequency and mode shapes..........................244
14.4 Discussion 247
14.5 Conclusions 251
References 252

Chapter 15: State-of-the-art of multiscale modeling of mechanical impacts to the human brain 255

Mark F. Horstemeyer

15.1 Introduction 255
15.2 Work to be completed..255
    15.2.1 Multiphysics aspects of the brain 255
    15.2.2 Multiscale structure–property relationships of the brain..................255
    15.2.3 Different biological effects on the brain......................................256
    15.2.4 The liquid–solid aspects of the brain...257
    15.2.5 Different human ages 257
15.3 Conclusions 258
    References 258

Index...259

Contributors

A. Alshareef   School of Engineering, University of Virginia, Charlottesville, VA, United States

A. Bakhtiarydavijani   Center for Advanced Vehicular Systems (CAVS), Mississippi State University, Mississippi State, MS, United States

M. Baskes   College of Engineering, University of North Texas, Denton, TX, United States

P.R. Berthelson   Center for Applied Biomechanics, University of Virginia, Charlottesville, VA, United States

D. Dickel   Bagley College of Engineering, Mississippi State University, Mississippi State, MS, United States

A. Dobbins   Biomedical Engineering Department, University of Alabama-Birmingham, Birmingham, AL, United States

T.R. Fonville   School of Engineering, Liberty University, Lynchburg, VA, United States

M. Gibson   Liberty University, Lynchburg, VA, United States

J.S. Giudice   School of Engineering, University of Virginia, Charlottesville, VA, United States

S. Gwaltney   Department of Chemistry, Mississippi State University, Mississippi State, MS, United States

Y. Hammi   Mississippi State University, Mississippi State, MS, United States

Mark F. Horstemeyer   School of Engineering, Liberty University, Lynchburg, VA, United States

M.D. Jones   Cardiff School of Engineering, Cardiff University, Cardiff, United Kingdom

C. Joseph   Liberty University, Lynchburg, VA, United States

G.A. Khalid   Electrical Engineering Technical College, Middle Technical University, Baghdad, Iraq

Amit Madhukar   Department of Mechanical Science & Engineering, University of Illinois at Urbana-Champaign, Urbana, IL, United States

R. Miralami   Center for Advanced Vehicular Systems (CAVS), Mississippi State University, Mississippi State, MS, United States

S. Mun   Center for Advanced Vehicular Systems (CAVS), Mississippi State University, Mississippi State, MS, United States

M.A. Murphy   Center for Advanced Vehicular Systems (CAVS), Mississippi State University, Mississippi State, MS, United States

A.B. Nellippallil   Department of Mechanical and Civil Engineering, Florida Institute of Technology, Melbourne, FL, United States

Contributors

Martin Ostoja-Starzewski Department of Mechanical Science & Engineering, University of Illinois at Urbana-Champaign, Urbana, IL, United States; Beckman Institute and Institute for Condensed Matter Theory, University of Illinois at Urbana-Champaign, Urbana, IL, United States

M.B. Panzer School of Engineering, University of Virginia, Charlottesville, VA, United States

L. Peterson Center for Advanced Vehicular Systems (CAVS), Mississippi State University, Mississippi State, MS, United States

Raj K. Prabhu USRA, NASA HRP CCMP, NASA Glenn Research Center, Cleveland, OH, United States

R. Prichard Liberty University, Lynchburg, VA, United States

K. Reynier School of Engineering, University of Virginia, Charlottesville, VA, United States

S.J. Scarola Liberty University, Lynchburg, VA, United States

D. Shedd School of Engineering, University of Virginia, Charlottesville, VA, United States

W. Strasser Liberty University, Lynchburg, VA, United States

A. Vo Center for Advanced Vehicular Systems (CAVS), Mississippi State University, Mississippi State, MS, United States; Department of Agricultural and Biological Engineering, Mississippi State University, Mississippi State, MS, United States

T. Wu School of Engineering, University of Virginia, Charlottesville, VA, United States

Preface

I have experienced about 12 concussions in my lifetime, and those are only the ones that I can recount at a much older age. All of my concussions came about because of sports until the last series, which came from a metal pipe striking my head. The first major concussion occurred from a football impact. I was a wide receiver and ran an end-around and had only the linebacker to beat for a touchdown, and he was very low, so I leaped trying to go over him. He barely touched my shoe, which was enough to completely flip me onto my head. I woke up with a bunch of players staring down at me. Later, I sandwiched between two defenders after I caught a ball down the middle of field, because of their perfect timing in hitting my chest and head on opposite sides, I woke up with both of those players staring down at me. Later, while playing third base on my Pony League baseball team, the center fielder collected a ball at the fence after the batter hit a ball into the right-center field gap. As the runner rounded second base, the center fielder heaved a through to me….a perfect throw in fact. Unfortunately, it was a night game, and I lost the ball in the lights. The ball struck me just under my left eye and to the left of my nose. It broke my zygomatic arch in three locations. I remember looking at the red third base bag from my bloody nose as I hovered over it in a kneeling position. The runner went back to second base, and we won the game by one run. The other concussions occurred from hockey and basketball, mainly from elbows to my head.

The most deleterious and recent concussion was the aforementioned metal pipe striking my head. My brother and I were out in the field of my farm and there was a 20 ft pipe that comprised two 10 foot sleeves. As we lifted it up, the sleeves slipped apart and one of pipes came down and struck me on the head. I did not pass out, but it hurt. I kept working. In the next 3 consecutive days, I slightly hit my head on the top freezer door, our van door, and a cupboard door, and each slight impact caused me to go unconscious. Although the pipe strike to my head did not cause me to go unconscious, it damage my brain enough so that even the slightest impacts afterward did, indeed, cause me to go unconscious. That lead to a 3-year process of trying to recover my brain and its function. My field of vision shrank; my hearing was impaired; my mental state was unstable; and the dizziness, headaches, and foggy thinking were prevalent. Thank God for Sharon Snider (Birmingham, AL) and Shannon Skelton (Starkville, MS) who helped me recover my brain conducting therapy on my brain, head, and neck as I have been able to recover my brain function.

Preface

From that point forward, I decided to study the brain. My research background is related to Integrated Computational Materials Engineering (ICME) of which I have written two books. ICME includes multiscale materials modeling from the smallest length scale at the electron level to the largest length scale like a car. Furthermore, ICME also includes modeling the chemistry–process–structure–property–performance sequence of a material. However, these past efforts focused on metals. The effort in this book is to employ the ICME methodologies to the brain to understand the multiscale mechanisms of traumatic brain injuries arising from mechanical impacts.

The different authors in this book have committed their research lives to understand traumatic brain injuries arising from mechanical impacts. For their dedication, I am very grateful and thankful that they would work toward using multiscale modeling techniques to help provide knowledge and understanding to help the next generation of people who may have incurred some sort of brain injury.

Sincerely,

Mark F. Horstemeyer
Dean of Engineering, Liberty University
ASME Fellow, ASM Fellow, SAE Fellow, AAAS Fellow
Member of the European Union National Academy of Science (EUNAS)

CHAPTER 1

The multiscale nature of the brain and traumatic brain injury

M.A. Murphy[a], A. Vo[a,b]

[a]Center for Advanced Vehicular Systems (CAVS), Mississippi State University, Mississippi State, MS, United States [b]Department of Agricultural and Biological Engineering, Mississippi State University, Mississippi State, MS, United States

1.1 Introduction

The human brain is a truly amazing structure that is complex both in its function and its anatomical multiscale hierarchy. While other organs and systems in the human body also exhibit unique and complex multiscale geometrical hierarchies, the brain's anatomical, physiological, and mechanical properties provide a very difficult problem for consideration. Scientists are still working to understand the brain's "normal" physiological responses and structures, and changes due to injuries can be even more confounding.

Generally, injuries to the brain are broadly classified under the term traumatic brain injury, which is commonly abbreviated as TBI. These injuries are one of the leading causes of mortality and disability globally (Humphreys et al., 2013; Taylor et al., 2017) and primarily due to accidents from sports and motor vehicles (Andriessen et al., 2010; Johnson et al., 2013). These and other injury events cause external mechanical forces to the head that can cause physical damage and dysfunctionality to the brain.

The question of what TBI is has been discussed and changed many times, including the initial distinction from the general classification of head injury (Menon et al., 2010). As a general definition, "TBI is defined as an alteration in brain function, or other evidence of brain pathology, caused by an external force" (Menon et al., 2010). In other words, TBI is any acceleration or impact that affects brain function or pathology.

While this description is very broad, TBI can be identified through observing an individual that has potentially sustained injury. A common way to identify an alteration in the brain's function is by observing apparent changes in an individual's mental or physical condition, such as loss of consciousness, memory loss, neurological deficits (weakness or sensory loss), or mental state alterations (confusion) (Menon et al., 2010). These changes can be short-lived or endure over longer periods of time. Alternatively, brain pathology can be viewed clinically using medical imaging, such as magnetic resonance imaging (MRI) or diffuse tensor imaging

Multiscale Biomechanical Modeling of the Brain.
DOI: https://doi.org/10.1016/B978-0-12-818144-7.00004-9
Copyright © 2022 Elsevier Inc. All rights reserved.

2 Chapter 1

(DTI) (Menon et al., 2010; Sabet et al., 2008). However, these images do not give information about the state of the brain during the process of injury, only the aftereffects of the injury. This limitation results in knowledge gaps regarding the injury process. Further, they may not allow an injury to be observed if it is not severe enough.

Mechanically speaking, when the head is subjected to a rapid acceleration or deceleration, TBI can result due to local strains within the brain, leading to various issues, including neuronal bilayer membrane deformation leading to mechanoporation, water molecule penetration, and the rearrangement of phospholipids (Prabhu et al., 2011; Murphy et al., 2016, 2018). This deformation causes membrane disruption and alters ion flow across the membrane, which can have a significant impact on the membrane's structural and dynamic properties as well as affect the transmembrane potential and cellular homeostasis (Murphy et al., 2018; Alaei, 2017). Eventually, these events can induce cell death, tissue damage, and brain dysfunction at higher length scales. Nonetheless, those cellular impairments, unlike visible macroscale structural and functional damages (e.g., bruises, bleeding, lacerations, etc.), are often too subtle to be detected or quantified in real-time studies through current imaging techniques or in vivo and in vitro measurements (Alaei, 2017; Montanino, 2019). These approaches are hindered by the time and spatial resolution as well as the seeming mismatch between cellular and subcellular scale physiological changes and damage observations at the organ scale (Murphy et al., 2018; Montanino, 2019; Rashid et al., 2013, 2014).

Determining each length scale's effect on the brain's mechanical properties and resulting pathology is of utmost importance because each can have profound effects on the macroscale brain's response. However, the multiscale nature of the brain and its injury mechanisms as well as the inability to easily observe injury mechanisms in real time can make this task unmanageable very quickly. Instead, each length scale must be considered separately to determine which boundary conditions must be scaled down from higher length scales and what information must be passed up from lower length scales. A graphical representation of this process is shown in Fig. 1.1. Then, experiments and simulation methods can be used to examine TBI at the length scale where each injury mechanism occurs.

1.2 The brain's multiscale structure

As mentioned above, the human brain has complex anatomical (geometrical) hierarchy. While the brain's complexity does make it more difficult to understand, being multiscale is not unique to the brain or even biological structures. Typically, any material or structure with properties from multiple length scales that can affect the overall properties can be described as multiscale, which necessitates that lower length scales be considered when considering mechanical responses and damage. Biological structures, like the brain, are often vastly different at the different length scales. For example, tendons have six distinct structural scales (tendon, fascicle, fiber, fibril, subfibril, and tropocollagen), which spans from the nanometer

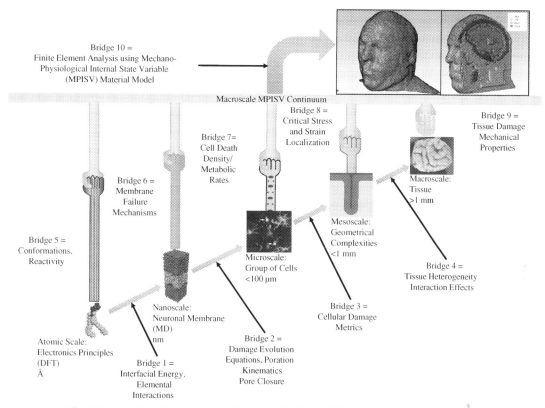

Fig. 1.1: Multiscale structure of the brain that will be used for the modeling and simulation discussed in the later chapters of this book.

to millimeter length scales (Harvey et al., 2009). For the brain, this complex multiscale structural hierarchy includes the brain, lobes, regions, sulci and gyri, groups of cells, individual cells, cellular organelles, and components. Therefore, a broad overview of the brain's multiscale anatomy has been included here.

1.2.1 Gross anatomy

The brain is covered with connective tissues that make up a system of membranes called meninges, which include the outer dura mater, the middle arachnoid mater, and the inner pia mater (Purves, 2012; Snell, 2010). Dura mater is a strong, thick, and nonelastic membrane that folds into septa, including the falx cerebri that separates the right and left half of the brain (Montanino, 2019; Kekere and Alsayouri, 2019). Arachnoid mater is a thin, web-like membrane that stretches between the dura and pia mater. Pia mater is another thin layer that tightly wraps the entire surface of the brain and aids in the production of cerebrospinal fluid (Patel and Kirmi, 2009; Decimo et al., 2012). The cerebrospinal fluid acts as a cushion in these protective layers, helping the brain preserve its shape and anchoring it in place within the

4 Chapter 1

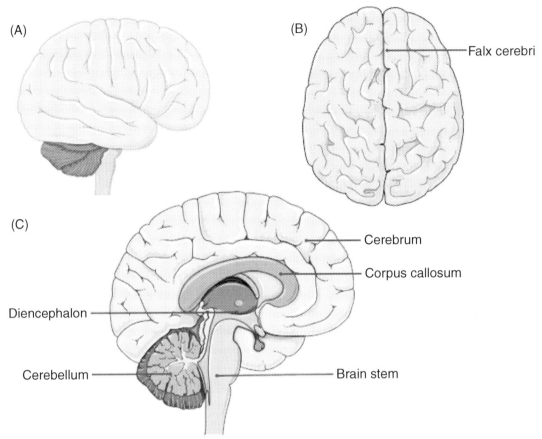

Fig. 1.2: Representative schematics of the brain (A) side and (B) top views and (C) brain structures. Images are modified and used with permission from Servier Medical Art - Creative Commons Attribution 3.0 Unported License.

skull (Patel and Kirmi, 2009; Decimo et al., 2012). If the brain is viewed in three dimensions and the meninges are removed, it contains four main divisions called cerebrum, cerebellum, brainstem, and diencephalon (Patestas and Gartner, 2016). The cerebrum, the cerebellum, and the brainstem are clearly visible, while the diencephalon is almost completely hidden from the brain surface (Snell, 2010; Patestas and Gartner, 2016) (Fig. 1.2).

1.2.1.1 Cerebrum

The cerebrum consists of two nearly symmetrical hemispheres and is covered by the cerebral cortex, which is 2–4 mm and highly folded (Patestas and Gartner, 2016; Kandel et al., 2000). Each fold or ridge is called a gyrus, and each groove between the folds is called a sulcus. Some large sulci are named according to their position or also called fissure, which divide the cerebrum into different regions (Snell, 2010; Patestas and Gartner, 2016; Squire et al., 2013). The cerebrum is incompletely separated into left and right hemispheres by a deep longitudinal fissure containing the falx cerebri, and joined again by the corpus callosum at the end of that fissure (Patestas and Gartner, 2016; Davey, 2011). Shown in Fig. 1.3, each hemisphere is subdivided into four main broad regions or lobes: the frontal, parietal, temporal, and occipital lobes (Snell, 2010; Patestas and Gartner, 2016; Gray and Standring, 2015). Besides, there are limbic and insular lobe (insula) hidden inside the hemisphere (Patestas and Gartner, 2016; Squire et al., 2013). The frontal lobe is the primary motor area, controlling movement, behaviors and personality, attention, and concentration (Snell, 2010; Patestas and Gartner, 2016; Kolb and Whishaw, 2009). The parietal lobe is the primary somesthetic area that integrates somatosensory information. The temporal lobe is the major processing center of auditory information (Patestas and Gartner, 2016; Freberg, 2009). The occipital lobe is the smallest lobe whose main functions are visual-spatial processing (Snell, 2010; Patestas and Gartner, 2016). The insula is associated with taste, visceral sensation and autonomic control (Patestas and Gartner, 2016). The limbic lobe is the cortical constituents of the limbic system, which is the center of emotions, learning and memory (Patestas and Gartner, 2016). Each lobe could be further divided into smaller regions serving very specific functions, making up approximately 50 functional areas in the cortex (Guyton and Hall, 2011).

Besides being divided into regions, the cerebral cortex is also organized in layers, which are characterized by different densities, sizes and morphology of 16 billion nerve cells or neurons (Bastiani and Roebroeck, 2015; Bigos et al., 2016; Azevedo et al., 2009). The cortex is classified into archicortex (allocortex), mesocortex (juxtallocortex), and neocortex (isocortex) (Patestas and Gartner, 2016). The archicortex contains three layers and is situated at the limbic system, while the mesocortex has three to six layers and is predominant in the insula and above the corpus callosum. The neocortex consists of six layers, which is known as the cytoarchitecture and comprises the bulk of the cerebral cortex (Patestas and Gartner, 2016; Bastiani and Roebroeck, 2015). Passing through this cytoarchitecture are cell columns, each of which is less

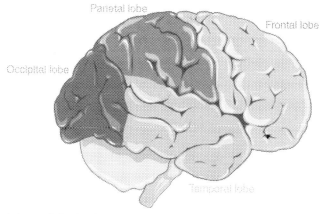

Fig. 1.3: Main lobes of the brain. Images are modified and used with permission from Servier Medical Art - Creative Commons Attribution 3.0 Unported License.

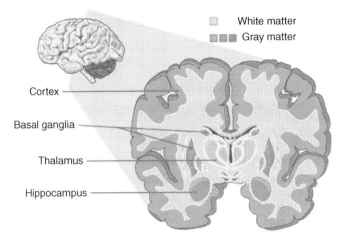

Fig. 1.4: Schematic representation of a brain section showing gray and white matter. Images are modified and used with permission from Servier Medical Art - Creative Commons Attribution 3.0 Unported License.

than 0.1 mm wide, perpendicular to the cortical surface and consists of neurons with similar functions (Patestas and Gartner, 2016). The cortical layers are made up of gray matter, which contains soma (nerve cell bodies or nuclei) and can be found at the brain regions involved in controlling muscular and sensory activity (Montanino, 2019). Beneath the cortex, deeper subcortical regions are made up of white matter, which includes axon (nerve fibers), connects different areas of the brain and the brain with other body parts (Patestas and Gartner, 2016; Bastiani and Roebroeck, 2015; Sampaio-Baptista and Johansen-Berg, 2017). Gray matter are also present within the white matter, in the deep structures of the cerebrum, diencephalon, cerebellum, and brainstem, such as the hippocampus, thalamus, hypothalamus, and basal ganglia (basal nuclei) (Montanino, 2019; Snell, 2010). A representative view of these structures shown in Fig. 1.4. Within the cerebrum, the basal ganglia acts alongside the cerebral cortex as part of the cognitive system (Purves, 2012; Patestas and Gartner, 2016; Stocco et al., 2010). The basal nuclei convey information to various brain regions and work with the cerebellum to control complex muscle movements (Patestas and Gartner, 2016; Guyton and Hall, 2011).

1.2.1.2 Cerebellum

The cerebellum is located below the cerebrum and consists of two cerebellar hemispheres connected by the narrow vermis (Snell, 2010). The vermis is subdivided into superior and inferior portion, which are respectively visible and hidden between the two hemisphere (Snell, 2010; Patestas and Gartner, 2016). The cerebellum has an outer surface made up of gray matter and an inner core composed of white matter (Patestas and Gartner, 2016; Gray and Standring, 2015). Certain masses of gray matter can also be found in the interior of the cerebellum and embedded in the white matter (Patestas and Gartner, 2016). The surface of the cerebellum is called cerebellar cortex, which is much thinner than the cerebral cortex and organized in three layers—the outermost molecular layer, the middle Purkinje layer, and the innermost granular layer (Patestas and Gartner, 2016; Gray and Standring, 2015). The cerebellar cortex has slender and parallel folds called folia, alternating with the grooves called sulci (Snell, 2010; Patestas and Gartner, 2016). Some deep sulci or fissures further subdivide each hemisphere into three lobes—anterior, posterior, and flocculonodular lobes (Patestas and Gartner, 2016; Guyton and Hall, 2011). The anterior and posterior lobes are separated by the transverse primary fissure and play an important role in coordinating complex muscle movements. The flocculonodular lobe is underneath these two lobes and is essential in maintaining balance (Patestas and Gartner, 2016; Guyton and Hall, 2011). Although the cerebellum is argued to be involved in some cognitive control, its main function is in motor control, maintenance of posture and balance (Squire et al., 2013; Guyton and Hall, 2011). It is linked to the cerebral motor strip and contributes to the precision of motor activity (Fine et al., 2002). The cerebellum and the basal ganglia, together with the thalamus in the diencephalon, work as the main movement coordination center that fine-tune motor functions.

1.2.1.3 Diencephalon

Besides the cerebrum and cerebellum, diencephalon is another structure that contains large collections of gray matter. It is hidden from the surface of the brain, interposed between the cerebrum and the brainstem, and is separated in two halves by the third ventricle—a narrow space filled with cerebrospinal fluid (Snell, 2010; Patestas and Gartner, 2016). The diencephalon consists of four components: the epithalamus, thalamus, hypothalamus, and subthalamus (Snell, 2010; Patestas and Gartner, 2016). The epithalamus is the dorsal posterior segment of the diencephalon, linking the limbic system with other parts of the brain (Caputo et al., 1998). It contains the pineal gland that modulates the body's internal clock, circadian rhythms, and sex hormones (Aulinas, 2000; Lowrey and Takahashi, 2000). The hypothalamus is located in the floor of the third ventricle and forms the ventral part of the diencephalon (Snell, 2010). It links the nervous system to the endocrine system through the pituitary gland, which secrets hormones for metabolism, growth and sexual development (Boron and Boulpaep, 2016). Located above the hypothalamus is the thalamus, both of which are part of the limbic system (Boeree, 2009). The thalamus is separated with the hypothalamus by the hypothalamic sulcus situated along the lateral walls of the third ventricle. The right and left thalami constitute the bulk of the diencephalon and are connected by a bridge of gray matter called interthalamic adhesion or massa intermedia. The thalamus functions as a relay station integrating and conveying information to the cortex (Gazzaniga et al., 2014), which is responsible for alertness, sensation and memory (Sherman, 2006; Sherman and Guillery, 2009; Aggleton et al., 2010). At the back of the thalamus is the brainstem (Higgins, 2006).

1.2.1.4 Brainstem

The brainstem is a stalk-like structure that connects the cerebrum, cerebellum, and diencephalon to the spinal cord (Snell, 2010; Patestas and Gartner, 2016). It is partly hidden

8   Chapter 1

by the cerebrum and cerebellum, linked to the cerebellum by the cerebellar peduncles, and consists of the midbrain, the pons, and the medulla oblongata (Patestas and Gartner, 2016; Gray and Standring, 2015). The brainstem controls the cardiovascular and respiratory system, consciousness, reflexes, and automatic processes (e.g., breathing, eye movements, swallowing, and digestion) (Gray and Standring, 2015; Guyton and Hall, 2011). It consists of many nerve tracts and nerve nuclei of the central as well as peripheral nervous system. Particularly, 10 of the 12 pairs of the nerves of the central nervous system (cranial nerves III through XII) directly emerge from the brainstem (Snell, 2010). Hence, it serves as a channel for transmitting information between the brain and different parts of the body, including the ascending tracts (the nerve pathways carrying sensory information from the body up the spinal cord to the brain) and descending tracts (the nerve pathways carrying motor information from the brain down the spinal cord to the body) (Snell, 2010; Gray and Standring, 2015).

1.2.2 Microanatomy

The cerebrum, the cerebellum, the diencephalon, and the brainstem are built from nervous tissues, which can be divided as gray and white matter (Montanino, 2019; Bastiani and Roebroeck, 2015). They are microscopically constituted from an abundance of cells, including two main categories: (1) non-neuronal cells (glia cells or neuroglia) consisting of microglia and macroglia (Montanino, 2019); (2) neuronal cells (nerve cells or neurons) comprising cell body or soma, branching dendrites and a longer projection called axon, in which the axon can be unmyelinated or myelinated (interruptedly wrapped by layers of a plasma membrane named myelin) (Alberts et al., 2014). Gray matter includes cell bodies in gray–brown color, dendrites, some glia cells and unmyelinated neurons (Montanino, 2019; Patestas and Gartner, 2016). White matter is made up of bundles of myelinated neurons, in which the myelin sheath covering the axon gives it the white color (Montanino, 2019; Gray and Standring, 2015). Overall, the human brain contains approximately 85 ± 10 billion of neuroglia and 86 ± 8 billion neurons, in which 16 billion neurons are in the cerebral cortex and 69 billion neurons are in the cerebellum (Bigos et al., 2016; Azevedo et al., 2009). Neurons are the key functional units of the brain, while the neuroglia provide them with nourishment, protection and structural support (Patestas and Gartner, 2016; Squire et al., 2013).

1.2.2.1 Neuroglia

The neuroglia can be divided into microglia and macroglia (Montanino, 2019), representative examples of which are shown in Fig. 1.5. Microglia are scattered throughout the brain and spinal cord (Snell, 2010; Ginhoux et al., 2013). From their cell bodies arise wavy branches or processes that give off multiple spine-like projections (Snell, 2010). They are responsible for the immune response and overall maintenance of the brain, continuously scavenging it for damaged neurons or infectious agents, and protecting it from invaders (Patestas and Gartner, 2016; Helmut et al., 2011). Macroglia consists of astrocytes, oligodendrocytes, and ependymal cells. Astrocytes are the largest neuroglia, consisting of fibrous and protoplasmic

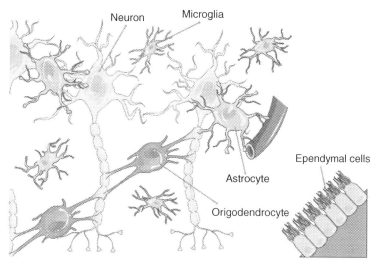

Fig. 1.5: Pictorial representation of neuroglia and neurons. Images are modified and used with permission from Servier Medical Art - Creative Commons Attribution 3.0 Unported License.

type, with small cell bodies and branching processes extending in all directions (Snell, 2010). Fibrous astrocytes are present mostly in white matter, where their long, slender and not much branched processes pass between the nerve fibers (Snell, 2010; Patestas and Gartner, 2016). Protoplasmic astrocytes are located mainly in gray matter, where their short, thick and branched processes pass between the nerve cell bodies (Snell, 2010; Patestas and Gartner, 2016). These astrocytes conduct metabolites, regulate blood flow, and form a supporting framework for the neurons (Snell, 2010). They regulate the uptake of neurotransmitters (electrical or chemical substances delivered between neurons), maintain ionic balance and modulate neural activity (Oberheim et al., 2012), as well as take up and replace dead neurons or degenerating axon terminals (Snell, 2010; Patestas and Gartner, 2016). Oligodendrocytes, with small cell bodies and a few delicate processes, are frequently found surrounding nerve cell bodies and along myelinated neurons (Snell, 2010). They can expand their membrane to form myelin sheaths that wrap around and insulate up to 50 axons, allowing electrical messages to travel faster (Saab and Nave, 2017). The third type of macroglia, ependymal cells, forms a single layer of cuboidal cells that line the brain ventricles, circulates cerebrospinal fluid and prevents it from leaking into underlying tissues (Snell, 2010; Patestas and Gartner, 2016). Altogether, the role of these neuroglia, although highly diverse and incompletely understood, is creating an environment where neurons can develop and function (Montanino, 2019; Herculano-Houzel, 2014).

1.2.2.2 Neurons

Neurons are electrically excitable nerve cells specializing in receiving, transmitting, and conducting electrochemical signals or nerve impulse (Snell, 2010; Patestas and Gartner, 2016).

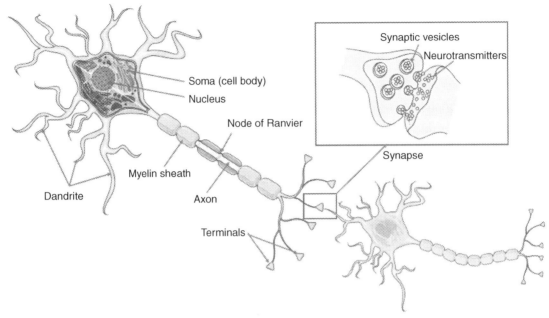

Fig. 1.6: Pictorial representation of neuron and intercellular communication. Images are modified and used with permission from Servier Medical Art - Creative Commons Attribution 3.0 Unported License.

A representative neuron is shown in Fig. 1.6. They vary in size, shape, location, connectivity, and neurochemical characteristics, but all have similar structures (Squire et al., 2013). Morphologically, each neuron consists of a cell body (soma or perikaryon) and multiple processes (neurites, composed of dendrites and axon) branching from the cell body, which facilitates intercellular communication through synapses (small gaps between the cells) (Snell, 2010; Patestas and Gartner, 2016; Alberts et al., 2014).

The nerve cell body, similar to other cells, contains a nucleus, cytoplasm and is enclosed by a plasma membrane (Montanino, 2019; Snell, 2010). The cytoplasm is rich in granular (rough) and agranular (smooth) endoplasmic reticulum (ER), which is an interconnected network of flattened, membrane-enclosed sacs (cisternae) that synthesize and transport cellular materials such as proteins, lipids and carbohydrates (Snell, 2010). It contains the nucleus, cytoplasmic organelles (including Nissl substance and Golgi complex), and the cytoskeleton needed for the functioning of the cell (Snell, 2010; Squire et al., 2013). The nucleus is typically rounded, centrally located within the cell body, and has an envelope made from two lipid bilayer membranes. It stores the genes and controls the cell activity (Snell, 2010). The Nissl substance contains stacks of rough ER interposed with ribosomes, distributed throughout the cytoplasm, extended into the proximal parts of the dendrites and absent from axon (Squire et al., 2013). Ribosomes containing RNA and protein are attached to the ER surface or insinuated between the cisterna, enabling the Nissl to synthesize proteins that flow along the neurites (Snell, 2010; Squire et al., 2013). The protein is then transferred to the Golgi complex in transport vesicles, where it is temporarily stored and maybe added with carbohydrate to form glycoproteins that contribute to the synthesis of cell membranes (Snell, 2010; Squire et al., 2013). The Golgi complex appears around the nucleus as clusters of cisternae made up of smooth ER, modifying and transporting vesicles of macromolecules to the nerve terminals (Snell, 2010).

Another feature of the cytoplasm, which extends throughout the cell body and its processes, is the cytoskeleton. It is composed of microfilaments (actin filaments), neurofilaments (intermediate filaments), and microtubules (Montanino, 2019; Squire et al., 2013). Microfilaments are helical polymers formed of actin (with a diameter of 3–5 nm in soma or 7–9 nm in axon), concentrated at the periphery of the cytoplasm (Snell, 2010; Fletcher and Mullins, 2010). They form a dense network beneath the plasma membrane, assist in cell transport, forming and retracting cell processes (Montanino, 2019; Snell, 2010). Larger than the microfilaments are the neurofilaments (with a diameter of about 10 nm). They are the main components of the cytoskeleton, which bundle together to form parallel neurofibrils running through the cell body into the neurites, determine the shape of neurons and axonal radial growth (Montanino, 2019; Snell, 2010). The biggest and stiffest components of the cytoskeleton are microtubules (hollow cylinders with an outer diameter of 25 nm). They are interspersed among the neurofilaments and composed of 13 protofilaments of tubulin heterodimers that are continuously added or removed at both ends (Snell, 2010; Fletcher and Mullins, 2010). In axons, they form dense bundles with uniform orientation—the fastest growing end pointing toward the axonal tip and the other end pointing to the cell body (Montanino, 2019). These microtubules, together with the microfilaments, provide a stationary track for specific organelles to move by molecular motors (Snell, 2010). On the whole, the filament network has two main roles: maintain the shape of the neuron and allow cellular transport along its structure (Montanino, 2019).

Besides the cytoskeleton, the plasma membrane is another component that extends from the soma into the axon and dendrites. The membrane is about 8 nm thick, mainly composed of lipids and proteins arranged in layers, in which the middle lipid bilayer (about 3 nm) separates the inner and outer layer of loosely arranged proteins (about 2.5 nm each) (Alaei, 2017; Montanino, 2019; Snell, 2010). The lipid bilayer is made up of phospholipids, each with two hydrophobic tails pointing inward and the polar headgroup pointing outward in contact with the protein layers (Alaei, 2017; Snell, 2010). Certain proteins span the entire width of the lipid bilayer, forming hydrophilic channels through which inorganic ions enter and leave the cell (Snell, 2010). Besides lipids and proteins, cholesterols (CHOL) are another important components (Alaei, 2017). Together they determine membrane properties such as stiffness, bending and permeability (Alaei, 2017; Harayama and Riezman, 2018). Additionally, carbohydrate molecules are attached outside the membrane and linked to the proteins or the lipids, forming the cell coat or glycocalyx (Snell, 2010). The plasma membrane and the cell

12   Chapter 1

coat form a semipermeable membrane, allowing diffusion of ions. When the neuron is resting (unstimulated), the passive efflux of K^+ (from the cytoplasm to the extracellular fluid) is greater than the influx of Na^+ (from the extracellular fluid into the cytoplasm), which makes the inside of the membrane more negative and results in the resting potential between -50 mV and -80 mV (Montanino, 2019; Snell, 2010). When the neuron is excited (simulated), the membrane permeability to Na^+ rapidly increases, progressively depolarizing the membrane and producing the action potential of +40 to +50 mV (Montanino, 2019; Snell, 2010). The increased permeability for Na^+ quickly stops after 5 milliseconds and the permeability for K^+ efflux increases again. The action potential spreads over the membrane and self-propagates along the neurites as nerve impulse (Snell, 2010). In general, the plasma membrane defines the intracellular-extracellular boundary, regulate the transport of materials and signals, and serves as the site for the initiation and conduction of the nerve impulse (Montanino, 2019; Giordano and Kleiven, 2014).

Although neurons resemble other cells in the body, they are distinguished by branching processes or neurites, including dendrites and axon (Montanino, 2019; Patestas and Gartner, 2016). Dendrites are short processes with tapered diameter and cytoplasm similar to the cell body (Snell, 2010). They branch profusely into dendritic shafts, which further protrude several smaller projections (about 2 µm) called dendritic spines (Snell, 2010; Squire et al., 2013). The spines regulate neurochemical events and increase the receptive surface for the cell body (Squire et al., 2013). Dendrites receive electrochemical stimulation or nerve impulse from other neurons via their synaptic terminals, then the inputs are conducted toward the cell body and transmitted along the axon (Montanino, 2019; Snell, 2010; Squire et al., 2013). Axon is a single long tubular neurite at the other pole of the neuron, extending farther from the cell body and containing more filaments than dendrites (Snell, 2010; Squire et al., 2013). Axon is the longest process (about 0.1 mm), with smooth surface and several branches (collaterals) along its length (Snell, 2010; Squire et al., 2013). Axon has a typical cellular architecture, including a cytoplasm (axoplasm) bounded by a plasma membrane (axolemma or plasmalemma). Axoplasm does not contain Nissl substance, Golgi complex, and the sites producing proteins (RNA and ribosomes) (Montanino, 2019; Snell, 2010). Meanwhile, axolemma is similar to the membrane of the cell body, containing ion channels that control the transmembrane ion flux and connect to the axonal cytoskeleton (Alberts et al., 2014; Leterrier et al., 2017). Most axons are surrounded by an insulating myelin sheath—an extended plasma membrane composed of lipoproteins and produced by oligodendrocytes (Montanino, 2019; Squire et al., 2013). Myelinated axonal portions (internodal segments) are interrupted by unmyelinated ones (nodes of Ranvier) clustered with sodium channels (Montanino, 2019). This configuration allows the action potential to propagate by jumping from node to node, facilitating efficient impulse conduction (Montanino, 2019; Squire et al., 2013). The axon leaves the cell body from a small elevated area called axon hillock, then the first 50–100 µm is the axon initial segment (AIS) rich in voltage-gated channels (Alberts et al., 2014; Kole et al., 2008). AIS is the most excitable axonal part, at which the action potential originates and the stimulation that a neuron receives is encoded for long-distance transmission (Snell, 2010; Alberts et al., 2014). After that the axon emits numerous collaterals with enlarged ends called terminals, forming synapses with other neurons (Montanino, 2019; Snell, 2010). An axon conducts nerve impulses away from the cell body and communicates with other neurons at synapses, where the neurotransmitters are released and bind to receptor molecules on the receiving neuron (Patestas and Gartner, 2016; Squire et al., 2013).

The synapse is a site of neuronal communication, in which an axon of one neuron makes contact with the dendrite or cell body, or rarely with the initial segment or terminals of another axon (Snell, 2010; Squire et al., 2013). Depending on the site of contact, synapses are referred to as axodendritic, axosomatic, or axoaxonic (Snell, 2010). Each synapse consists of a presynaptic element (a portion of the axon making the contact), a narrow synaptic cleft, and a postsynaptic element (a portion of the receiving dendrite, soma, or axon) (Snell, 2010). The portion of the presynaptic element proximate to the postsynaptic element is the active zone, where the synaptic vesicles are concentrated or anchored for fusion. The active zone is enriched with voltage-gated calcium channels that aid in the vesicle fusion and the release of neurotransmitter into the synaptic cleft (Squire et al., 2013). The synaptic cleft is around 20 µm, determining the volume where each synaptic vesicle releases its contents and the maximum concentration of released neurotransmitters (Squire et al., 2013). Several neurotransmitters could be released, one of which is the key activator functioning directly on the postsynaptic membrane, while others modulate its activities (Snell, 2010). The synapses excites or inhibits the neuron when the released neurotransmitters cause a small depolarization (excitatory postsynaptic potential) or hyperpolarization (inhibitory postsynaptic potential) in the postsynaptic membrane (Alberts et al., 2014). Overall, a single neuron receive inputs from thousands of other neurons through neurotransmission, and can form synapses with thousands of other neurons (Snell, 2010; Alberts et al., 2014). This interneuronal communication forms tree-like structures called neuronal arbors, which are responsible for distributing information over a region of the brain (Teeter and Stevens, 2011). Furthermore, the neuronal arbors connect together to form local circuits that constitute the structural and functional basis for the brain activity (Snell, 2010; Squire et al., 2013).

1.3 The multiscale nature of TBI

While TBI can be the result of an externally visible injury with a clear cause, such as a puncture wound or blast wave, TBIs often occur due to less apparent causes as well and range in severity from mild to severe. Importantly, injury severity greatly influences both the ability to detect TBI cases as well as a patient's expected short-term and long-term outcomes, which can include death due to injury. To understand this problem better, a broad overview of injury mechanisms at each length scale, injury types and examples, neurobehavioral sequelae, and research methods are explored here.

14   Chapter 1

1.3.1 Multiscale injury mechanisms

The brain's multiscale properties also affect how the brain is damaged during injury, which ranges from macroscale brain tissue death to nanoscale cellular organelle and biochemical changes. While it may be tempting to simply look at the macroscale tissue death as the "injury," injury actually spans across each length scale. Therefore, TBI must also be considered as a multiscale problem to identify essential brain injury mechanisms at each length scale. Specifically, consider the brain on five distinct scales: macroscale, mesoscale, microscale, nanoscale, and atomic scale, as shown in Fig. 1.1.

At the highest length scale, the macroscale is likely what individuals are most familiar with when considering the brain (and head). This scale is typically what someone will see when learning about the brain's gross anatomy. At this scale, severe injuries are typically apparent during medical imaging due to displaced tissue, tissue contusions, and hematomas, but less severe injuries that primarily affect lower length scales may be less visible. Further, damage can cause swelling, which can result in increased intracranial pressure and cause additional brain tissue damage.

The mesoscale contains the sulci and gyri geometric features of the brain's surface. These structures serve the purpose of increasing the brain's surface area, so much of the damage here will be similar to that of the macroscale. However, this complex geometry can affect the way stress waves propagate through the brain tissue and be the site of local damage resulting from macroscale deformations.

Moving down another length scale gives the microscale, which starts to be smaller than what is easily observable with the human eye. At this scale, the primary focus is how many cells are dying in a local area. Unfortunately, even if there are only a few cells killed in an initial injury, it can result in biological cascades that cause additional cells in the local area to die. This cell death happens via a combination of apoptosis (programmed cell death using ATP) and necrosis (sudden cell death that does not require ATP). with the potential for local tissue death.

At the nanoscale, changes to local cellular structures are the primary focus. This includes mechanical changes, such as mechanoporation in the membrane that can cause a loss of cell homeostasis, the activation of proteins that trigger cell death, and other direct damage to cellular structures. There are many different proteins that can be triggered here, so readers are referred to more dedicated resources for more information regarding the proteins and cascades related to these processes (Whalen et al., 2009; Farkas and Povlishock, 2007; Friedlander, 2003; Raghupathi, 2004; Wong et al., 2005).

Lastly, the atomic scale includes particle interactions, which affects the atomic interactions. Injury at this scale includes the effects of excitotoxic particles, such as calcium ions (Ca^{2+}).

1.3.2 Types of injury

TBIs vary greatly in severity as well as the type of injury they cause, which can be largely classified as primary or secondary injury. Primary injury is manifested by localized damage in relation to a central focal point with injuries that cause direct damage, whereas secondary injury is the result of ongoing physiological response after the initial injury. Due to primary injury being a direct result of a brain insult, damage from these injuries is often observed more easily in brain imaging than secondary injury. Conversely, secondary injuries are at least partially caused by biochemical pathways and can lead to damage that exceeds the initial focal injury over a period of time (Meythaler et al., 2001) and considered by many to be the primary determinant of adverse outcomes (Farkas and Povlishock, 2007; Meythaler et al., 2001). However, secondary injury may not be as apparent in brain imaging, especially just after injury.

Diffuse axonal injury (DAI) is one form of secondary injury that can result in damage that exceeds a focal injury (Meythaler et al., 2001). DAI has been previously reported to be present in 40–50% of the TBI injuries requiring hospitalization in the United States (Meythaler et al., 2001). More importantly, it can be located in a remote location from any focal injury or lack a focal injury while being microscopic in nature, meaning that it is not easily noticed on CT and MRI imaging methods (Farkas and Povlishock, 2007; Meythaler et al., 2001). This creates a large threat that is not easily detectable despite causing damage effects throughout the brain. Axonal bulbs that result from biochemical factors will develop due to neuron damage and may result in continued myelin degeneration for years afterward (Meythaler et al., 2001). DAI has also been linked to cognitive deficits, and neurons that are affected by DAI often suffer from axon degeneration indicating membrane integrity loss (Farkas and Povlishock, 2007; Meythaler et al., 2001).

While primary and secondary injuries can both result in permanent damage to the brain through the death of neurons (Farkas and Povlishock, 2007; Raghupathi, 2004; Wong et al., 2005; Meythaler et al., 2001; DePalma et al., 2005; Maiden, 2009; Wolf et al., 2009; Shukla and Devi, 2010), injury severity will largely determine the final outcome and how long recovery will take, if possible. TBI is classified into severity categories of mild, moderate, and severe, with mild TBI often being referred to as a concussion. To determine injury severity, the Glasgow coma scale (GCS) can be used (Andriessen et al., 2010; Shukla and Devi, 2010; Haddad and Arabi, 2012). This scale is based on visual, verbal, and motor responsiveness with a higher score being indicative of less severe injury (Andriessen et al., 2010). For mild TBI, there is a loss of consciousness for less than 30 minutes, memory loss around the time of injury, and disorientation or confusion, and the GCS should be greater than 13 (Andriessen et al., 2010; Shukla and Devi, 2010). In comparison, severe TBI has a score of less than eight indicating the individual is much less responsive, which indicates greater damage. This score is a predictor of mortality risk, but it does not provide information regarding the underlying brain damage (Andriessen et al., 2010), such as if the injury was primary or secondary.

16   Chapter 1

1.3.3 Examples of injuries

TBI events also vary greatly, with the type of event greatly influencing whether the damage will be predominantly primary or secondary in nature. Some common examples of TBI events include the head being struck by or striking an object and a projectile, such as a bullet, penetrating the skull. Pressure waves from blasts and movements without external impact that cause rapid accelerations to the head can also result in TBI (Menon et al., 2010).

Injuries sustained from the motor vehicle accident, and falls, are often due to the brain coming in contact with the inside of the skull after contact with an object and are often known as closed-head, or nonpenetrating, injuries. When this occurs, large inertial motions of the brain tissue with respect to the skull can cause brain tissue to deform excessively. These injury events can lead to both immediate primary damage and secondary damage (Pfister et al., 2003). Primary injury in this context takes the form of cerebral contusions, a form of tissue bruising where the brain contacts the skull, and secondary injury often takes the form of DAI. Depending on severity, these injuries may require immediate surgical intervention (Shukla and Devi, 2010).

Injuries sustained from skull penetration events, known as open-head or penetrating brain injury, cause a focal injury where the brain is directly damaged by an external source. This penetration can leave a cavity, cause significant primary injury, and can greatly increase the risk of detrimental effects. Further, this damage can be higher for high energy ballistics (i.e., fast-moving projectiles), which create a temporarily larger cavity through pressure waves and cause further damage, which may be increased by designs that cause a projectile to flatten or otherwise change form after entering the body (Maiden, 2009). Pressure waves resulting from these ballistics cause secondary diffuse damage that extends beyond the focal injury (Meythaler et al., 2001; Maiden, 2009).

Injuries sustained from blast waves have several factors that affect the amount of damage done by a blast, including the blast medium, the distance from the blast epicenter, and the size of the blast (Wolf et al., 2009; Moss et al., 2009). Additionally, the problem is compounded by the various types of blast injuries that occur. In a single event, a person can receive primary, secondary, tertiary, and quaternary blast injuries. Primary blast injuries are direct tissue damage from the blast overpressure, secondary blast injuries are caused by items displaced by the blast pressure wave becoming projectiles resulting in blunt force trauma or penetration, tertiary blast injuries are caused by the person being displaced by the pressure wave and injured by blunt trauma after hitting an object (e.g., a wall), and quaternary blast injuries are any injuries caused directly by the explosion but not covered by the first three injuries such as radiation, burns, or psychological trauma (Wolf et al., 2009). This assortment of injuries results in a variety of potential damages to the brain, including directly damaged tissue from skull penetrations and blast overpressure, contusions from blunt force trauma, hemorrhaging from penetration, shear damages, and possible skull flexure (Wolf et al., 2009; Moss et al., 2009; Kato et al., 2007).

Another important mode of injury that should be considered is rapid acceleration or deceleration of the head. This type of injury often occurs in conjunction with injuries when the head is exposed to a sudden acceleration change due to a force on the body (Sabet et al., 2008). For example, the above examples of motor vehicle accidents and blast injuries would both result in significant acceleration changes. Other examples would include athletes, such as fighters or football players. In many of these cases, brain injury can occur if the brain excessively deforms due to accelerations (Sabet et al., 2008). Often, this type of injury is diffuse in nature and may not have a large focal point of damage if no other type of injury occurs as well.

1.3.4 Neurobehavioral sequelae

In addition to the directly observable injury, TBI often results in neurobehavioral sequeulae. A person may experience multiple neurobehavioral sequelae following an injury. These TBI sequelae can be grouped into two overarching categories: somatic and neuropsychiatric, which correspond to the physical effects and mental changes following injury (Riggio and Wong, 2009). While different in nature, both types of sequelae can be temporary or chronic in duration and can significantly impact a person's quality of life when severe, especially when chronic.

Somatic sequelae can include loss of consciousness, headache, dizziness, fatigue, sleep disturbances, and seizures as well as other effects (Riggio and Wong, 2009). While these sequelae can be acute or chronic, they often go away on their own, especially in cases of mild TBI. For example, any loss of consciousness during a mild injury typically lasts less than 30 minutes. However, some somatic sequelae, such as headaches, can last multiple months even with mild injury and can be debilitating during more severe injury (Riggio and Wong, 2009).

Neuropsychiatric sequelae include both cognitive deficits and behavioral disorders (Riggio and Wong, 2009). Cognitive deficits include impairments in attention/concentration, memory, or executive function and can interfere in performing everyday tasks or routines and a direct result of injury (Riggio and Wong, 2009; McAllister, 2011). These impairments may also affect the ability to plan or organize. Because of an inability to perform as well as before the injury, cognitive disorders may also result in depression, substance abuse, irritability, or anxiousness (Riggio and Wong, 2009; McAllister, 2011). While cognitive deficits rarely last more than 3 months, there is a correlation between loss of consciousness or post trauma amnesia and severity of the deficit (Riggio and Wong, 2009). In comparison, behavioral disorders include changes to a person's behavior postinjury, such as changes in personality, major depression, and anxiety disorders like post-traumatic stress disorder (Riggio and Wong, 2009; McAllister, 2011). These changes can be directly rooted in underlying neural damage from the injury or in cognitive problems resulting from the injury (Riggio and Wong, 2009). Changes due to physical damage to different parts of the brain can often be correlated to different behavioral disorders (McAllister, 2011). Further, fear and emotional responses brought

18   Chapter 1

about by weak defense mechanisms, poor social support, or medication can be the cause of behavioral disorders (Riggio and Wong, 2009). Regardless of the cause, behavioral disorders that manifest after TBI can range in severity and be long-lasting, which can lead to significant changes to interpersonal interactions and interfere with daily life.

1.3.5 TBI research methods

TBI cannot be explored in a controlled setting using human subjects, so a combination of experimental and computational methods must be used. These methods cover an extremely wide range of techniques, so a very broad overview is covered here. Regardless of the method, one should keep in mind that the goal of these research methods is to elucidate the injury mechanisms and the resulting changes due to TBI. These complex interconnections, therefore, require extensive study in morphology, stress–strain response, and other biomechanical cues behind TBI.

1.3.5.1 Experiments

Experiments examining injury include in vivo, ex vivo, and in vitro methods, which refer to experiments taking place inside an organism's body, outside of an organism's body after being taken from an organism, and outside a body altogether in an artificial setting. Each of these have advantages and disadvantages, including differences in the type of information they can provide.

One common in vivo method includes using animal models as representative proxies for human TBI. These models allow specific TBI aspects and outcomes can be investigated and correlated back observations in humans. For example, rats have been used with blunt impact and blast devices to study the immediate and ongoing effects of TBI (Bayly et al., 2006; Begonia et al., 2014; Shultz et al., 2017). AI studies have used rats for these studies. Examining injury in this kind of study includes a mixture of medical imaging, cell (tissue) staining, biomarkers, and observation. These observations can be made both before and following injury but can be difficult to capture during the actual injury. The primary benefits of this method are experiment repeatability in a known model, number of samples, and the ability to observe injury effects over time, including changes to behavior and biomarker concentrations. However, animal studies are limited by available resources, including equipment and laboratory space, time, and personnel, and must abide by procedural restraints as established by governing bodies. Additionally, even though there are often challenges relating common models, such as rodents, to human TBI (Agoston et al., 2019), the cost and related restrictions increase dramatically when using larger animal models or nonhuman primates.

Another in vivo method includes medically imaging, observing human subjects postinjury, and postmortem findings. This has led to many discoveries related to TBI, but each injury is unique (i.e., not controlled) and observations are limited to only postinjury in most cases, meaning they lack a preinjury state for comparison. One exception to this are studies that include individuals participating in activities with a high-risk of TBI. In these cases, an individual's brain can be scanned prior to injury to create an individualized brain model (Garimella et al., 2019). By wearing sensors that detect potential injury events (e.g., impacts or blasts), these boundary conditions can be used to generate individualized simulations (Garimella et al., 2019). If necessary, the person in question could also be rescanned following injury to confirm whether injury has occurred and to what severity.

Ex vivo experiments include mechanical testing of tissues removed from the body. Because brain tissue has rate-dependent (Arbogast et al., 1995; Begonia et al., 2010; Prabhu et al., 2019) and anisotropic (Arbogast and Margulies, 1997) mechanical properties, both high- and low-rate and different stress state testing are needed. For low-rate testing, micromechanical devices are used for testing (Begonia et al., 2010; Prabhu et al., 2019). Note that the load cells with these devices must be matched with the material being tested, so a load cell of ~1 kg is often used with for brain tissue (Begonia et al., 2010; Prabhu et al., 2019). For high-rate testing, the split Hopkinson pressure bar is more appropriate (Prabhu et al., 2019; Pervin and Chen, 2009). Rather than measuring load directly with a load cell as typically used for low-rate devices, split Hopkinson pressure bar measure the change in pressure along the bars using strain gauges. Therefore, the bar material is important because an overly stiff bar will not be sensitive enough to identify deforming soft materials compared to changes in the bar. While traditional bars are made of steel for testing metals, softer bars with lower impedance, such as hollow aluminum (Pervin and Chen, 2009) and polycarbonate (Prabhu et al., 2019), are used for soft materials to better match impedance between bar and material. While these mechanical experiments provide a better understanding of tissue properties, testing is often destructive to the sample, especially during high-rate testing. Hence, they require interruption-testing (stopping a test mid-way) to investigate intermediate deformation states. Additionally, these tests do not provide any insight into ongoing injury effects as occurs in the body, such as biomarkers.

In vitro experiments differ from in vivo and ex vivo in that the samples are completely maintained outside the body, such as cell cultures (Geddes et al., 2003; Farkas et al., 2006; Kilinc et al., 2008). For example, in vitro cell culture experiments have shown that membrane mechanoporation results in membrane disruptions and increases membrane permeability postinjury (Geddes et al., 2003; Farkas et al., 2006; Kilinc et al., 2008), which can cause a loss of cell homeostasis. While many of these culture studies are performed on a single plane, some researchers have implemented 3D cell culture to better replicate their natural structure (LaPlaca and Thibault, 1997; Cullen et al., 2007, 2011). Many properties, such as physiological changes, mechanical properties, or fluorescence expression, can be examined using cultures and should be chosen as appropriate for each problem. However, the difference between local forces and macroscale forces as well as culture neuron heterogeneities, orientation, and processes in comparison to natural tissue must be considered when performing mechanical tests (Cullen and LaPlaca, 2006).

20 Chapter 1

One final branch of experiments is TBI equipment efficacy testing. These experiments differ from the others mentioned here in that they often do not include a biological component. Rather, they use a testing apparatus with a helmeted headform, such as a drop tower (NOC-SAE, 2019; Rush et al., 2017; Bliven et al., 2019). This type of experiment allows for equipment to be tested for minimum safety requirements and graded based on standardized rubrics.

1.3.5.2 Computational models (simulations)

While experiments help to identify potential injury locations of interest, they can be complemented by in silico computational methods. Researchers have developed computational models at different length scales to develop a more comprehensive picture of TBI and establish brain injury thresholds (Prabhu et al., 2011; Murphy et al., 2016, 2018; Alaei, 2017; Montanino, 2019; Rashid et al., 2013, 2014; Bakhtiarydavijani et al., 2019b, 2019a). Multiscale in silico models help to investigate potential injury mechanisms across length and time scales, which ultimately serve as a noninvasive predictor in primary responses to TBI (Montanino, 2019; Bakhtiarydavijani et al., 2019b, 2019a). Additionally, models have the benefit of no ongoing consumable costs as seen in many experiments and may require less personnel to run. However, initial model development and software licenses can be expensive upfront costs.

As shown in Fig. 1.1, each length scale has both information it needs and what information it can provide—similarly, computational models are limited in what they can include. Unlike real-world experiments, which inherently include information on lower-length scales, even though it may not be easily accessed, many computational models only include details for the scale being examined. In this case, any information needed from other length scales must be passed into the model through boundary conditions, material models, or some other method.

Take, for example, finite element analysis (FEA), which is often used to examine TBI. Because FEA is a continuum method, FEA simulations require constitutive material models to better replicate the brain's morphology for more accurate stress responses that general material models do not capture (Prabhu et al., 2011; Cloots et al., 2008; Yang et al., 2014; Colgan et al., 2010). Specifically, the constitutive material models better capture effects from lower length scale changes and related changes to the brain. Without these details, the model cannot capture history effects and dependencies on rate and stress state. Similarly, a constitutive model is required for injury effects to be implemented, which some studies have done through damage criteria (El Sayed et al., 2008; Prevost et al., 2011; McElhaney et al., 1973; Mihai et al., 2017; de Rooij and Kuhl, 2018; Weickenmeier et al., 2017). However, a true rubric based on changes to the lower length scales is needed.

Due to small length and time scales making it difficult to experimentally capture sufficient information for lower length scale injury effects, lower-length scale simulations must be considered. For example, molecular dynamics can be used to explore the mechanical properties of the cellular membrane structure and deformation-related mechanoporation damage (Murphy et al., 2016, 2018; Koshiyama and Wada, 2011; Shigematsu et al., 2014, 2015). In addition to providing a better understanding of cell membrane mechanoporation, results from the simulations can be used to estimate cell ion permeation (Bakhtiarydavijani et al., 2019a). Creating a more robust model of TBI with these underlying details will aid in determining cellular death and injury.

This brief description is intended to provide a general description of the multiscale modeling process. More detailed descriptions of methods used at each length scale will be covered in the chapters included herein.

1.4 Summary

The human brain is a complex multiscale system comprised of several elements, which are highly interconnected in a structural hierarchy and concurrently work at different levels of organization (Bastiani and Roebroeck, 2015; van den Heuvel and Yeo, 2017). From the macroscale brain cortical structures to the microscale cellular structures and the nanoscale cellular components, the brain's length scales span nine orders of magnitude, each with a distinct organization. Is it any wonder then that TBI is also complex in its mechanisms and effects? A result of direct impacts and accelerations to the head, TBI causes a combination of focal and diffuse injuries that lead to cell death in the brain and a variety of injury sequelae that can be debilitating when severe or chronic. Due to the severity and frequency of TBI, a variety of methods must be considered to study the mechanisms behind the injury. While real-world observations of TBI and experiments can provide many insights into injury, computer models can help elucidate TBI mechanisms that cannot be readily observed.

References

Aggleton, J.P., O'Mara, S.M., Vann, S.D., Wright, N.F., Tsanov, M., Erichsen, J.T., 2010. Hippocampal-anterior thalamic pathways for memory: uncovering a network of direct and indirect actions. Eur. J. Neurosci. 31, 2292–2307. https://doi.org/10.1111/j.1460-9568.2010.07251.x.

Agoston, D.V., Vink, R., Helmy, A., Risling, M., Nelson, D., Prins, M., 2019. How to translate time: the temporal aspects of rodent and human pathobiological processes in traumatic brain injury. J. Neurotrauma 36, 1724–1737. https://doi.org/10.1089/neu.2018.6261.

Alaei, Z., 2017. Molecular dynamics simulations of axonal membrane in traumatic brain injury. Master thesis, KTH Royal Institute of Technology, School of Technology and Health (STH), Stockholm, Sweden. https://kth. diva-portal.org/smash/get/diva2:1127564/FULLTEXT01.pdf.

Alberts, B., Johnson, A., Lewis, J., Morgan, D., Raff, M., Roberts, K., et al., 2014. Molecular Biology of the Cell. W. W. Norton & Company, New York, NY.

Andriessen, T., Jacobs, B., Vos, P.E., 2010. Clinical characteristics and pathophysiological mechanisms of focal and diffuse traumatic brain injury. J. Cell. Mol. Med. 14, 2381–2392. https://doi.org/10.1111/j.1582-4934.2010.01164.x.

Arbogast, K.B., Margulies, S.S., 1997. Regional Differences in Mechanical Properties of the Porcine Central Nervous System. 41st Stapp Car Crash Conference 106, 3807–3814. https://doi.org/10.4271/973336.

Arbogast, K.B., Meaney, D.F., Thibault, L.E., 1995. Biomechanical Characterization of the Constitutive Relationship for the Brainstem. SAE International. https://doi.org/10.4271/952716.

Aulinas, A., 2000. Physiology of the Pineal Gland and Melatonin. MDText.com, Inc, South Dartmouth, MA.

Azevedo, F.A.C., Carvalho, L.R.B., Grinberg, L.T., Farfel, J.M., Ferretti, R.E.L., Leite, R.E.P., et al., 2009. Equal numbers of neuronal and nonneuronal cells make the human brain an isometrically scaled-up primate brain. J. Comp. Neurol. 513, 532–541. https://doi.org/10.1002/cne.21974.

Bakhtiarydavijani, A., Murphy, M.A., Mun, S., Jones, M.D., Bammann, D.J., LaPlaca, M.C., et al., 2019a. Damage biomechanics for neuronal membrane mechanoporation. Model Simul. Mater. Sci. Eng. 27, 065004. https://doi.org/10.1088/1361-651X/ab1efe.

Bakhtiarydavijani, A.H., Murphy, M.A., Mun, S., Jones, M.D., Horstemeyer, M.F., Prabhu, R.K., 2019b. Multiscale modeling of the damage biomechanics of traumatic brain injury. Biophys. J. 116, 322a. https://doi.org/10.1016/j.bpj.2018.11.1748.

Bastiani, M., Roebroeck, A., 2015. Unraveling the multiscale structural organization and connectivity of the human brain: the role of diffusion MRI. Front. Neuroanat 9, 77. https://doi.org/10.3389/fnana.2015.00077.

Bayly, P.V., Black, E.E., Pedersen, R.C., Leister, E.P., Genin, G.M., 2006. In vivo imaging of rapid deformation and strain in an animal model of traumatic brain injury. J. Biomech. 39, 1086–1095. https://doi.org/10.1016/j.jbiomech.2005.02.014.

Begonia, M.T., Prabhu, R., Liao, J., Horstemeyer, M.F., Williams, L.N., 2010. The influence of strain rate dependency on the structure–property relations of porcine brain. Ann. Biomed. Eng. 38, 3043–3057. https://doi.org/10.1007/s10439-010-0072-9.

Begonia, M.T., Prabhu, R., Liao, J., Whittington, W.R., Claude, A., Willeford, B., et al., 2014. Quantitative analysis of brain microstructure following mild blunt and blast trauma. J. Biomech. 47, 3704–3711. https://doi.org/10.1016/j.jbiomech.2014.09.026.

Bigos, K., Hariri, A., Weinberger, D., 2016. Neuroimaging Genetics: Principles and Practices. Oxford University Press, Oxford, UK.

Bliven, E., Rouhier, A., Tsai, S., Willinger, R., Bourdet, N., Deck, C., et al., 2019. Evaluation of a novel bicycle helmet concept in oblique impact testing. Accid. Anal. Prev. 124, 58–65. https://doi.org/10.1016/j.aap.2018.12.017.

Boron, W.F., Boulpaep, E.L., 2016. Medical Physiology. Elsevier, Philadelphia, PA. https://doi.org/10.1016/s0033-3182(64)72477-2.

Caputo, A., Ghiringhelli, L., Dieci, M., Giobbio, G.M., Tenconi, F., Ferrari, L., et al., 1998. Epithalamus calcifications in schizophrenia. Eur. Arch. Psychiatry Clin. Neurosci. 248, 272–276. https://doi.org/10.1007/s004060050049.

Catani, M., Dell'Acqua, F., De Schotten, M.T., 2013. A revised limbic system model for memory, emotion and behaviour. Neurosci. Biobehav. Rev. 37 (8), 1724–1737. https://doi.org/10.1016/j.neubiorev.2013.07.001.

Cloots, R.J.H., Gervaise, H.M.T., van Dommelen, J.A.W., Geers, M.G.D., 2008. Biomechanics of traumatic brain injury: influences of the morphologic heterogeneities of the cerebral cortex. Ann. Biomed. Eng. 36, 1203–1215. https://doi.org/10.1007/s10439-008-9510-3.

Colgan, N.C., Gilchrist, M.D., Curran, K.M., 2010. Applying DTI white matter orientations to finite element head models to examine diffuse TBI under high rotational accelerations. Prog. Biophys. Mol. Biol. 103, 304–309. https://doi.org/10.1016/j.pbiomolbio.2010.09.008.

Cullen, D.K., LaPlaca, M.C., 2006. Neuronal response to high rate shear deformation depends on heterogeneity of the local strain field. J. Neurotrauma 23, 1304–1319. https://doi.org/10.1089/neu.2006.23.1304.

Cullen, D.K., Simon, C.M., LaPlaca, M.C., 2007. Strain rate-dependent induction of reactive astrogliosis and cell death in three-dimensional neuronal–astrocytic co-cultures. Brain Res. 1158, 103–115. https://doi.org/10.1016/j.brainres.2007.04.070.

Cullen, D.K., Vernekar, V.N., LaPlaca, M.C., 2011. Trauma-induced plasmalemma disruptions in three-dimensional neural cultures are dependent on strain modality and rate. J. Neurotrauma 28, 2219–2233. https://doi.org/10.1089/neu.2011.1841.

Davey, G., 2011. Applied Psychology. Wiley, Hoboken, NJ.

de Rooij, R., Kuhl, E., 2018. A physical multifield model predicts the development of volume and structure in the human brain. J. Mech. Phys. Solids 112, 563–576. https://doi.org/10.1016/j.jmps.2017.12.011.

Decimo, I., Fumagalli, G., Berton, V., Krampera, M., Bifari, F., 2012. Meninges: from protective membrane to stem cell niche. Am. J. Stem Cells 1, 92–105.

DePalma, R.G., Burris, D.G., Champion, H.R., Hodgson, M.J., 2005. Blast Injuries. N. Engl. J. Med. 352, 1335–1342. https://doi.org/10.1056/NEJMra042083.

El Sayed, T., Mota, A., Fraternali, F., Ortiz, M., 2008. A variational constitutive model for soft biological tissues. J. Biomech. 41, 1458–1466. https://doi.org/10.1016/j.jbiomech.2008.02.023.

Farkas, O., Lifshitz, J., Povlishock, J.T., 2006. Mechanoporation induced by diffuse traumatic brain injury: an irreversible or reversible response to injury? J. Neurosci. 26, 3130–3140. https://doi.org/10.1523/JNEUROSCI.5119-05.2006.

Farkas, O., Povlishock, J.T., 2007. Cellular and subcellular change evoked by diffuse traumatic brain injury: a complex web of change extending far beyond focal damage. In: Weber, J.T., Maas, A.I.R. (Eds.). Progress in Brain Research, 161. Elsevier, Amsterdam; Boston, pp. 43–59. https://doi.org/10.1016/S0079-6123(06)61004-2.

Fine, E.J., Ionita, C.C., Lohr, L., 2002. The history of the development of the cerebellar examination. Semin. Neurol. 22, 375–384. https://doi.org/10.1055/s-2002-36759.

Fletcher, D.A., Mullins, R.D., 2010. Cell mechanics and the cytoskeleton. Nature 463, 485–492. https://doi.org/10.1038/nature08908.

Freberg, L., 2009. Discovering Biological Psychology. Cengage Learning, Boston, MA.

Friedlander, R.M., 2003. Apoptosis and caspases in neurodegenerative diseases. N. Engl. J. Med. 348, 1365–1375. https://doi.org/10.1056/NEJMra022366.

Garimella, H.T., Menghani, R.R., Gerber, J.I., Sridhar, S., Kraft, R.H., 2019. Embedded finite elements for modeling axonal injury. Ann. Biomed. Eng. 47, 1889–1907. https://doi.org/10.1007/s10439-018-02166-0.

Gazzaniga, M.S., Ivry, R.B., Mangun, G.R., 2014. Cognitive Neuroscience - The Biology of The Mind. New York: W. W. Norton.

Geddes, D.M., Cargill 2nd, R.S., LaPlaca, M.C., 2003. Mechanical stretch to neurons results in a strain rate and magnitude-dependent increase in plasma membrane permeability. J. Neurotrauma 20, 1039–1049. https://doi.org/10.1089/089771503770195885.

Ginhoux, F., Lim, S., Hoeffel, G., Low, D., Huber, T., 2013. Origin and differentiation of microglia. Front. Cell Neurosci. 7, 1–14. https://doi.org/10.3389/fncel.2013.00045.

Giordano, C., Kleiven, S., 2014. Evaluation of axonal strain as a predictor for mild traumatic brain injuries using finite element modeling. Stapp Car Crash 58, 29–61.

Gray, H., Standring, S., 2015. Gray's Anatomy: The Anatomical Basis of Clinical Practice. Elsevier Health Sciences, New York, NY.

Guyton, A.C., Hall, J.E., 2011. Guyton and Hall Textbook of Medical Physiology, 12th ed. Saunders, Philadelphia, PA.

Haddad, S.H., Arabi, Y.M., 2012. Critical care management of severe traumatic brain injury in adults. Scand. J. Trauma Resusc. Emerg. Med. 20, 12. https://doi.org/10.1186/1757-7241-20-12.

Harayama, T., Riezman, H., 2018. Understanding the diversity of membrane lipid composition. Nat. Rev. Mol. Cell. Biol. 19, 281–296. https://doi.org/10.1038/nrm.2017.138.

Harvey, A.K., Thompson, M.S., Cochlin, L.E., Raju, P.A., Cui, Z., Cornell, H.R., et al., 2009. Functional imaging of tendon. Ann. BMVA 2009, 1–11.

Helmut, K., Hanisch, U.K., Noda, M., Verkhratsky, A., 2011. Physiology of microglia. Physiol. Rev. 91, 461–553. https://doi.org/10.1152/physrev.00011.2010.

Herculano-Houzel, S., 2014. The glia/neuron ratio: how it varies uniformly across brain structures and species and what that means for brain physiology and evolution. Glia 62, 1377–1391. https://doi.org/10.1002/glia.22683.

Higgins, V., 2006. Human physiology: the basis of medicine. Oxford University Press, Oxford, UK.

Humphreys, I., Wood, R.L., Phillips, C.J., Macey, S., 2013. The costs of traumatic brain injury: a literature review. Clinicoecon. Outcomes Res. 5, 281–287. https://doi.org/10.2147/CEOR.S44625.

Johnson, V.E., Stewart, W., Smith, D.H., 2013. Axonal pathology in traumatic brain injury. Exp. Neurol. 246, 35–43. https://doi.org/10.1016/j.expneurol.2012.01.013.

24   Chapter 1

Kandel, E., Schwartz, J., Jessell, T., 2000. Principles of neural science. McGraw-hill, New York.

Kato, K., Fujimura, M., Nakagawa, A., Saito, A., Ohki, T., Takayama, K., et al., 2007. Pressure-dependent effect of shock waves on rat brain: induction of neuronal apoptosis mediated by a caspase-dependent pathway. J. Neurosurg. 106, 667–676. https://doi.org/10.3171/jns.2007.106.4.667.

Kekere, V., Alsayouri, K., 2019. Anatomy, Head and Neck, Dura Mater. StatPearls, Treasure Island, FL.

Kilinc, D., Gallo, G., Barbee, K.A., 2008. Mechanically-induced membrane poration causes axonal beading and localized cytoskeletal damage. Exp. Neurol. 212, 422–430. https://doi.org/10.1016/j.expneurol.2008.04.025.

Kolb, B., Whishaw, I.Q., 2009. Fundamentals of Human Neuropsychology. Macmillan, New York, NY.

Kole, M.H.P., Ilschner, S.U., Kampa, B.M., Williams, S.R., Ruben, P.C., Stuart, G.J., 2008. Action potential generation requires a high sodium channel density in the axon initial segment. Nature 11, 178–186. https://doi.org/10.1038/nn2040.

Koshiyama, K., Wada, S., 2011. Molecular dynamics simulations of pore formation dynamics during the rupture process of a phospholipid bilayer caused by high-speed equibiaxial stretching. J. Biomech. 44, 2053–2058. https://doi.org/10.1016/j.jbiomech.2011.05.014.

LaPlaca, M., Thibault, L., 1997. An in vitro traumatic injury model to examine the response of neurons to a hydrodynamically-induced deformation. Ann. Biomed. Eng. 25, 665–677. https://doi.org/10.1007/BF02684844.

Leterrier, C., Dubey, P., Roy, S., 2017. The nano-architecture of the axonal cytoskeleton. Nat. Rev. Neurosci. 18, 713–726. https://doi.org/10.1038/nrn.2017.129.

Lowrey, P.L., Takahashi, J.S., 2000. Genetics of the mammalian circadian system: photic entrainment, circadian pacemaker mechanisms, and posttranslational regulation. Annu. Rev. Genet. 34, 533–562. https://doi.org/10.1146/annurev.genet.34.1.533.

Maiden, N., 2009. Ballistics reviews: mechanisms of bullet wound trauma. Forensic Sci. Med. Pathol. 5, 204–209. https://doi.org/10.1007/s12024-009-9096-6.

McAllister, T.W., 2011. Neurobiological consequenses of traumatic brain injury. Dialog. Clin. Neurosci. 13, 287–300.

McElhaney, J.H., Melvin, J.W., Roberts, V.L., Portnoy, H.D., 1973. Dynamic Characteristics of the Tissue of the Head. Palgrave Macmillan, London, 215–222.

Menon, D.K., Schwab, K., Wright, D.W., Maas, A.I., 2010. Position statement: definition of traumatic brain injury. Arch. Phys. Med. Rehabil. 91, 1637–1640.

Meythaler, J.M., Peduzzi, J.D., Eleftheriou, E., Novack, T.A., 2001. Current concepts: diffuse axonal injury-associated traumatic brain injury. Arch. Phys. Med. Rehabil. 82, 1461–1471.

Mihai, L.A., Budday, S., Holzapfel, G.A., Kuhl, E., Goriely, A., 2017. A family of hyperelastic models for human brain tissue. J. Mech. Phys. Solids 106, 60–79. https://doi.org/10.1016/j.jmps.2017.05.015.

Montanino, A., 2019. Definition of Axonal Injury Tolerances Across Scales. PhD Thesis, KTH Royal Institute of Technology, Stockholm, Sweden. http://kth.diva-portal.org/smash/get/diva2:1387578/FULLTEXT01.pdf.

Moss, W.C., King, M.J., Blackman, E.G., 2009. Skull flexure from blast waves: a mechanism for brain injury with implications for helmet design. Phys. Rev. Lett. 103, 108702.

Murphy, M.A., Horstemeyer, M.F.F., Gwaltney, S.R., Stone, T., Laplaca, M.C., Liao, J., et al., 2016. Nanomechanics of phospholipid bilayer failure under strip biaxial stretching using molecular dynamics. Model Simul. Mater. Sci. Eng. 24, 055008. https://doi.org/10.1088/0965-0393/24/5/055008.

Murphy, M.A., Mun, S., Horstemeyer, M.F., Baskes, M.I., Bakhtiary, A., LaPlaca, M.C., et al., 2018. Molecular dynamics simulations showing 1-palmitoyl-2-oleoyl-phosphatidylcholine (POPC) membrane mechanoporation damage under different strain paths. J. Biomol. Struct. Dyn. 37, 1–14. https://doi.org/10.1080/07391102.2018.1453376.

NOCSAE. Standard Performance Specification For Newly Manufactured Football Helmets NOCSAE DOC (ND)002-17m19. National Operating Committee on Standards For Athletic Equipment. 2019. https://doi.org/NOCSAEDOC(ND)002-17m19.

Oberheim, N.A., Goldman, S.A., Nedergaard, M., 2012. Heterogeneity of astrocytic form and function. Methods Mol. Biol. 814, 23–45. https://doi.org/10.1007/978-1-61779-452-0_3.

Patel, N., Kirmi, O., 2009. Anatomy and imaging of the normal meninges. Semin. Ultrasound CT MRI 30, 559–564. https://doi.org/10.1053/j.sult.2009.08.006.

Patestas, M., Gartner, L., 2016. A Textbook of Neuroanatomy. Wiley, Hoboken, NJ.

Pervin, F., Chen, W.W., 2009. Dynamic mechanical response of bovine gray matter and white matter brain tissues under compression. J. Biomech. 42, 731–735. https://doi.org/10.1016/j.jbiomech.2009.01.023.

Pfister, B.J., Weihs, T.P., Betenbaugh, M., Bao, G., 2003. An in vitro uniaxial stretch model for axonal injury. Ann. Biomed. Eng. 31, 589–598. https://doi.org/10.1114/1.1566445.

Prabhu, R., Horstemeyer, M.F., Tucker, M.T., Marin, E.B., Bouvard, J.L., Sherburn, J.A., et al., 2011. Coupled experiment/finite element analysis on the mechanical response of porcine brain under high strain rates. J. Mech. Behav. Biomed. Mater. 4, 1067–1080. https://doi.org/10.1016/j.jmbbm.2011.03.015.

Prabhu, R.K., Begonia, M.T., Whittington, W.R., Murphy, M.A., Mao, Y., Liao, J., et al., 2019. Compressive mechanical properties of porcine brain: experimentation and modeling of the tissue hydration effects. Bioengineering 6, 40. https://doi.org/10.3390/bioengineering6020040.

Prevost, T.P., Balakrishnan, A., Suresh, S., Socrate, S., 2011. Biomechanics of brain tissue. Acta Biomater. 7, 83–95. https://doi.org/10.1016/j.actbio.2010.06.035.

Purves, D., 2012. Neuroscience, fifth ed. Sinauer Associates Inc, Sunderland, MA.

Raghupathi, R., 2004. Cell death mechanisms following traumatic brain injury. Brain Pathol. 14, 215–222. https://doi.org/10.1111/j.1750-3639.2004.tb00056.x.

Rashid, B., Destrade, M., Gilchrist, M.D., 2013. Mechanical characterization of brain tissue in simple shear at dynamic strain rates. J. Mech. Behav. Biomed. Mater. 28, 71–85. https://doi.org/10.1016/j.jmbbm.2013.07.017.

Rashid, B., Destrade, M., Gilchrist, M.D., 2014. Mechanical characterization of brain tissue in tension at dynamic strain rates. Forensic Biomech. 33, 43–54. https://doi.org/10.1016/j.jmbbm.2012.07.015.

Riggio, S., Wong, M., 2009. Neurobehavioral sequelae of traumatic brain injury. Mt Sinai J. Med. A J. Transl. Pers. Med. 76, 163–172. https://doi.org/10.1002/msj.20097.

Rush, G.A., Rush, G.A., Sbravati, N., Prabhu, R., Williams, L.N., DuBien, J.L., et al., 2017. Comparison of shell-facemask responses in American football helmets during NOCSAE drop tests. Sport Eng. 20, 199–211. https://doi.org/10.1007/s12283-017-0233-2.

Saab, A.S., Nave, K.A., 2017. Myelin dynamics: protecting and shaping neuronal functions. Curr. Opin. Neurobiol. 47, 104–112. https://doi.org/10.1016/j.conb.2017.09.013.

Sabet, A.A., Christoforou, E., Zatlin, B., Genin, G.M., Bayly, P.V., 2008. Deformation of the human brain induced by mild angular head acceleration. J. Biomech. 41, 307–315. https://doi.org/10.1016/j.jbiomech.2007.09.016.

Sampaio-Baptista, C., Johansen-Berg, H., 2017. White matter plasticity in the adult brain. Neuron 96, 1239–1251. https://doi.org/10.1016/j.neuron.2017.11.026.

Sherman, S., 2006. Thalamus. Scholarpedia 1, 1583. https://doi.org/10.4249/scholarpedia.1583.

Sherman, S.M., Guillery, R.W., 2009. Exploring the Thalamus and Its Role in Cortical Function. The MIT Press, Cambridge, MA.

Shigematsu, T., Koshiyama, K., Wada, S., 2014. Molecular dynamics simulations of pore formation in stretched phospholipid/cholesterol bilayers. Chem. Phys. Lipids 183, 43–49. https://doi.org/10.1016/j.chemphyslip.2014.05.005.

Shigematsu, T., Koshiyama, K., Wada, S., 2015. Effects of stretching speed on mechanical rupture of phospholipid/cholesterol bilayers: molecular dynamics simulation. Sci. Rep. 5, 15369. https://doi.org/10.1038/srep15369.

Shukla, D., Devi, B.I., 2010. Mild traumatic brain injuries in adults. J. Neurosci. Rural Prac.t 01, 082–088. https://doi.org/10.4103/0976-3147.71723.

Shultz, S.R., McDonald, S.J., Vonder Haar, C., Meconi, A., Vink, R., van Donkelaar, P., et al., 2017. The potential for animal models to provide insight into mild traumatic brain injury: translational challenges and strategies. Neurosci. Biobehav. Rev. 76, 396–414. https://doi.org/10.1016/j.neubiorev.2016.09.014.

Snell, R.S., 2010. Clinical Neuroanatomy. Lippincott Williams & Wilkins, Philadelphia, PA.

Squire, L.R., Berg, D., Bloom, F.E., Du Lac, S., Ghosh, A., Spitzer, N.C., 2013. Fundamental Neuroscience. Elsevier, Cambridge, MA. https://doi.org/10.1016/C2010-0-65035-8.

Stocco, A., Lebiere, C., Anderson, J.R., 2010. Conditional routing of information to the cortex: a model of the basal ganglia's role in cognitive coordination. Psychol. Rev. 117, 541–574. https://doi.org/10.1037/a0019077.

Taylor, C.A., Bell, J.M., Breiding, M.J., Xu, L., 2017. Traumatic brain injury-related emergency department visits, hospitalizations, and deaths -- United States, 2007 and 2013. MMWR Surveill. Summ. 66, 1–16. https://doi.org/10.15585/mmwr.ss6609a1.

Teeter, C.M., Stevens, C.F., 2011. A general principle of neural arbor branch density. Curr. Biol. 21, 2105–2108. https://doi.org/10.1016/j.cub.2011.11.013.

van den Heuvel, M.P., Yeo, B.T.T., 2017. A spotlight on bridging microscale and macroscale human brain architecture. Neuron 93, 1248–1251. https://doi.org/10.1016/j.neuron.2017.02.048.

Weickenmeier, J., Saez, P., Butler, C.A.M., Young, P.G., Goriely, A., Kuhl, E., 2017. Bulging brains. J. Elast. 129, 197–212. https://doi.org/10.1007/s10659-016-9606-1.

Whalen, M.J., Yager, P., Lo, E.H., Lok, J., Noviski, N., 2009. XXXXX. In: Wheeler, D.S., Wong, H.R., Shanley, T.P. (Eds.), Molecular Biology of Brain Injury: The Central Nervous System in Pediatric Critical Illness and Injury. Springer, London, pp. 1–12. https://doi.org/10.1007/978-1-84800-993-6_2.

Wolf, S.J., Bebarta, V.S., Bonnett, C.J., Pons, P.T., Cantrill, S.V., 2009. Blast injuries. Lancet 374, 405–415. https://doi.org/10.1016/s0140-6736(09)60257-9.

Wong, J., Hoe, N., Zhiwei, F., Ng, I., 2005. Apoptosis and traumatic brain injury. Neurocrit. Care 3, 177–182. https://doi.org/10.1385/ncc:3:2:177.

Yang, B., Tse, K.-.M., Chen, N., Tan, L.-.B., Zheng, Q.-.Q., Yang, H.-.M., et al., 2014. Development of a finite element head model for the study of impact head injury. Biomed. Res. Int. 2014, 1–14. https://doi.org/10.1155/2014/408278.

CHAPTER 2

Introduction to multiscale modeling of the human brain

Raj K. Prabhu[a], Mark F. Horstemeyer[b]

[a]USRA, NASA HRP CCMP, NASA Glenn Research Center, Cleveland, OH, United States [b]School of Engineering, Liberty University, Lynchburg, VA, United States

2.1 Introduction

Now that we have described the mechanical multiscale structures of the human brain in Chapter 1, here we discuss a brief history of the material modeling and simulation methods that capture the human brains behavior under mechanical impacts. Motivated from the 1.7 million traumatic brain injury (TBI) cases that occur in the United States every year (Faul et al., 2010) arising from mechanical impacts, updated tools to comprehend the cause-effect relationships are warranted. However, due to the multiscale and complex nature of TBI, it is currently not well understood. One modern tool for examining TBI employs the use of a simulation method called finite element analysis (FEA). Although FEA started in the early 1960s when computers were first developed, the application of FEA to the brain did not occur until Shugar and Katona (1974) presented macroscale continuum model of the brain in a skull at a conference. Shortly afterward, Ward and Thompson (1975) published probably the first FEA of a macroscale continuum brain model related to vibrations. From these simple FEA simulations, brain modeling simulations have moved to much greater details. Recently, FEA has been used to for design optimization in an effort to mitigate TBI as per Chatelin et al. (2011), Johnson et al. (2016), and Giordano et al. (2017), which all provided a better understanding of the brain's local deformation behavior and helped produce safety criteria for human-centric designs. Major benefits of using FEA for TBI analysis are the following: (1) physical humans do not have to be used or injured to garner information about the level of TBI, (2) analysts and designers can make decisions for mitigating TBI regarding helmets, (3) physical experiments of human cadavers cannot give a comprehensive state of strain or stress within the brain, and (4) human cadavers cannot give "live" tissue responses to a mechanical impact.

2.2 Constitutive modeling of the brain

A recent brief review by Horstemeyer et al. (2019) summarized the current state of brain modeling and simulation from a macroscale continuum perspective as they led a group of

Multiscale Biomechanical Modeling of the Brain.
DOI: https://doi.org/10.1016/B978-0-12-818144-7.00012-8
Copyright © 2022 Elsevier Inc. All rights reserved.

28   Chapter 2

brain modelers produce their works in a special issue of the *Ann. Biomed. Eng*. One major conclusion of the researchers' work was the need for an accurate constitutive model for the brain. A constitutive model is a continuum mathematical model that represents the constitution of the material as exhibited by its mechanical properties. Some researchers loosely call this a material model or a rheological model. Constitutive models (Madhukar and Ostoja-Starzewski, 2019) are typically defined as a function of observable state variables such as time and strain, while other models also include internal state variables (Prabhu et al., 2011) defined per the models needed to capture viscoelasticity, hyperelasticity, viscoplasticity, the Mullins effect, material hysteresis, and damage in brain tissue. However, these mechanical properties arise because of the multiscale hierarchical structures that interplay with each other as described in Chapter 1. The difficulty is to quantify the structure–property relationships at each length scale and then determine the cause-effect relationships than can be upscaled into the continuum framework.

Initial mechanical properties studies of brain tissue focused on continuum viscoelasticity associated with the time rate of change of the elastic modulus. Assuming the brain as a fluid, a Maxwell–Kelvin model was first used to describe the viscoelastic brain behavior under creep and relaxation in compression tests (Galford and McElhaney, 1970). Shuck and Advani (1972) later analytically solved the equation of motion for cylindrical brain samples under torsion to calculate the viscoelastic storage and loss moduli.

Later constitutive models were used to describe the viscoelastic and hyperelastic properties of brain tissue. The hyperelasticity of nonlinear behavior of brain tissue or soft tissue can be calculated from a derivation of a strain energy function (SEF). A list of the associated energy-based constitutive models is presented in Table 2.1. These SEFs were mostly phenomenological equations based on a mathematical formalism that used empirical data. These SEF models lack microstructural details, include no history effects, and have not been too predictive. Most of the SEFs were first developed for polymers and rubber materials and later extended to soft tissue to explain their hyperelastic properties. The first of these models dates back to the generalized Mooney–Rivlin model developed by Mooney (1940) and the neo-Hookean model developed by Rivlin (1948),

$$\psi\left(J_1,J_2\right) = \sum_{i+j=1}^{N} C_{ij}\left(J_1 - 3\right)^i \cdot \left(J_2 - 3\right)^j. \tag{2.1}$$

Mendis et al. (1995) used a first-order Mooney–Rivlin SEF with viscoelasticity implemented with a linear (first-order) decay function acting on the strain energy. Their model was calibrated to quasi-static uniaxial compression tests of brain tissue. Donnelly and Medige (1997) used a Kelvin–Voigt model to define the nonlinear viscoelastic model with data coming from quasi-static shear tests on human brain tissue. Arbogast and Margulies (1999) also looked at the viscoelastic properties of the brain stem using a fiber-reinforced composite model. Miller and Chinzei (1997) then used a second-order Mooney–Rivlin SEF to define the hyperelastic

Table 2.1: A list of experimental tests carried out on brain tissue.

Author	Tissue sample	Test type	Strain	Strain rate
Fallenstein et al. (1969)	Human brain tissue	Shear/compression	0.37	9–10 Hz
Metz et al. (1970); Estes and McElhaney (1970)	Human brain tissue	Compression		
Miller and Chinzei (1997)	Porcine brain tissue	Compression		
Arbogast et al. (1997)	Sylgard gel mixture (similar to soft tissue)	Shear	Up to 20%	20–200 Hz
Darvish and Crandall (2001)	Bovine brain tissue	Shear	0.15	0.5–2000
Brands et al. (2004)	Porcine brain tissue	Shear		
Hrapko et al. (2006)	Porcine brain tissue	Shear		
Francheschini et al. (2006)	Human brain tissue	Tension/compression		
Prabhu et al. (2011)	Porcine brain tissue	High strain rate compression		
Prevost et al. (2011)	Swine brain tissue	Compression		
Prevost et al. (2011)	Swine brain tissue	Indentation		
Javid et al. (2014)	Porcine brain	Tension (stress relaxation)		
Lee et al. (2014)	Rat brain tissue	Indentation (measuring viscoelasticity		

behavior of the brain in quasi-static compression, and later calibrated their material model to indentation tests performed in vivo swine tissue for comparison with surgical procedures. Bilston et al. (2001) proposed a viscoelastic, hyperelastic, and strain rate dependent material model for large strains that was calibrated to oscillatory shear and required 28 constants. Their model allowed for softening of tissue through mathematically capturing the stress strain behavior. Later, Miller and Chinzei (2002) replaced the hyperelastic Mooney–Rivlin SEF with an Ogden (1972) SEF to properly capture the tension and compression behavior with one SEF. Prange and Margulies (2002) looked at regional, directional, and age dependent properties of brain, where they used first order Ogden (1972) with viscoelasticity calibrated to shear tests in different regions of the white matter for different directions. Age dependence of porcine brain properties was also evaluated. It was found that adult human tissue is slightly stiffer than porcine and that directional anisotropy exists in the corona radiata and corpus callosum, but the directional anisotropy does not change significantly with time. They used five time points. They also use isochrone measurements to show that shear modulus relaxation is independent of stress for their viscoelastic modulus. Shear tests on brain sections with oriented axon orientation to capture shear modulus differences.

Meaney (2003) produced a hyperelastic viscoelastic model of optic tracts that included the hyperelastic properties and accounted for the undulations of axons in the white matter tract with anisotropy arising from the fiber direction. This was accomplished by using a SEF that included I_4 and I_5. Brands et al. (2004) expanded this work through multiplicatively decomposing the deformation gradient into elastic and plastic parts, and capturing viscoelasticity by implementing plastic flow. The validation of their model overestimated the shear experimental results by 31%. Bischoff et al. (2004) and Velardi et al. (2006) performed tension and compression tests on white matter to capture anisotropy in their SEF. This was done by adding a function of the fourth invariant with axonal stiffening to the SEF. Hrapko et al. (2006) presented a pseudo-elastic viscoelastic model for the brain where the stress was additively decomposed into the volumetric, viscoelastic, and deviatoric part. The deformation gradient was multiplicatively decomposed into the elastic and plastic parts, where plastic flow, or viscoplasticity, was defined as a function of the deviatoric stress and the viscosity parameter that allows for capturing finite strains.Franceschini et al. (2006) attempted to capture permanent deformations and the Mullins effect of human brain tissue using a pseudo-elastic SEFs developed for polymers by Ogden and Roxburgh (1999) and Dorfmann and Ogden (2004). The first captured Mullins effect but not permanent deformation, while the second model captured permanent deformation, but could not capture cyclic loading. El Sayed et al. (2008a, 2008b) presented their own SEF to be the sum of an elastic, viscoelastic, and a plastic part. The plastic SEF is broken up into a deviatoric (function of shear strain) and a volumetric (function of volumetric strain) parts. The deviatoric is defined with a power law, and the volumetric plasticity defined as a function of void volume fraction in the body. Strain rate sensitivity is also considered through using strains as internal variables of the brain. Chafi (2009) then produced a geometrically detailed brain model combined with a simpler hyperviscoelastic model to look at the effects of blast trauma. Prabhu et al. (2011) used an internal state variable model presented by Bouvard et al. (2010) to capture the high rate behavior of brain under high rate compression tests where three internal state variables describe the entanglement of fibers, their hardening, and their alignment. Prevost et al. (2011) combined an elastic eight chain polymer model with a short-term nonlinear and long-term linear viscosity to present their strain rate dependent model that captured cyclic softening, hyperelasticity, and viscoelasticity fairly well. Laksari et al. (2012) developed their quasi-linear viscoelastic model for compression strains up to 35% with stress–strain being measured 0.06 seconds after loading to remove inertia effects. The hyperelastic and viscoelastic properties of porcine brain were also measured in compression (Rashid et al., 2012), shear (Rashid et al., 2013), and tension (Rashid et al., 2014) for a range of strain rates and fitted with Fung, Gent, and Ogden SEFs. Haldar and Pal (2018) used internal state variables to introduce strain rate dependence

Some models have been proposed to capture the structure–property relationships related to constitutive models. While tracking axonal paths using diffusion tensor imaging (DTI) date back to as early as 1996 (Pierpaoli and Basser, 1996; Basser et al., 2000), anisotropic mechanical properties of the white matter dependent on the axonal orientation was not presented until much later. Colgan et al. (2010) defined their SEF as that of the matrix, defined by a neo-Hookean model, and the fiber, defined by the axonal bundle behavior. DTI coupled with the magnetic resonance imaging was then used to describe the strengthening effects of axonal bundles to implement axonal orientations into their SEF. Chatelin et al. (2011) took a different approach and mapped fiber orientations of 12 human patients using DTI to create a FE mesh of a human brain to measure the axonal stretch resulting from such impacts. After that, Cloots et al. (2013) also used the Colgan et al. (2010) model when looking at mesoscale brain structures. Sahoo et al. (2014) also introduced DTI and fractional anisotropy into their brain material model to look at axonal tensile elongation. Giordano et al. (2014) further developed the Cloots et al. (2013) model by implementing it on the whole head to compare simulation results for isotropic and anisotropic models and found the anisotropic model can produce increased strains. Giordano et al. (2017) further expanded the model by adding two strains as internal variables: $\{\mathbb{Z}\} = \{C_{v\alpha}, C_{v\beta}^{f}\}$. These were the viscoelastic part of the deformation gradient and the viscoelastic part of the fiber deformation tensor, that are Cauchy–Green type strain tensors, as internal variables so that the history of the model can be captured. Haldar and Pal (2018) used viscoelastic deformation Cauchy strains as internal state variables to produce strain rate dependence. Ganpule et al. (2018) used an anisotropic Holzapfel–Gasser–Ogden SEF with internal strain variables in the viscoelastic strain energy to capture strain rate dependence while looking at the effect of varying the bulk modulus.

The hyperelastic part of material models used in brain tissue has also been studied in comparison to one another. Mihai et al. (2015) looked at various hyperelasticity models and found that the Ogden (1972) hyperelastic model gives the best results for finite element computations by addressing three unusual mechanical properties of brain tissue, that is, sharp increase in shear modulus as compression in the orthogonal direction to the shear direction increases, the shear modulus remaining constant or decreasing as tension increases in the orthogonal direction to shear, the elastic modulus increases or remains constant when compression increases. Madireddy et al. (2015) focused on using a Bayesian approach to compare the first-order Mooney–Rivlin model, exponential model and the Ogden model ($N = 1, 2, 3$) for tensile testing of porcine brain and found none of the models satisfied their criteria for a parsimonious model that were the qualitative fit of the model to the experimental data, evidence values, maximum likelihood values, and landscape of the likelihood function. Budday et al. (2017) also look at the goodness of fit of hyperelastic models to shear, compression, and tension properties of human brain and find that the modified first-order Ogden method produces the best fit compared to the neo-Hookean, Mooney–Rivlin, Demiray, and Gent model. However, the range of mechanical properties is extremely high between different brain samples and may not be the best data considering the age of the samples (55–81 years for 10 subjects). Mihai et al. (2017) proposed a SEF to capture tension, compression, or shear by combining three modified Ogden SEFs. That would allow the effect of tensile and compression before secondary deformation to be captured in a model that is not history dependent.

Through the last few decades, advances in mechanical testing of soft tissue have provided more insight into mechanical properties of brain tissue. This has paved the way for more accurate models to be developed. These models can now capture the viscoelastic,

32 Chapter 2

hyperelastic, Mullins effect, hysteresis, and anisotropy of white matter to be captured. Some aspects of mechanical properties such as strain rate dependence and damage still require further study. From another standpoint, mechanical properties of brain tissue have been seen to vary spatially and person-to-person. Use of microstructurally based brain constitutive models can help address such variations. Specifically, internal state variables can be used introduced spatially dependent properties that characterize the mechanical properties through capturing the microstructure effects such as neuronal distribution and concentration. Such an approach not only captures the variations in mechanical properties, but also helps in identifying location dependent neuronal cell death.

2.3 Brain tissue experiments used for constitutive modeling calibration

Various tests are used to define the mechanical properties of the brain. Over the past decades, several research groups investigated the mechanical properties of brain tissue, which has given us a range of different loading conditions to establish constitutive relationships. In general, four tests have been used to determine the mechanical properties of the brain, which are tensile tests, compression tests, shear tests, and indentation tests.

Mechanical property testing has included shear, compression, and tension of brain tissue. Mostly dynamic oscillatory shear tests were conducted (Arbogast et al., 1997; Bilston et al., 2001; Brands et al., 2004; Darvish and Crandall, 2001; Fallenstein et al., 1969; Hrapko et al., 2006; Nicolle et al., 2004; Nicolle et al., 2005; Prange and Margulies, 2002; Shuck and Advani, 1972; Thibault and Margulies, 1998) and unconfined compression tests (Cheng and Bilston, 2007; Estes and McElhaney, 1970; Gilchrist, 2004; Miller and Chinzei, 1997; Pervin and Chen, 2009; Prange and Margulies, 2002; Rashid et al., 2012; Tamura et al., 2007). Only a limited number of tensile tests were conducted (Miller and Chinzei, 2002; Tamura et al., 2008; Velardi et al., 2006). Experimental data for brain tissue in tension at dynamic strain rates are available from Tamura et al. (2008) who performed tests at 0.9, 4.3, and 25/s and Rashid et al. (2014) who conducted tests at 30, 60, and 90/s strain rates.

Considering the difficulty of obtaining human brain tissue for in vitro testing, experiments are usually performed on animal brain samples (monkey, porcine, bovine, rabbit, calf, rat, or mouse). Galford and McElhaney (1970) showed that shear, storage, and loss moduli are 1.5, 1.4, and 2 times greater, respectively, for monkeys than for humans. Similarly, Estes and McElhaney (1970) performed tests on human and Rhesus monkey tissue and found that the response of the Rhesus monkey tissue was slightly greater than the response of human brain tissue at comparable compression rates. Differences between human and porcine brain properties were also pointed out by Prange et al. (2000), who demonstrated that human brain tissue stiffness was 1.3 times greater than that of porcine brain. However, Nicolle et al. (2004) observed no significant difference between the mechanical properties of human and porcine brain matter. Pervin and Chen (2009) found no difference between the in vitro dynamic mechanical response of brain matter in different animals (porcine, bovine, and caprine), different breeds, and different genders. Because of the similarities of porcine and human brain tissue, it is convenient to use porcine brain tissue for material characterization and to use these material parameters in human finite element head models.

2.4 Modeling summary of upcoming chapters in the book

Now that we have described the different length scale heterogeneities in the brain in Chapter 1 and given a brief summary of the constitutive modeling and associated experiments in Chapter 2, the reader now has a frame of reference for the different modeling and simulation examples related to the different length scales associated with mechanical impacts. From Horstemeyer's (2012), (2018), integrated computational materials engineering (ICME) perspective of downscaling first and then upscaling the results to solve the structural scale boundary value problem, we first describe the downscaling from Fig. 2.1.

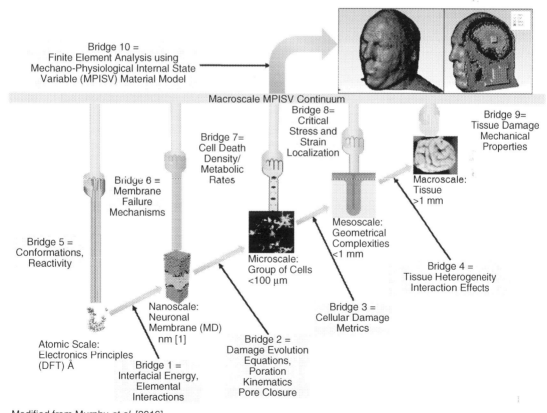

Modified from Murphy et al. [2016]

Fig. 2.1: Multiscale schematic of the different bridges of information required to garner a comprehensive constitutive model at the continuum macroscale in order to use the finite element analysis for the structural scale simulations.

34 Chapter 2

Because in the end we are focused on developing a macroscale mechano-physiological internal state variable (MPISV) constitutive model that can be used in FEA and still be able to incorporate different structures from the subscales, downscaling starts at the macroscale level. Bridges 5–9 in Fig. 2.1 describe the different effects that the subscale simulations need to provide for the macroscale MPISV set of equations. Also, each length below the macroscale also needs information from the subscale simulations. Bridges 1–4 in Fig. 2.1 define the pertinent effects between the lower length scales that need to be upscaled. Now that the downscaling has defined the pertinent upscaling information, the simulations starting at the lowest length need to be conducted to pass on the effects the next higher length scale and to the macroscale MPISV equations. Each of the following chapters discusses the different length scale simulations and what information is passed from there to the next higher length scale or to the macroscale MPISV model.

2.5 Summary

We provide a review of the different length scales for the modeling and simulations, the major constitutive models at the macroscale, and the associated experiments that can help calibrate the macroscale models. As the latter chapters show, once the macroscale models are calibrated, then they can be used for different simulations related to impacting the brain.

References

Arbogast, K.B., Margulies, S.S., 1999. A fiber-reinforced composite model of the viscoelastic behavior of the brainstem in shear. J. Biomech. 32 (8), 865–870.

Arbogast, K.B., Thibault, K.L., Pinheiro, B.S., Winey, K.I., Margulies, S.S., 1997. A high-frequency shear device for testing soft biological tissues. J. Biomech. 30 (7), 757–759.

Basser, P.J., Pajevic, S., Pierpaoli, C., Duda, J., Aldroubi, A., 2000. In vivo fiber tractography using DT-MRI data. Magn. Reson. Med. 44, 625–632.

Bilston, L.E., Liu, Z., Phan-Thien, N., 2001. Large strain behaviour of brain tissue in shear: some experimental data and differential constitutive model. Biorheology 38, 335–345.

Bischoff, J.E., Arruda, E.M., Grosh, K., 2004. A rheological network model for the continuum anisotropic and viscoelastic behavior of soft tissue. Biomech. Model. Mechanobiol. 3, 56–65. https://doi.org/10.1007/s10237-004-0049-4.

Bouvard, J.L., Ward, D.K., Hossain, D., Marin, E.B., Bammann, D.J., Horstemeyer, M.F., 2010. A general inelastic internal state variable model for amorphous glassy polymers. Acta Mech. 213, 71–96. https://doi.org/10.1007/s00707-010-0349-y.

Brands, D.W.A., Peters, G.W.M., Bovendeerd, P.H.M., 2004. Design and numerical implementation of a 3-D non-linear viscoelastic constitutive model for brain tissue during impact. J. Biomech. 37, 127–134. https://doi.org/10.1016/S0021-9290(03)00243-4.

Budday, S., Sommer, G., Holzapfel, G.A.A., Steinmann, P., Kuhl, E., 2017. Viscoelastic parameter identification of human brain tissue. J. Mech. Behav. Biomed. Mater. 74, 463–476.

Chafi, M.S., Karami, G., Ziejewski, M., 2009. Biomechanical assessment of brain dynamic responses due to blast pressure waves. Ann. Biomed. Eng. 38, 490–504. https://doi.org/10.1007/s10439-009-9813-z.

Chatelin, S., Deck, C., Renard, F., Kremer, S., Heinrich, C., Armspach, J.-P., Willinger, R., 2011. Computation of axonal elongation in head trauma finite element simulation. J. Mech. Behav. Biomed. Mater. 4, 1905–1919. https://doi.org/10.1016/j.jmbbm.2011.06.007.

Cheng, S., Bilston, L.E., 2007. Unconfined compression of white matter. J. Biomech. 40 (1), 117–124.

Cloots, R.J.H., Van Dommelen, J.A.W., Kleiven, S., Geers, M.G.D., Kleiven, S., Geers, M.G.D., Kleiven, S., 2013. Multi-scale mechanics of traumatic brain injury: predicting axonal strains from head loads. Biomech. Model. Mechanobiol. 12, 137–150. https://doi.org/10.1007/s10237-012-0387-6.

Colgan, N.C., Gilchrist, M.D., Curran, K.M., 2010. Applying DTI white matter orientations to finite element head models to examine diffuse TBI under high rotational accelerations. Prog. Biophys. Mol. Biol. 103, 304–309. https://doi.org/10.1016/j.pbiomolbio.2010.09.008.

Darvish, K.K., Crandall, J.R., 2001. Nonlinear viscoelastic effects in oscillatory shear deformation of brain tissue. Med. Eng. Phys. 23 (9), 633–645.

Donnelly, B.R., Medige, J., 1997. Shear properties of human brain tissue. J. Biomech. Eng. 119, 423–432. https://doi.org/10.1115/1.2798289.

Dorfmann, A., Ogden, R.W., 2004. A constitutive model for the Mullins effect with permanent set in particle-reinforced rubber. Int. J. Solids Struct. 41, 1855–1878. https://doi.org/10.1016/J.IJSOLSTR.2003.11.014.

El Sayed, T., Mota, A., Fraternali, F., Ortiz, M., 2008a. A variational constitutive model for soft biological tissues. J. Biomech. 41, 1458–1466. https://doi.org/10.1016/j.jbiomech.2008.02.023.

El Sayed, T., Mota, A., Fraternali, F., Ortiz, M., 2008b. Biomechanics of traumatic brain injury. Comput. Methods Appl. Mech. Eng. 197 (51–52), 4692–4701.

Estes, M.S., McElhaney, J.H., 1970. Response of brain tissue of compressive loading. ASME, Paper no. 70-BHF-13.

Fallenstein, G.T., Hulce, V.D., Melvin, J.W., 1969. Dynamic mechanical properties of human brain tissue. J. Biomech. 2 (3), 217–226.

Faul, M., Xu, L., Wald, M.M., Coronado, V., Dellinger, A.M., 2010. Traumatic brain injury in the United States: national estimates of prevalence and incidence, 2002–2006. Inj. Prev. 16 (Suppl 1), A268–A268.

Franceschini, G., Bigoni, D., Regitnig, P., Holzapfel, G.A.A., 2006. Brain tissue deforms similarly to filled elastomers and follows consolidation theory. J. Mech. Phys. Solids 54, 2592–2620. https://doi.org/10.1016/j.jmps.2006.05.004.

Galford, J.E., McElhaney, J.H., 1970. A viscoelastic study of scalp, brain, and dura. J. Biomech. 3, 211–221. https://doi.org/10.1016/0021-9290(70)90007-2.

Ganpule, S., Daphalapurkar, N.P., Cetingul, M.P., Ramesh, K.T., 2018. Effect of bulk modulus on deformation of the brain under rotational accelerations. Shock Waves 28, 127–139. https://doi.org/10.1007/s00193-017-0791-z.

Gilchrist, M.D., 2004. Experimental device for simulating traumatic brain injury resulting from linear accelerations. Strain 40 (4), 180–192.

Giordano, C., Cloots, R.J.H., van Dommelen, J.A.W., Kleiven, S., 2014. The influence of anisotropy on brain injury prediction. J. Biomech. 47, 1052–1059.

Giordano, C., Zappalà, S., Kleiven, S., 2017. Anisotropic finite element models for brain injury prediction: the sensitivity of axonal strain to white matter tract inter-subject variability. Biomech. Model. Mechanobiol. 16, 1–25. https://doi.org/10.1007/s10237-017-0887-5.

Haldar, K., Pal, C., 2018. Rate dependent anisotropic constitutive modeling of brain tissue undergoing large deformation. J. Mech. Behav. Biomed. Mater. 81, 178–194.

Horstemeyer, M.F., 2012. Integrated Computational Materials Engineering (ICME) for Metals: Reinvigorating Engineering Design with Science. Wiley Press, Hoboken, New Jersey.

Horstemeyer, M.F., 2018. Integrated Computational Materials Engineering (ICME) for Metals: Concepts and Case Studies. Wiley Press, Hoboken, New Jersey.

Horstemeyer, M.F., Panzer, M.B., Prabhu, R.K., 2019. State-of-the-art modeling and simulation of the brain's response to mechanical loads. Ann. Biomed. Eng. 47 (9), 1829–1831.

Hrapko, M., van Dommelen, J.a.W., Peters, G.W.M., Wismans, J.S.H.M., 2006. The mechanical behaviour of brain tissue: large strain response and constitutive modelling. Biorheology 43, 623–636.

Javid, S., Rezaei, A., Karami, G., 2014. A micromechanical procedure for viscoelastic characterization of the axons and ECM of the brainstem. J. Mech. Behav. Biomed. Mater. 30, 290–299.

Johnson, K.L.L., Chowdhury, S., Lawrimore, W.B.B., Mao, Y., Mehmani, A., Prabhu, R., Rush, G.A.A., Horstemeyer, M.F.F., 2016. Constrained topological optimization of a football helmet facemask based on brain response. Mater. Des. 111, 108–118. https://doi.org/10.1016/j.matdes.2016.08.064.

36 Chapter 2

Laksari, K., Shafieian, M., Darvish, K., 2012. Constitutive model for brain tissue under finite compression. J. Biomech. 45, 642–646. https://doi.org/10.1016/j.jbiomech.2011.12.023.

Lee, S.J., King, M.A., Sun, J., Xie, H.K., Subhash, G., Sarntinoranont, M., 2014. Measurement of viscoelastic properties in multiple anatomical regions of acute rat brain tissue slices. J. Mech. Behav. Biomed. Mater. 29, 213–224.

Madhukar, A., Ostoja-Starzewski, M., 2019. Finite element methods in human head impact simulations: a review. Ann. Biomed. Eng. 47 (9), 1832–1854.

Madireddy, S., Sista, B., Vemaganti, K., 2015. A Bayesian approach to selecting hyperelastic constitutive models of soft tissue. Comput. Methods Appl. Mech. Eng. 291, 102–122. https://doi.org/10.1016/J.CMA.2015.03.012.

Meaney, F.D., 2003. Relationship between structural modeling and hyperelastic material behavior: application to CNS white matter. Biomech. Model. Mechanobiol. 1, 279–293. https://doi.org/10.1007/s10237-002-0020-1.

Mendis, K.K., Stalnaker, R.L., Advani, S.H., 1995. A constitutive relationship for large deformation finite element modeling of brain tissue. J. Biomech. Eng. 117, 279–285. https://doi.org/10.1115/1.2794182.

Metz, H., McElhaney, J., Ommaya, A.K., 1970. A comparison of the elasticity of live, dead, and fixed brain tissue. J. Biomech. 3 (4), 453–458.

Mihai, L.A., Budday, S., Holzapfel, G.A., Kuhl, E., Goriely, A., 2017. A family of hyperelastic models for human brain tissue. J. Mech. Phys. Solids 106, 60–79. https://doi.org/10.1016/j.jmps.2017.05.015.

Mihai, L.A., Chin, L., Janmey, P.A., Goriely, A., 2015. A comparison of hyperelastic constitutive models applicable to brain and fat tissues. J. R. Soc. Interface 12, 0486. https://doi.org/10.1098/rsif.2015.0486.

Miller, K., Chinzei, K., 1997. Constitutive modelling of brain tissue: experiment and theory. J. Biomech. 30, 1115–1121. https://doi.org/10.1016/S0021-9290(97)00092-4.

Miller, K., Chinzei, K., 2002. Mechanical properties of brain tissue in tension. J. Biomech. 35, 483–490. https://doi.org/10.1016/S0021-9290(01)00234-2.

Mooney, M., 1940. A theory of large elastic deformation. J. Appl. Phys. 11 (9), 582–592.

Nicolle, S., Lounis, M. and Willinger, R., 2004. *Shear properties of brain tissue over a frequency range relevant for automotive impact situations: new experimental results (No. 2004-22-0011).* SAE Technical Paper.

Nicolle, S., Lounis, M., Willinger, R., Palierne, J.F., 2005. Shear linear behavior of brain tissue over a large frequency range. Biorheology 42 (3), 209–223.

Ogden, R.W., 1972. Large deformation isotropic elasticity—on the correlation of theory and experiment for incompressible rubberlike solids. Proc. R. Soc. Lond. A Math. Phys. Sci. *326* (1567), 565–584.

Ogden, R.W., Roxburgh, D.G., 1999. A pseudo-elastic model for the Mullins effect in filled rubber. Proc. R. Soc. A Math. Phys. Eng. Sci. 455, 2861–2877. https://doi.org/10.1098/rspa.1999.0431.

Pervin, F., Chen, W.W., 2009. Dynamic mechanical response of bovine gray matter and white matter brain tissues under compression. J. Biomech. 42 (6), 731–735.

Pierpaoli, C., Basser, P.J., 1996. Toward a quantitative assessment of diffusion anisotropy. Magn. Reson. Med. 36, 893–906. https://doi.org/10.1002/mrm.1910360612.

Prabhu, R., Horstemeyer, M.F.F., Tucker, M.T.T., Marin, E.B.B., Bouvard, J.L.L., Sherburn, J.A.A., Liao, J., Williams, L.N., 2011. Coupled experiment/finite element analysis on the mechanical response of porcine brain under high strain rates. J. Mech. Behav. Biomed. Mater. 4, 1067–1080. https://doi.org/10.1016/j.jmbbm.2011.03.015.

Prange, M.T., Margulies, S.S., 2002. Regional, directional, and age-dependent properties of the brain undergoing large deformation. J. Biomech. Eng. 124, 244. https://doi.org/10.1115/1.1449907.

Prange, M.T., et al., 2000. Defining brain mechanical properties: effects of region, direction, and species. J. Stapp Car Crash 44, 205–213.

Prevost, T.P., Balakrishnan, A., Suresh, S., Socrate, S., 2011. Biomechanics of brain tissue. Acta Biomater. 7, 83–95. https://doi.org/10.1016/j.actbio.2010.06.035.

Rashid, B., Destrade, M., Gilchrist, M.D., 2012. Mechanical characterization of brain tissue in compression at dynamic strain rates. J. Mech. Behav. Biomed. Mater. 10, 23–38. https://doi.org/10.1016/j.jmbbm.2012.01.022.

Rashid, B., Destrade, M., Gilchrist, M.D., 2013. Mechanical characterization of brain tissue in simple shear at dynamic strain rates. J. Mech. Behav. Biomed. Mater. 28, 71–85. https://doi.org/10.1016/j.jmbbm.2013.07.017.

Rashid, B., Destrade, M., Gilchrist, M.D., 2014. Mechanical characterization of brain tissue in tension at dynamic strain rates. Forensic Biomech. 33, 43–54. https://doi.org/10.1016/j.jmbbm.2012.07.015.

Rivlen, R.S., 1948. Large elastic deformations of isotropic materials IV. Further developments of the general theory. Philos. Trans. Royal Soc. London. Series A, Math. Phys. Sci. 241 (835), 379–397.

Sahoo, D., Deck, C., Willinger, R., 2014. Development and validation of an advanced anisotropic visco-hyperelastic human brain FE model. Forensic Biomech. 33, 24–42. https://doi.org/10.1016/j.jmbbm.2013.08.022.

Shuck, L.Z., Advani, S.H., 1972. Rheological response of human brain tissue in shear. J. Basic Eng. 94, 905–911. https://doi.org/10.1115/1.3425588.

Shugar, T., Katona, M., 1974. Development of a finite element head injury model, ASCE National Structural Engineering Meeting. Cincinnati, OH April.

Tamura, A., Hayashi, S., Nagayama, K., Matsumoto, T., 2008. Mechanical characterization of brain tissue in high-rate extension. J. Biomech. Sci. Eng. 3 (2), 263–274.

Tamura, A., Hayashi, S., Watanabe, I., Nagayama, K., Matsumoto, T., 2007. Mechanical characterization of brain tissue in high-rate compression. J. Biomech. Sci. Eng. 2 (3), 115–126.

Thibault, K.L., Margulies, S.S., 1998. Age-dependent material properties of the porcine cerebrum: effect on pediatric inertial head injury criteria. J. Biomech. 31 (12), 1119–1126.

Velardi, F., Fraternali, F., Angelillo, M., 2006. Anisotropic constitutive equations and experimental tensile behavior of brain tissue. Biomech. Model. Mechanobiol. 5, 53–61. https://doi.org/10.1007/s10237-005-0007-9.

Ward, C.C., Thompson, R.B., 1975. The development of a detailed finite element brain model. SAE Trans., 3238–3252.

CHAPTER 3

Density functional theory and bridging to classical interatomic force fields

D. Dickel[a], S. Mun[b], M. Baskes[c], S. Gwaltney[d], Raj K. Prabhu[e], Mark F. Horstemeyer[f]

[a]*Bagley College of Engineering, Mississippi State University, Mississippi State, MS, United States*
[b]*Center for Advanced Vehicular Systems (CAVS), Mississippi State University, Mississippi State, MS, United States* [c]*College of Engineering, University of North Texas, Denton, TX, United States*
[d]*Department of Chemistry, Mississippi State University, Mississippi State, MS, United States* [e]*USRA, NASA HRP CCMP, NASA Glenn Research Center, Cleveland, OH, United States* [f]*School of Engineering, Liberty University, Lynchburg, VA, United States*

3.1 Introduction

The electronic length scale is the lowest length scale to date that has accomplished any modeling of the brain. Fig. 3.1 focuses our attention on the electronics scale where density functional theory (DFT) is conducted in the context of the multiscale framework for the brain. Information from this length scale will be upscaled (Bridge 5) to the macroscale mechano-physiological internal state variable model and into the nanoscale molecular dynamics (MD) simulations (Bridge 1). The information related to the conformations and reactivity is used for the mechano-physiological internal state variable model in Bridge 5, and the interfacial energies, elastic moduli, and lattice parameters are used in the Bridge 1 for MD simulations.

3.1.1 Why quantum mechanics?

A common refrain from both modelers and experimentalists alike when presented with techniques or data that use individual atoms as their primary object of interest is the question, "Why do I need atomic physics to study biomechanics?" Setting aside the related but distinct fact that many of the instruments used in biomechanics rely on a deep knowledge of fundamental physics both for their operation and the interpretation of the data they provide, the question of the connection between these lowest length-scale processes and what is or can be observed at even the cellular level is a deep one. To extend the question further, if a macroscale model can be informed and validated by experimental results, and experimental work can be extended and guided by robust, validated models, why then is there a need for any lower length-scale modeling? The answer of course is a large part of the subject matter of this text. Multiscale modeling provides deep insights into macroscopic behavior and can

Multiscale Biomechanical Modeling of the Brain.
DOI: https://doi.org/10.1016/B978-0-12-818144-7.00007-4
Copyright © 2022 Elsevier Inc. All rights reserved.

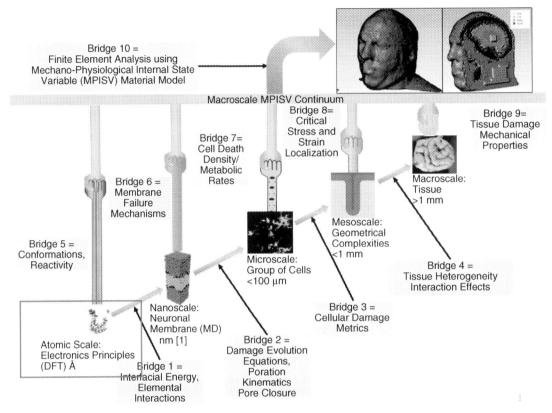

Modified from Murphy et al. [2016]

Fig. 3.1: Multiscale modeling of the brain requires modeling and analysis at different length scales where the bridges of information of clearly defined by downscaling but the cause–effect relations are determined by upscaling. In this chapter, we focus on the electronic scale, sometimes called the "quantum" scale.

help develop, refine, and validate higher length-scale models through the use of lower ones at a lower cost and faster than experimental results. Often, lower length-scale models are even able to provide information and detail that is not directly available to experiment.

The question remains though: Why quantum physics? While the use of multiscale modeling may be self-evident, why must it reach down all the way to the level of individual atoms and electrons? The answer is that all models—at least, all useful models—must eventually make contact with the reality of physics. For example, macroscale material models have both a form and a parameterization that are empirically motivated and validated by experimental results. Although the macroscale models are based on the mechanics laws of physics, as they are currently understood, they can only be derived based on the results of precise experimental measurements taken in the real world. In the case of multiscale modeling where greater connection with the equations of physical reality are made as we move down to lower length scales and remove simplifying assumptions and the coarse-graining essential to pose solvable problems, we are adding physics into the macroscale models. Hence, the most accurately verified science we have yet to know is quantum mechanics.[1] If the equations of quantum mechanics are correct, which we know to extremely high precision they are, we are given firm ground to stand on as we upscale to higher and higher length scales. Care must still be taken in connecting back to these fundamental equations but without this basis from which to begin, the reliability of our models would always be in question no matter how well connected to the model below them.

3.1.2 Physical chemistry of biomechanical systems

While the physical laws that determine the structure and interaction of systems of atoms are universal, their particular manifestation will vary greatly for different types of materials. To wit, bulk metallic systems or semiconductors will behave very differently from organic molecules despite the same governing equations dictating their behavior. The reason for this is due to the atomic species involved in these molecules and the nature of their bonding. An even cursory overview of quantum mechanics behind these relations is beyond the scope of this work, so it will be left to other more detailed sources (Ballentine, 1998; Feynman, 1971; Griffiths, 1994). However, we will briefly summarize some of the key features that make biological materials unique at the atomistic scale.

We first note that the bonding of individual atoms is driven by the interaction between a positively charged ion and the negatively charged electrons, which surround them. Quantum mechanics shows that structures will form in which electrons can occupy the lowest energy states. In the case of metal systems, this results in large numbers of ions sharing electrons and extreme delocalization of s- and p-electrons, which give metals their regular crystal structure, conductive character, and relatively weak bonding. This delocalization of electrons also means that the electron density is relatively uniform within metal crystals except in the immediate vicinity of metal nuclei. In direct contrast, the elements of organic chemistry tend to form local, covalent bonds where electrons are shared only be neighboring atoms in, for example, a saturated hydrocarbon. This limits the number of bonded neighbors as the number of shared electrons for any one atom is limited due to the Pauli exclusion principle (Pauli, 1925). It also increases the intramolecular bond strength versus metals and limits their conductivity. Because electrons occupy particular local orbitals, the resulting charge density is also less uniform for organic molecules like those for the brain.

---

[1] The value of the dimensionless gyromagnetic ratio for a free electron is predicted by quantum field theory and can be measured experimentally. The two values agree to better than one part in one billion. This has often been compared to correctly predicting cross-country distances to within the width of a human hair.

42 Chapter 3

3.2 Density functional theory

While the quantum mechanical equations governing the behavior of, for example, methane are known exactly, they cannot be solved analytically, and even brute force computational methods are unable to produce accurate solutions. Methane, by the way, has recently been observed to help brain function as an antioxidant and healer of brain trauma (Jia et al., 2018). In order to model something like methane, numerical methods of approximately solving these equations have been developed which can capture all the salient features of the interactions among atomic species. The most prominent of these methods is DFT. Capable of describing the ground-state electronic configuration of bulk solids as well as organic molecules, DFT utilizes the total charge density as the object of study rather than the wavefunction of individual electrons, greatly simplifying the calculation. However, in writing the Kohn-Sham Equations (1965), which must be solved for a given system, an expression must be given for the exchange-correlation (XC) energy functional. However, this functional is not known in general with an analytical form only being known for the model homogenous electron gas. As such, a number of different approximations describe the XC functional.

The simplest of these is the local density approximation (LDA) (Lee et al., 1988), which assumes the XC functional depends only on the local charge density:

$$E_{XC}[\rho] = \int \rho(r) \in_{XC} (\rho(r)) dr E_{XC}[\rho] = \int \rho(r) \in_{XC} (\rho(r)) dr \tag{3.1}$$

where $\rho(r)$ is the charge density at r, and ϵ_{XC} is the XC energy of a homogenous electron gas. Highly accurate formula for the XC functional under the LDA has been derived from quantum Monte Carlo simulations.

One of the limitations of the LDA is the implicit assumption that the charge density does not vary greatly and, because of this, only the local density needs to be considered. For systems where the electron density is less homogenous, such as covalently bonded systems, additional terms are included in the XC functional related to the spatial derivatives of the density. When the first derivative, the gradient, is included, the functional is said to follow the generalized gradient approximation (GGA) while inclusion of higher order derivatives comprise the so-called meta-GGA functionals.

While more complex functionals can correct many of the errors of simpler approximations, they come with a pair of important drawbacks. First, the simplest forms of the LDA functional can be derived analytically from theoretical considerations of the homogenous electron gas, tying them directly to physical theory, an important consideration as discussion in Section 3.1. More complicated functional forms require a larger parameter space and these parameters must be fit using known values, often from experiment. The other, probably more significant consequence is that these more sophisticated methods require more and more expensive computational cost. DFT suffers from costly calculation times with poor scaling.

Density Functional Theory and Bridging to Classical Interatomic Force Fields   43

Calculations of energies and forces for just a few electron system might require minutes on modern processors. And because DFT calculations scale approximately with the cube of the number of electrons, if a system of 10 atoms requires, say, half an hour of computer time, a similar system with 100 atoms will require almost 3 weeks, and a system of 1000 atoms would require decades. Parallelization can, of course, greatly reduce the total time required, but the severe limitations on system size should be clear. And, as methods become more complicated, the scaling becomes even poorer, to the point where a single calculation of a relatively small molecule becomes a prohibitive computational endeavor.

These computational limitations, even in the easiest cases, are why multiscale modeling is employed to begin with. If computers were powerful enough to provide DFT solutions to problems at the mesoscale or macroscale, there would be no need for further approximations and the development of independent models at these scales. Such a need does exist, however, and for DFT, the most direct link is to classical force field models. For a truly bottom-up approach to multiscale modeling, these force fields must be motivated based on the results of DFT and their outputs be further used by even higher length-scale models.

3.3 Downscaling requirements of classical force field atomistic models

The simplifying assumption underlying classical atomistic models is that individual electrons no longer need to be considered. Entire atoms, implicitly including electromagnetic interactions, become the primary object of study representing a length scale higher than the electronic scale. These atoms follow classical laws of motion, responding to a position dependent force field exerted by other atoms in the system. The work of developing models for classical force fields is determining the functional form of these fields and parameterizing them based on experimental or DFT data. While ultimately the dynamics are entirely determined by the force acting on the atoms for any given molecular configuration, a number of more physically meaningful values are used to encode information about the force field. Force fields are typically defined on a per atom basis, with an energy being assigned to each atom that is a function of its local environment. The parameterization of this function will be different for different atomic species.

Of fundamental importance when modeling interactions, particularly for organic molecules, are the type and variety of molecules considered. Because of the wide variety of even simple structures that can be created, it is important that the properties of several molecules be considered. However, since the energy is determined by the local environment, it is typically sufficient to reproduce the behavior of a number of small molecules, with the dynamics of larger ones reflecting the composite of many short-range interactions. For example, in developing a force fields model for saturated hydrocarbons, it would be important to consider the short, saturated molecules of at least methane, ethane, and propane, as well as isobutane and neopentane in order to independently account for all the possible local saturated environments

44 Chapter 3

for carbon. However, it may not be necessary to consider longer chains such as octane in the fitting procedure as its properties should be fully described by the environments present in the smaller molecules. For the brain, we would need to consider sugars (glucose made of $C_6H_{12}O_6$) (Mergenthaler et al., 2013), water (H_2O), and different proteins (for example, tryptophan is $C_{11}H_{12}N_2O_2$). In these cases for the brain, we would need to run DFT simulations combining C, H, O, and N, and the atomistic potential would need to be developed for the C, H, O, and N interactions.

3.3.1 Upscaling properties

When determining which properties should be used in the creation or parameterization of a higher length-scale model, there are a few important considerations. Of the utmost importance is the relevance of the property to the phenomena under investigation at that particular length scale. In molecular chemistry, one of the most important considerations is whether the formalism is reactive or not. That is, whether bonds can be spontaneously broken during simulation or if the chemistry of individual species is fixed. It is also important to consider, however, the ease with which the property can be calculated by both lower and higher length-scale models, particularly when parameterization of a model is considered. The rate at which a given parameterization can be evaluated is limited by the slowest calculation used in that evaluation so in the case where rapid optimization is desired, simpler properties should take precedence. It can also be the case that a difficult to calculate property, say the melting temperature of a solid, can be highly correlated with simpler properties like the cohesive energy. In this case it is much more efficient to parameterize to satisfy the simpler calculation and use the more rigorous one for validation.

The simplest properties to consider in developing a classical force fields are those that can be evaluated statically, without dynamic simulation. First among these is the cohesive energy, or heat of formation at 0 K for the reference structures considered. These values are well known experimentally for most simple molecules and are a standard output of DFT calculations. We note that nothing really exists at 0 K but it is a great computational starting point.

A discrepancy should be noted here that is particularly important for the study of small molecules as are seen in biological systems. Due to the stochastic nature of quantum mechanics, real molecules are never observed to have potential energy in agreement with the minimum of the energy well, even at vanishingly small temperature. The excess energy about the minimum, called the "zero-point" energy will not be explicitly calculated by DFT and is ignored in classical MD calculations. Typically, the known zero-point energy is added to DFT or classical force fields values to compare to the experimentally observed cohesive energy.

The next straight-forward target for development of a force fields is the bond lengths of the reference molecules. These require energy minimization of the reference structures as the force fields model may give the same cohesive energy as DFT results but with different bond lengths or angles.

Once the energy and structure of the reference structures are confirmed to agree with experiment or DFT, the elastic properties are the next logical consideration. In bulk solids, this takes the form of elastic moduli, whereas for smaller molecules, they are revealed by their vibrational modes. There are many equivalent ways to approach this data, but vibrational frequencies are the most common value examined as these can be accessed experimentally by various spectroscopic techniques. The full vibrational spectrum of a molecule will reveal the behavior of both local bond stretching and rotation thus ensuring the atomic forces are correct and therefore the behavior of individual molecules at low to moderate temperatures. Note that these modes are normally only considered for lower amplitude vibrations and, as such, while their consistency between force fields and either DFT or experiment will ensure reliable behavior at low temperature, the inherent anharmonicity of these vibrational modes will limit the predictive power at higher temperatures, certainly above temperatures where the molecule is expected to break down or react chemically with other species.

In addition to intramolecular interactions, another upscaling property to consider is the interaction between or among molecules. This can include electrostatic interactions in the case of charged or polar species or dispersion forces like the van der Waals force (Dzyaloshinskii et al., 1961). While weaker than intramolecular bonding, their consideration is essential for correctly describing complex systems of molecules. Dispersion forces, such as those between mutually induced dipoles, can be quite difficult to model as these interactions are not considered by, and do not appear in, DFT results. This necessarily requires the use of other models or methods to determine the strength of the interaction. These models are usually represented by an additional energy term in the potential and can take the form of pairwise power laws like the Lennard-Jones potential (Lennard-Jones, 1924). More complicated forms, which include three-body interactions, such as the D3 potential (Grimme et al., 2010) can also be used.

Finally, the formation or dissolution of molecules and their reactions should also be considered for reactive potentials. While reaction pathways can be quite complicated and not necessarily well known, simple pathways, such as the linear separation of a hydrogen atom from a hydrocarbon, can be modeled directly with DFT and reproduced. Particular reaction kinetics may require advanced methods and robust experiment or DFT results.

With the development of force fields that satisfactorily reproduces the above properties for all relevant molecules, it is crucial to validate the force fields. While a robust formalism should provide a great deal of transferability. More complicated properties or phenomena which are only distantly related to the properties used to design the force fields should be validated using experimental method or other means to ensure reliability.

46   Chapter 3

3.4 Sample atomistic force fields formalism and development of an interatomic potential for hydrocarbons

Before one can start with the whole set of C, H, O, and N required for the brain materials, we start only with just C and H for hydrocarbons. As a demonstration of the above method for using DFT to provide information about organic molecules and to develop classical force fields models capable of simulating dynamic systems of millions or billions of atoms, we will outline the development of a reactive chemo-thermomechanical hydrocarbon potential which should be capable of reproducing correctly the behavior and interaction of the full suite of hydrocarbon materials. For this exercise, we use the Modified Embedded Atom Method with Bond Order (MEAMBO) and the hydrocarbon parameterization of Mun et al. (2017).

3.4.1 MEAMBO

The modified embedded atom method (MEAM) (Baskes et al., 1989) was developed in an attempt to improve on the embedded atom method (Daw and Baskes, 1984). These two formalisms have shown exceptional success in bulk metals and semiconductors, but limited application in biological materials, although a model for saturated hydrocarbon systems has been fairly recently produced using MEAM (Nouranian et al., 2014). The bond-order extension allows for varying degrees of covalent bonding (Mun et al., 2017). Essential for organic molecules, the bond-order contribution allows for the explicit consideration of double and triple bonds in addition to the single bonds successfully characterized by MEAM.

The fundamental energy equation of MEAM assumes atomic energies is due to a pairwise interaction with its neighbors and an embedding function which results from the local electron density

$$E = \sum_i F_{\tau_i}\left(\rho_i\right) + \sum_{i,j} S_{i,j}\Phi_{\tau_i,\tau_j}\left(R_{ij}\right) \tag{3.2}$$

where F_τ is the embedded energy required to insert an atom of type τ into a local electron density of ρ, and $\Phi_{\tau\tau}$ is the pair potential between atoms of type i and j separated by distance R. The bond order additionally considers a bond energy for configurations where the geometry of the molecule indicates the presence of a double or triple bond, with linear structure indicating a triple bond for carbon and locally planar structures indicating a double bond.

MEAM and MEAMBO explicitly use reference structures to determine a subset of their parameters. A single reference structure exists for each possible pair of elements considered. These reference structures should be simple with readily available information on their properties from experiment or first principles results. For the hydrocarbon systems, these reference structures are diamond for the carbon–carbon interaction, the H_2 molecule for hydrogen–hydrogen, and methane for carbon–hydrogen.

The intermolecular forces are modeled using a simple 9-6 pair potential with a radial cutoff included to truncate the distance over which intermolecular forces act. This cutoff serves a role in reducing the computation time of calculations as it limits the number of pairwise interactions which must be considered, most of which would be vanishingly small if included.

3.4.2 Calibration of the MEAMBO potential

While the reference structures are necessary for the parameterization of the MEAMBO potential, they do not exhaust the range of possible interactions in hydrocarbon systems. As such, a number of other structures are used to generate data for fitting.

As mentioned above, longer chains of saturated hydrocarbons provide additional information about the local environment for the carbon atom, so ethane, propane, and isobutane are included to help exhaust these possibilities. The methyl radical CH_3 is also included as a reaction intermediary and to correctly predict the coordination of both carbon and hydrogen. Additionally, double and triple bonding between carbons is possible and must be considered. This adds ethylene for doubly bonded environments, as well as acetylene for triply bonded environments. The remaining intramolecular bonding environment available to hydrocarbons is those found in resonance structures such as benzene. The MEAMBO formalism treats these as fractional bonds according to the number of electrons shared among carbon atoms. To make sure these fractional bonds are correctly modeled, benzene and graphene are included in the model calibration.

While the tendency can exist to increase the fitting database without bounds and attempt to reproduce all available data from first principles and experiments, this approach suffers from a number of limitations. The main one being that if unlimited data are available for generation of the potential, there is nothing left for the potential to do; all problems it would be used to solve are available, implicitly or explicitly, in the available training data. Particularly in the case of first principles calculations where, in principle, the target dataset can be expanded indefinitely, the researcher will at some point have sacrificed the primary benefit of these atomic scale models, namely the decreased computational time and higher length-scale calculations which can be simulated. A second limitation is that the number of free parameters in any force field model is limited and so the ability of any particular formalism to reproduce an arbitrary amount of target data is limited. Formalisms better suited to particular problems or material systems will be more transferable with accuracy for a well-defined fitting database implying accuracy for a wide variety of configurations outside of the targeted data. Ensuring this accuracy is the purpose of validation which will be discussed below.

Due to the structure of the MEAMBO formalism, the parameters for the saturated hydrocarbon potential are determined first, so these structures are used exclusively in the fitting of base MEAM part of the force fields. With these in place, the unsaturated structures and their properties are used to determine the bond-order parameters. This leaves the intermolecular dispersion interaction to be fit to fully characterize the potential. As mentioned above, DFT is not an

48 Chapter 3

effective means of determining these forces as they are necessarily excluded from DFT calculations. As such the second-order Møller–Plesset perturbation (MP2) method (Grimme, 2003) is used to determine the intensity of the energy and force contribution. For the simple hydrocarbon force fields being developed, we will consider the binding curve of two symmetric methane molecules to parameterize this interaction. Note that the term binding is used loosely here as the force involved is quite small compared to the intramolecular ones. There is, however, a well-defined minimum potential energy as the separation between the molecules is changed.

3.4.3 Parameterization of the interatomic potential

The primary targets for calibration of the MEAMBO potential are listed in Tables 3.1 and 3.2. Table 3.1 gives the targets for saturated hydrocarbons, while Table 3.2 shows the unsaturated. While all of these values can be calculated using first principles, experimental values are shown in the tables as well. As mentioned above, due to the numerous approximations used in any first principles method, systemic errors occur. While there is value in simply reproducing these results in the higher length-scale method, being aware of the discrepancy is invaluable for evaluating the quality of results from these length scales. Additionally, the opportunity presents itself to incorporate the known experimental results into the force fields model, correcting the errors of the first principles results while retaining its valuable contributions.

Table 3.1: Experimental values used as targets for the development of a MEAMBO potential for saturated hydrocarbons and the values predicted by the resulting potential.

Molecule/structure	Property	Experiment	MEAMBO
Diamond	E_{corr} (eV/atom)	7.346	7.346
	Lattice constant (Å)	3.567	3.567
	Bulk modulus (GPa)	443.0	442.2
Graphene	E_{corr} (eV/atom)	7.315	7.304
	Lattice constant (Å)	2.462	2.460
Methane	E_{corr} (eV)	17.018	17.018
	C–H bond length (Å)	1.087	1.087
Ethane	E_{corr} (eV)	28.885	30.941
	C–H bond length (Å)	1.094	1.114
	C–C bond length (Å)	1.535	1.533
Propane	E_{corr} (eV)	40.880	43.723
	C–H bond length (Å)	1.107	1.120
	C–C bond length (Å)	1.532	1.537
Isobutane	E_{corr} (eV)	52.977	56.559
	C–H bond length (Å)	1.113	1.119
	C–C bond length (Å)	1.535	1.541
Neopentane	E_{corr} (eV)	65.123	69.416
	C–H bond length (Å)	1.114	1.113
	C–C bond length (Å)	1.537	1.548

Density Functional Theory and Bridging to Classical Interatomic Force Fields 49

Table 3.2: Experimental values used as targets for the development of a MEAMBO potential for unsaturated hydrocarbons and free radicals as well as the values predicted by the resulting potential.

Molecule/structure	Property	Experiment	MEAMBO
Ethene	E_{corr} (eV/atom)	23.066	23.070
	C–H bond length (Å)	1.086	1.110
	C–C bond length (Å)	1.339	1.337
Acetylene	E_{corr} (eV/atom)	16.857	16.856
	C–H bond length (Å)	1.063	1.111
	C–C bond length (Å)	1.203	1.202
Benzene	E_{corr} (eV)	56.619	56.637
	C–H bond length (Å)	1.084	1.140
	C–C bond length (Å)	1.397	1.396
CH radical	E_{corr} (eV/atom)	3.469	6.484
	C–H bond length (Å)	1.120	0.967
CH_2 radical	E_{corr} (eV/atom)	7.410	8.002
	C–H bond length (Å)	1.085	1.010
CH_3 radical	E_{corr} (eV/atom)	12.534	13.694
	C–H bond length (Å)	1.091	1.050

While there need be no single systemic method for parameterizing a given potential, the formalism may lend itself more directly to particular methods. This can occur when particular parameters have an explicit relationship to known experimental or first principles values. In the case of MEAMBO, this is the case, by design, for a number of parameters. For this reason, as discussed above, the MEAM parameters are determined first using the values for the saturated hydrocarbon molecules. As the bond-order parameters should not affect these values, no reparameterization of these parameters is necessary.

Fitting of the MEAM parameters produces the potential values presented in Table 3.1 on for individual saturated hydrocarbons. With these parameters set, the bond-order parameters can be determined using the unsaturated targets in the same way. And finally, the intermolecular contributions to the total energy can be found using the intramolecular interactions. This separation is possible because of the construction of the MEAMBO formalism containing separate energy terms some of which are explicitly zero in the case of saturated coordination.

3.4.4 Validation of the interatomic force fields

While at this point, the force field has been fully parameterized and could be used for dynamic simulations, it is imperative that it undergo some validation measures to ensure reliability. While the transferability of the MEAMBO formalism is intended to require minimal validation, it is likely, particularly with a relatively small number of fitting targets that multiple parameter sets could reproduce the same level of fit. Distinguishing these parameters requires extending the fitting to more complicated systems and interactions.

50 Chapter 3

Table 3.3: Experimental results for the pressure of simple hydrocarbon gases at various temperatures and densities and the prediction made by the MEAMBO potential developed by Mun et al. (2017).

System	Temperature (K)	Density (g/cm^3)	Experimental pressure (MPa)	MEAMBO pressure (MPa)
Methane	400	0.0246	5.005	4.70
	298	0.1185	14.994	10.6
	450	0.2021	59.975	49.4
	373	0.5534	1000.000	959.1
Ethane	308	0.0371	2.550	2.2
	308	0.2726	5.387	7.4

In order to validate the full potential, including intramolecular interactions, as well as to demonstrate the ability of the force fields to properly reproduce dynamic behavior, systems of homogenous hydrocarbon gases were created and allowed to equilibrate at some fixed density and temperature. The pressure exerted by the gas is them measured (computationally) and compared to experimental values. Table 3.3 shows these results and we again see agreement between the *prediction* of the force fields simulation and actual experimental values.

3.5 Summary

In this chapter, we focused on the lowest length-scale modeling and simulation methods for the brain. In short, we presented a brief overview of DFT methods that rely on fundamental physical equations to predict the energy and forces on molecular systems. We further showed how these computationally expensive methods can be used to generate a fitting database for the next higher length-scale semiempirical interatomic force fields, which are faster. As an illustrative example, we presented a sample database of simple hydrocarbons and showed the steps involved in generating a useful potential from this data using the MEAMBO. This force field was able to reproduce the cohesive energy, bond lengths, and intermolecular interactions for both saturated and unsaturated hydrocarbons. Simple validation using experimental results was then shown, demonstrating the predictive power of the potential. This potential, among a range of other candidates developed in similar ways, is then useful for large-scale MD simulation.

For a thorough analysis of modeling and simulations for the brain at this length scale, many more electronics principles and atomistic calculations will need to be conducted. Each structure in the brain with its associated chemical make-up will need to be defined beyond the hydrocarbons discussed herein in order to upscale the needed the information for a macroscale analysis of the brain. Hence, many the elements of the periodic table and their interactions are necessary to quantify the structure–property relations at this small length scale.

Finally, at the atomic level, one could use other interatomic potentials that were first start for polymers as opposed to the MEAMBO method, which was modified from metals. However, in order to capture all of the features for the brain, like blood which includes iron, the authors can only foresee that MEAMBO will be able to capture the complete set of cause–effect relations in the brain, while the other methods will run into their limits of application.

References

Ballentine, L.E., 1998. Quantum Mechanics: A Modern Development. World Scientific Publishing Company, River Edge, NJ.

Baskes, M.I., Nelson, J.S., Wright, A.F., 1989. Semiempirical modified embedded-atom potentials for silicon and germanium. Phys. Rev. B 40, 6085.

Daw, M.S., Baskes, M.I., 1984. Embedded-atom method: derivation and application to impurities, surfaces, and other defects in metals. Phys. Rev. B 29, 6443.

Dzyaloshinskii, I.E., Lifshitz, E.M., Pitaevskii, L.P., 1961. The general theory of van der Waals forces. Adv. Phys. 10, 165.

Feynman, R.P., 1971. The Feynman Lecture of Physics, Vol. 3, Quantum Mechanics, Basic Books, New York, NY.

Griffiths, D.J., 1994. Introduction to Quantum Mechanics. Prentice Hall, Upper Saddle River, NJ.

Grimme, S., 2003. Improved second-order Møller–Plesset perturbation theory by separate scaling of parallel- and antiparallel-spin pair correlation energies. J. Chem. Phys. 118, 9095–9102.

Grimme, S., Antony, J., Ehrlich, S., Krieg, H., 2010. A consistent and accurate ab initio parametrization of density functional dispersion correction (DFT-D) for the 94 elements H–Pu. J. Chem. Phys. 132, 154104.

Jia, Y., Li, Z., Liu, C., Zhang, J., 2018. Methane medicine: a rising star gas with powerful anti-inflammation, antioxidant, and antiapoptosis properties. Oxid. Med. Cell. Longevity 2018, 1912746. doi: 10.1155/2018/1912746.

Kohn, W., Sham, L.J., 1965. Self-consistent equations including exchange and correlation effects. Phys. Rev. 140, A1133–A1138.

Lee, C., Yang, W., Parr, R.G., 1988. Development of Colle-Salvetti correlation-energy formula into a functional of the electron density. Phys. Rev. B 37, 785.

Lennard-Jones, J.E., 1924. On the determination of molecular fields. II. From the equation of state of gas. Proc. R. Soc. A. 106, 463.

Mergenthaler, P., Lindauer, U., Dienel, G.A., Meisel, A., 2013. Sugar for the brain: the role of glucose in physiological and pathological brain function. Trends Neurosci. 36 (10), 587–597.

Mun, S., Bowman, A.L., Nouranian, S., Gwaltney, S.R., Baskes, M.I., Horstemeyer, M.F., 2017. Interatomic potential for hydrocarbons on the basis of the modified embedded-atom method with bond order (MEAM-BO). J. Phys. Chem. A 121, 1502–1524.

Murphy, M.A., Horstemeyer, M.F., Gwaltney, S.R., Stone, T., LaPlaca, M., Liao, J., Williams, L., Prabhu, R., 2016. Nanomechanics of phospholipid bilayer failure under strip biaxial stretching using molecular dynamics. Model. Simul. Mater. Sci. Eng. 24 (5), 055008.

Nouranian, S., Tschopp, M.A., Gwaltney, S.R., Baskes, M.I., Horstemeyer, M.F., 2014. An interatomic potential for saturated hydrocarbons based on the modified embedded-atom method. Phys. Chem. Chem. Phys. 16, 6233–6249.

Pauli, W., 1925. Über den Zusammenhang des Abschlusses der Elektronengruppen im Atom mit der Komplexstruktur der Spektren. Z. Phys. 31, 373.

CHAPTER 4

Modeling nanoscale cellular structures using molecular dynamics

M.A. Murphy[a], Mark F. Horstemeyer[b], Raj K. Prabhu[c]

[a]*Center for Advanced Vehicular Systems (CAVS), Mississippi State University, Mississippi State, MS, United States* [b]*School of Engineering, Liberty University, Lynchburg, VA, United States* [c]*USRA, NASA HRP CCMP, NASA Glenn Research Center, Cleveland, OH, United States*

4.1 Introduction

In the previous chapter, we discussed the electronic and atomistic length scales related to modeling and simulations of features related to the brain. In this chapter, we focus on the next higher length scale as illustrated in Fig. 4.1. The brain also has a complex multi-scale structural hierarchy (brain, lobe, region, sulci and gyri, group of cells, individual cells, cellular organelles, and components). The consideration of these scales and their effects on the mechanical properties of the brain's pathology and mechanical properties is of utmost importance because effects at the nanoscale can have profound effects on the brain at the macroscale. Bridge 6 information illustrated in Fig. 4.1 focuses on brain damage that the mechano-physiological internal state variable (MPISV) model would be able to use from these lower length nanoscale simulations. Bridge 2 information also illustrated in Fig. 4.1 upscales the damage progression equations and physics-basis for the microscale simulations.

At the macroscale, typical simulations of the brain include limited details from lower length scales. The most accurate brain material models employ internal state variables (ISVs) that focus on the brain's mechanical properties but lack lower length scale physiological effects (El Sayed et al., 2008; Mendis et al., 1995; Miller, 1999; Prabhu et al., 2011; Rashid et al., 2014). Some studies have also included some heterogeneities due to axonal tracts through embedded elements (Garimella et al., 2019). The lack of physiological aspects means the models are only telling a portion of the story. By relating physiological injury mechanisms into the mechanical response and adding injury mechanisms to ISV models, in effect creating MPISV models, traumatic brain injuries (TBIs) can be more accurately modeled.

It is currently not feasible, and likely unnecessary, to incorporate all of the physiological injury mechanisms into the models. Rather, specific mechanisms that are known to have a

Multiscale Biomechanical Modeling of the Brain.
DOI: https://doi.org/10.1016/B978-0-12-818144-7.00001-3
Copyright © 2022 Elsevier Inc. All rights reserved.

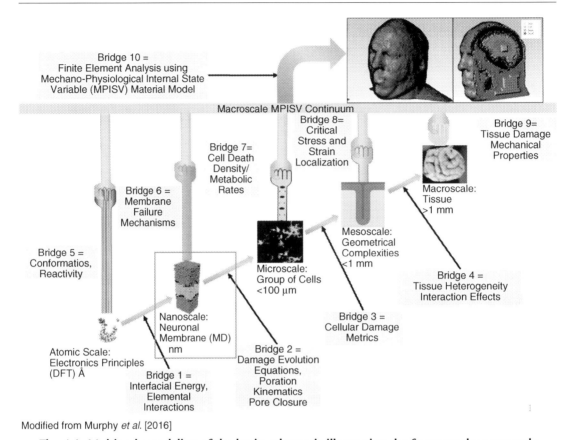

Fig. 4.1: Multiscale modeling of the brain schematic illustrating the focus on the nanoscale neuronal membrane in this chapter.

relatively more significant impact related to TBI injury and thus should be selected, quantified, and incorporated into the MPISV model. One such example of importance is the perturbance of the nanoscale neuronal membrane arising from an external mechanical impact, that is, membrane mechanoporation, which produces significant effects on higher length scales, such as neuron death during TBI (Farkas et al., 2007; Geddes et al., 2003). Fig. 4.2 shows how the membrane comprises the outside layer of a neuron, or nerve cell in the brain. Due to the membrane's important roles in regulating diffusion and maintaining homeostasis (Gwag et al., 1999; Kampfl et al., 1996), a loss of integrity in the neuron's membrane due to mechanoporation can cause multiple biological pathways to be activated (Gwag et al., 1999; Kampfl et al., 1996; Cooper and McNeil, 2015; Terasaki et al., 1997) in a negative manner. For example, disrupting intracellular ion concentrations, like calcium ions, triggers a number of pathways resulting in the benefit of membrane repair (vesicle–plasma membrane fusion) or the harmful effect of apoptosis or necrosis based on the level of disruption (Gwag et al., 1999; Kampfl et al., 1996; Cooper and McNeil, 2015; Terasaki et al., 1997).

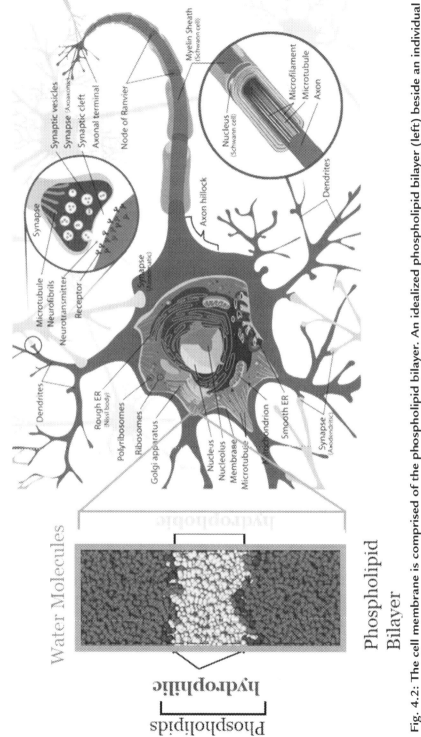

Fig. 4.2: The cell membrane is comprised of the phospholipid bilayer (left) beside an individual neuron cell diagram (right).

56 **Chapter 4**

The molecules that comprise the cell membrane are fats with phosphorus called phospholipids. These are made of compact, phosphate "heads" that are **hydrophilic** ("water-seeking") and point toward the exterior of the membrane on each side, and a pair of long fatty acid chains that are **hydrophobic** ("water-fearing") and face each other. This arrangement means that these heads face the exterior of the cell on one side and the cytoplasm on the other as shown in Fig. 4.2.

The lipid layer on the interior is the only true double layer in the cell membrane mix, because here, there are two consecutive membrane sections consisting almost solely of lipid tails. One set of tails from the phospholipids on one half of the bilayer, and one set of tails from the phospholipids on the other half of the bilayer. One lipid bilayer function is to protect the cell from threats from the outside so if it is weakened or torn, the whole neuron is vulnerable to death. The membrane is **semipermeable**, meaning that some substances can pass through while others are denied entry or exit outright. Molecules like water and oxygen can easily diffuse through the membrane, but other ions, nucleic acids (DNA or RNA), and sugars require the help of membrane transport proteins to pass through the membrane. Without a way to transfer substances into and out of the cell, the cell would rapidly run out of energy and not be able to expel metabolic waste products thus inducing neuron death.

Studying the mechanical behavior of the neuron membrane at the nanoscale introduces a host of difficulties. Experimentally, the length and time scales needed for examining phenomena at this small scale can make capturing data arduous and sometimes impossible. For example, experiments that deform cell cultures can capture cell death deformation effects by tracking biomarkers at length scales greater than neuronal membranes (LaPlaca and Thibault, 1997; LaPlaca et al., 2005; LaPlaca and Prado, 2010). However, these studies cannot easily quantify the local membrane strain, strain rate, or stress state caused by deformations (Cullen et al., 2011). Some methods examine individual cells or specific parts of the cell as well, such as investigating lipid structures with atomic force microscopy (Ovalle-García et al., 2011; Picas et al., 2012) or micropipette aspiration (Evans et al., 2003; Hochmuth, 2000; Needham and Nunn, 1990; Rawicz et al., 2008). While these methods allow more localized deformations to be examined and provide useful information, such as rupture strength and material constants, they are limited by applied rates and imaging methods. Furthermore, while the human body and other biological structures appear to be relatively in equilibrium at the macroscale level, the nanoscale is highly transient, because the local structures are always changing due to physiological, chemical, and mechanical processes. Cellular membranes are often modeled as a fluid, because the individual components are free to flow.

At the nanoscale molecular dynamics (MD) is used to analyze molecules such as the phospholipid bilayer that is prevalent in the brain. We note that this length scale is above that of electrons and atoms. Herein, we start with atoms and build the molecules. With a focus of membrane mechanoporation at the nanoscale due to TBI, deformation boundary conditions are downscaled from macroscale conditions. Then high-rate mechanisms, pore nucleation, pore growth, stress–strain behavior, and diffusion rates can be upscaled. The current chapter will explore the methods related to performing MD simulations of simplified membranes to examine mechanoporation and representative results.

4.2 Methods

4.2.1 Molecular dynamics simulation method

The MD method is a versatile simulation technique for investigating structures in the nanoscale to microscale length scales that, at its base, is a stochastic method based on Newton's equations of motion and astomistic force fields describing particle interactions. Together, they are used to calculate the energy in the system for a given thermodynamic ensemble and set of boundary conditions. Based on Newton's equations of motion, the energy is then used to compute new forces, velocities, and positions for every atom. Due to the time-average nature of this method, any properties of interest must be averaged over a suitable time interval to obtain useful information.

This method allows nanoscale TBI phenomena to be examined that cannot be easily imaged experimentally and are often not included in macroscale simulations. Unfortunately, these models can be relatively expensive computationally to run, but they are significantly less expensive than their quantum counterparts. A brief overview of some relevant concepts is provided here, but readers should seek out more dedicated resources for additional details, such as the book *Molecular Modelling Principles and Applications* by Leach (Leach, 2001).

4.2.2 Atomic force fields

Atomic force fields, which are numerically implemented as a combination of functional equations and constants, allow the calculation of the potential energy for a set of atoms and describe how strongly each atom is affected by changes in the system in relation to its neighboring atoms. Three common categories are classical, reactive, and coarse-grained force field, but there are many other force field, including those specialized for select types of molecules (e.g., water models; Jorgensen et al., 1983; Price and Brooks, 2004; Horn et al., 2004; Khalak et al., 2018; Izadi et al., 2014).

Although the focus of Chapter 3 was on the modified embedded atom method with bond order (MEAMBO) (Marrink et al., 2007), more classical atomistic force fields such as AMBER (Wang et al., 2004) and CHARMM (Klauda et al., 2010) can simulate polymers and biological materials. For example, force fields designed with a focus on organic molecules tend to have terms for bonds, angles, dihedral (torsion) angles, and nonbonded (van der Waals and electrostatic) interactions. Although MEAMBO (Marrink et al., 2007) can account for

58 Chapter 4

these features, its novelness has admitted just a small group of biological materials to date when compared to CHARMM for example. For the membrane analysis used in the study in this chapter, the CHARMM force field is used and is given by the following equation:

$$V\left(\hat{R}\right) = \sum_{bonds} K_b \left(b - b_0\right)^2 + \sum_{angles} K_\theta \left(\theta - \theta_0\right)^2$$
$$+ \sum_{dihedrals} \sum_j K_\varphi \left(1 + \cos\left(n\varphi - \delta\right)\right)$$
$$+ \sum_{nonbonded\ pairs\ i,j} \varepsilon_{ij} \left[\left(\frac{R_{min,ij}}{r_{ij}}\right)^{12} - \left(\frac{R_{min,ij}}{r_{ij}}\right)^6\right] \tag{4.1}$$
$$+ \sum_{nonbonded\ pairs\ i,j} \frac{q_i q_j}{\varepsilon_D r_{ij}}$$

Here, V is the potential energy; K_b, K_θ, and K_ϕ are the bond, angle, and dihedral force constants, respectively; b and b_0 are the current and equilibrium bond distances; θ and θ_0 are the current and equilibrium bond angles; n is the multiplicity of the function; ϕ is the dihedral angle; δ is the phase shift; εij is the well depth; $Rmin_{ij}$ is the radius for the Lennard-Jones 6–12 term used for van der Waals interactions; r_{ij} is the distance between atoms i and j; q_i, q_j are point charges; and εD is the relative permittivity of free space. Together, these terms allow for the potential energy to be determined, which can then be used to determine the change in atomic positions using Newton's laws of motion combined with the existing atomic motions. One limitation of classical force fields, however, is that they cannot simulate chemical changes (bond breaking and forming) because they typically define each bond explicitly. For example, if you have two carbon atoms that are bonded, this bond would be explicitly defined when using a classical force field and would not break even if the two atoms were moved very far apart. Rather, moving them apart by some excessive distance would result in increased energy in the bond because any deviation from the equilibrium distance increases the amount of energy. Barring any algorithms to prevent it, this can lead to a cycle of increasing energy due to the simulation overcorrecting (i.e., atoms too close → too far → too close… etc.) and simulation instability. This phenomenon is often observable when starting from a bad configuration where atoms are too close to one another.

To address the aforementioned limitations, reactive atomic force fields like ReaxFF (Senftle et al., 2016; Russo and Van Duin, 2011) and MEAMBO (Mun et al., 2017) were developed. These force fields introduce the ability to break and form chemical bonds by describing bonds in terms of bond order (Senftle et al., 2016; Russo and Van Duin, 2011; Mun et al., 2017) rather than defining explicit bonds as used in classical force fields. This difference allows the modeling of chemical reactions without reverting to quantum methods. Furthermore, these force fields may not explicitly include certain energy terms, such as the van der Waals contribution, or may only include the interactions for certain types of molecules (Mun et al., 2017).

However, a model like MEAMBO will allow the admission of all of the elements of periodic table like iron that is on blood.

The computational costs of the aforementioned force fields are great which limits their application. Coarse-grained models are designed to allow larger models that can run much faster for a given system to reduce the computational cost, often drastically so, and thus increase the system size of molecules. Examples of this type of force field include the United Atom (Lee et al., 2014), Martini (Marrink et al., 2007), and Dry Martini (Arnarez et al., 2015) force fields, each of which offers a tradeoff of model resolution versus simulation speedup. Out of the examples listed, the United Atom force field groups hydrogen atoms with heavier atoms and contains the lowest level of coarse graining. This coarse graining allows the simulation to be sped up in two ways: (1) there are fewer particles to simulate and (2) the time step can be increased because it is based on the fastest bond in the simulation—note: a simulation with explicit hydrogen requires a time step of 0.5 femtoseconds unless bond constraint algorithms (e.g., SHAKE; Ciccotti and Ryckaert, 1986, RATTLE; Andersen, 1983, and SETTLE; Miyamoto and Kollman, 1992) are used. The Martini force field further increases the level of coarse graining by replacing multiple heavy atoms in each simulated particle, which allows for larger systems and additional performance boosts. Rather than increasing the level of coarse graining compared to the Martini force field, the Dry Martini force field replaces the explicit coarse-grained water molecules with implicit water. This change allows the simulation to keep water effect while minimizing the overhead simulation cost of water molecules, which can be considerable. Further, while these models are nonreactive, some recent coarse-grained models have begun implementing reactive capabilities as well (Dannenhoffer-Lafage and Voth, 2020).

Because higher levels of coarse graining reduce the model resolution, coarse-grained force fields may not reproduce some properties as well as equivalent atomic force fields. For example, for some systems, coarse-grained models have been shown to affect properties like self-diffusion (Yoon et al., 1993), packing density (Yoon et al., 1993), interchain interactions (Yoon et al., 1993), polymer chain folding (Li et al., 2010) as well as specific heat and thermal properties of polymers (Yoon et al., 1993). Additionally, coarse-grained models may give significantly different free volume calculations during stretching (Hossain et al., 2010). Hence, the ability to use any coarse-grained force field will depend on the properties being investigated. In any case, a thorough literature review should be performed to determine which force field(s), whether atomistic or coarse-grained, is appropriate for the problem being investigated.

4.2.3 Simulation ensembles of atoms

The thermodynamic ensemble(s) chosen for a simulation must also be chosen with care depending on the problem being investigated. These statistical ensembles represent a set of conditions

60   Chapter 4

where the system can be simulated in a way that is in equilibrium with observable parameters. The way this equilibrium is accomplished depends on each ensemble. However, they each provide a way to simulate systems at equilibrium while maintaining energy conservation.

While there are many different ensembles, the microcanonical (*NVE*), canonical (*NVT*), and isothermal–isobaric (*NPT*) are common in MD simulations. Note the abbreviated names (i.e., *NVE*, *NVT*, and *NPT*) help to quickly identify what is being directly defined in the simulations, where N is the number of particles, V is the volume, E is the energy, P is the pressure, and T is the temperature.

There are also *NPT* variations specific to membranes, including the *NPAT* and *NPγT* ensembles, where the A represents the area and γ represents the surface tension. These are specialized ensembles that control the membrane in a way that varies from the typical isotropic pressure control used by the default *NPT* ensemble. Specifically, the *NPAT* ensemble controls the lateral membrane area and the pressure normal to them membrane, and the *NPγT* ensemble controls the lateral surface tension and the pressure normal to them membrane (Ikeguchi, 2004). These ensembles allow for specific membrane properties to be controlled to match physically observed properties (e.g., area per lipid). This restriction is needed for many older force fields, which did not accurately reproduce surface tension and area per lipid properties when using *NPT*. However, many modern force fields should accurately replicate membrane physical properties using the standard *NPT* ensemble (Klauda et al., 2010; Dickson et al., 2014).

4.2.4 Boundary conditions

Boundary conditions describe how the outer boundaries of a system (i.e., the system box of atoms in the ensemble) are defined. Two common boundary condition types, periodic and nonperiodic boundary conditions can be used for the MD simulations. Systems with periodic boundary conditions replicate the primary unit cell in every direction. This boundary condition means that every periodic cell (or image) is identical, but it effectively allows macroscopic properties to be examined by making the structure seemingly much larger with a small computational cost. An example of periodic boundary conditions is shown in Fig. 4.3, which shows a unit cell with its adjacent periodic images in a 2D view for a lipid that can be found in the brain. In reality, additional periodic images "exist" in this system in all dimensions, which can allow particles to travel through an indefinite number of periodic images (within computational abilities). This particle travel is handled by tracking particle coordinates in relation to the unit cell combined with additional values recording which periodic image a particle is currently located. In effect, each atom that leaves one side of a cell re-enters the other side of the cell to maintain the same number of atoms, meaning the unit cell has all atoms in the system even though the "original" atom may no longer be there. Considering the unit cell and periodic cells are all the same, the exact cell an atom is located in may seem

Modeling nanoscale cellular structures using molecular dynamics 61

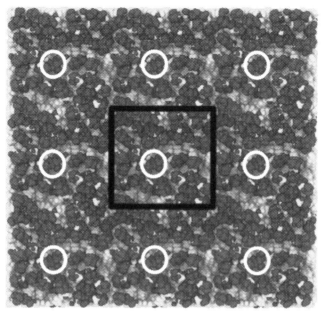

Fig. 4.3: Periodic image example showing the top (perpendicular) view of a phospholipid bilayer with water hidden with its eight identical periodic images. The primary unit cell is marked with a black box, and nine white circles mark the same spot in each bilayer structure.

unimportant—and that may indeed be true, depending on the problem. For some problems, however, it may be very important to identify which atoms belong to which cells. For example, some properties of interest, for example, diffusion characteristics, may require this information. It may also be visually preferable to view a system with atoms in their periodic images, which could mean the difference of having N chains bunched into the unit cell or a single chain stretching across the corresponding number of cells.

Limitations of periodicity include computational cost, box shape, minimum image convention, and system size variability. While there is some computational cost associated with periodic systems due to interactions with atoms in periodic images, this limitation is slowly becoming less of a concern with increasing computational capabilities. Another limitation is the shape of the system box, which must be able to fit into a periodic system in a way that leaves no empty space. A rectangular system box is often used and will work for most problems, but more complex shapes also exist that meet this requirement that may be more desirable (e.g., may fit a particular system better with less wasted space). Another significant limitation is known as the minimum image convention, which states that an atom should not be interacting with its own image. In practice, this means that the atomic interaction distances should be considered with regards to the system box size. It also means that care must be taken to ensure simulated systems do not thin too much during deformations, which can violate the minimum image convention. To counteract this effect, an extra thick layer of particles

62 Chapter 4

(e.g., water molecules) may be needed to ensure the system size does not decrease too much during deformations. The system size variability can also be an issue in periodic systems when a small unit cell is used. This limitation basically means that, due to all cells being the same, you cannot simulate a property that is larger than your unit cell. For example, if you simulate a small membrane (e.g., 10 nm square) using periodic boundary conditions, you will not see some membrane properties, such as undulations, that require a larger system. In this case, the small membrane system would not accurately be simulating properties of a larger membrane. This fact may or may not be important, depending on the problem being examined, but should be considered when setting up the simulation.

Many of the periodicity limitations mentioned above can be avoided by using nonperiodic boundary conditions, which only considers atoms in the unit cell (i.e., there are no periodic images). This change eliminates any interactions outside the box, imbuing lower computational costs for a system of the same size and no potential for an atom to interact with itself. Furthermore, the box shape is not restricted as it does not need to fit into a grid where empty space is a concern. However, it introduces additional options for how the system's boundaries will be implemented. Depending on the problem being examined, the walls can be allowed to change with the system or held constant while reflecting or deleting atoms when they encounter the wall. Alternatively, the walls can be lined with particles that may or may not interact with (repel or attract) other atoms in the system. These are just a few of the tactics for implementing nonperiodic boundary conditions. The chosen implementation should be considered carefully as it can affect simulation results and even cause a loss of energy conservation (e.g., the example listed above that deletes atoms would result in a loss of energy conservation with no other changes). The biggest nonperiodic boundary condition limitation (or possibly advantage, depending on the problem), however, is that the system becomes isolated. This isolation makes macroscale properties more difficult to replicate and can require much larger systems to examine a particular property, which may be more computationally expensive than a smaller system with periodic images.

A combination of periodic and nonperiodic systems may also be considered for some problems. For example, for equilibrium membrane systems, it may make sense to make the in-plane dimensions periodic and the out-of-plane dimension nonperiodic. This can allow a smaller out-of-plane dimension, with a correspondingly smaller water layer, and restrict particle movement between above and below the membrane while still allowing macroscale properties in the in-plane dimensions. When feasible, this setup could potentially allow for a simulation that is less computationally expensive than a fully periodic counterpart.

In addition to these boundary conditions, the system boundaries can also be used to simulate system-wide deformation conditions. This ability is of particular importance when considering injury biomechanics. A variety of deformation boundary conditions have been used previously in MD studies for deforming phospholipid bilayer structures (Tieleman et al.,

2003; Leontiadou et al., 2004; Tolpekina et al., 2004; Tomasini et al., 2010; Koshiyama and Wada, 2011; Shigematsu et al., 2014, 2015). These will be considered as two primary groups here: simulations where structure deformations were controlled through box pressures versus directly controlling the box dimensions.

The first group of simulations deformed the bilayer under a state of tension based on the lateral in-plane pressures. For example, Tieleman et al. (2003) varied the lateral plane pressure to deform the lipid bilayer. This applied lateral pressure resulted in pore nucleation at greater pressures and led to phospholipid bilayer destabilization thus destroying the membrane and in time the neuron. They noted that during the mechanical deformations the phospholipid bilayer exhibited considerable thinning before rupture. Leontiadou et al. (2004) used this method to demonstrate that pores were stable at low pressures when the pore was inserted prior to simulations but ruptures occurred at greater lateral pressures. One other point to consider is that controlling deformations using pressure can result in tearing the bilayer apart due to barostat feedback rather than due to pore nucleation (Tolpekina et al., 2004).

Alternatively, the phospholipid bilayer can be deformed by changing the dimensions of the in-plane box dimensions. Tomasini et al. (2010) used incremental in-plane boundary stretches followed by relaxation periods where the bilayer was allowed to adjust. While this method avoids the problem introduced by pressure controls, these increments did not represent a continuous deformation. A more continuous method was implemented by using unsteady state stretching in the in-plane direction through coordinate scaling (Koshiyama and Wada, 2011; Shigematsu et al., 2015). A variation of this unsteady stretching method allowed the molecules to flow with the deforming system box rather than coordinate scaling to better simulate the fluid nature of the membrane (Murphy et al., 2016, 2018). These unsteady stretching methods provided a continuous deformation method that allowed the phospholipid bilayer structure to form pores without allowing the structure to relax and partially recover while being deformed.

4.2.5 Current simulation details

The simulations used as examples here have been previously explored in Murphy et al. (2018). In summary, a simplified membrane structure containing 72 1-palmitoyl-2-oleoylphosphatidylcholine phospholipid and 9070 TIP3P water molecules (original structure sourced from Klauda et al., 2010 and 6828 TIP3P water molecules were added) was equilibrated for 10 nanoseconds and then deformed under different stress states. To maintain comparability, deformation velocities for each deformation were determined so that each strain state resulted in a constant von Mises strain rate. Furthermore, a constant temperature of 310 K, chosen to reflect the human body temperature, was also maintained for both equilibration and deformation simulations, and all simulations were performed using the MD code LAMMPS (Plimpton, 1995) paired with the CHARMM36 all-atom lipid force field (Klauda et al., 2010).

64   Chapter 4

Table 4.1: Summary of average deformation velocities and coefficients of variation (CV) correlating to a von Mises strain rate ($\dot{\varepsilon}_{VM}$) of 5.5×10^8 s^{-1}.

Strain loading state	$\dot{\varepsilon}_y/\dot{\varepsilon}_x$	v_x (m/s) ± CV	v_y (m/s) ± CV
Equibiaxial tension	1	3.9 ± 4.4%	3.8 ± 4.0%
2:1 Nonequibiaxial tension	½	4.5 ± 3.8%	2.2 ± 4.5%
4:1 Nonequibiaxial tension	¼	4.4 ± 1.3%	1.0 ± 5.6%
Strip biaxial tension	0	3.9 ± 4.4%	0.0 ± 0.0%
Uniaxial tension	n/a	3.9 ± 4.4%	Pressure of 1 atm

Deformations are focused on the in-plane (x and y) dimensions because of the membrane's planar structure and were applied by deforming the system box. Averaged deformation velocities for the current simulations are shown in Table 4.1. To allow comparison between strain states, velocities were chosen so that all strain states were deformed at the same von Mises strain rate ($\dot{\varepsilon}_{VM}$) of 5.5×10^8 s^{-1}. In all cases, the out-of-plane z dimension was allowed to adjust freely under a pressure of 1 atm while the system deformed.

Because the MD method employs stochasticity for the thermal vibrations, results are averaged. Therefore, each set of simulations were performed three times. This gave three different initial starting structures (dimensions of approximately 4.8 nm, 4.6 nm, and 14 nm for the x, y, and z dimensions, respectively).

Due to the membrane having a fluid nature, it did not experience plastic deformation and material failure as observed in solid materials under mechanical loading. Therefore, failure was presumed to be when initial water penetration occurred, which is the event where water penetrates through both phospholipid bilayer leaflets and forms a solid water bridge through them because the membrane has been compromised. Hence, the initial water penetration is the straightforward and simple way to define failure of the membrane throughout this chapter.

4.2.6 Molecular dynamics analysis methods for the phospholipid bilayer (neuron membrane)

When determining what information to pass on to higher length scales, one must consider the downscaling requisites as discussed earlier in Bridges 2 and 6 in Fig. 4.1. For the current simulations, analysis revolved two distinct avenues: calculating the mechanical response via stress and strain behavior and using 2D image analysis to calculate the stereological properties of damage. Many studies use the areal stretch or areal strain rather than von Mises strain to describe the change in membrane size. However, while areal stretch and areal strain can describe the change in area for a single strain state, they do not allow the different strain states to be directly compared. Hence, the methods described herein serve only as a few representative examples for the types of analyses that were performed. Some additional methods include

Modeling nanoscale cellular structures using molecular dynamics   65

headgroup mean nearest neighbor distance (Murphy et al., 2018), lipid chain order parameter (Shigematsu et al., 2015), molecule tilt angle (Shigematsu et al., 2015), and membrane rupture rate (Koshiyama and Wada, 2011), electrostatic potential (Koshiyama and Wada, 2011), and mass density profile (Koshiyama and Wada, 2011). Therefore, the reader will have to determine what methods and metrics will best enable them to explore their problem, and we encourage them to keep in mind that sometimes inspiration can be taken from seemingly unrelated fields.

4.2.6.1 Stress–strain behavior of the neuron membrane

In addition to image analysis, mechanical properties can be quantified from the membrane simulations to directly investigate other system properties, such as the stress response. Many membrane computational studies have used membrane surface tension as the stress metric (Tomasini et al., 2010; Koshiyama and Wada, 2011), which is given by the following,

$$\gamma = \left(\sigma_z - \frac{\sigma_x + \sigma_y}{2} \right) * l_z \tag{4.2}$$

where γ is the surface tension; σ_x, σ_y, and σ_z are the x, y, and z pressures terms, respectively; and l_z is the atomic box height. The surface tension property is typically sufficient for computational studies examining membranes at equilibrium, because it allows a comparison to theoretical and experimental models. However, this surface tension equation has an assumption of a thin water layer (Zhang et al., 1995), which is not true when performing membrane deformation simulations because these simulations require a thick water layer that is two-plus times the bilayer height. Without this additional water, there is not enough water to maintain periodic boundary conditions because the water layer thins drastically during the applied deformation. Therefore, two problems arise with using surface tension for deformation simulations. The first issue is that the surface tension result becomes dependent on how much water is in the system, and the second issue is that the water layer reduction results in a corresponding drop in the surface tension when using the box dimensions that is not reflected by the true stress response. Therefore, l_z must be estimated based on the bilayer height or some other factor. Alternatively, the von Mises stress can be used, which can be calculated using only the pressure terms, which is defined as the following,

$$\sigma_{VM} = \left(\frac{1}{2} * \left[\left(\sigma_x - \sigma_y \right)^2 + \left(\sigma_y - \sigma_z \right)^2 + \left(\sigma_z - \sigma_x \right)^2 \right] + 3 * \left[\sigma_{xy}^2 + \sigma_{yz}^2 + \sigma_{xz}^2 \right] \right)^{0.5} \tag{4.3}$$

As shown in Fig. 4.4, the surface tension dependence on system height can result in an artificial drop in stress compared to von Mises stress when deforming at high strain rates. Therefore, the von Mises stress was chosen as the stress metric for the current deformation simulations. Additional details of this difference can be found in the supplemental materials of Murphy et al. (2016).

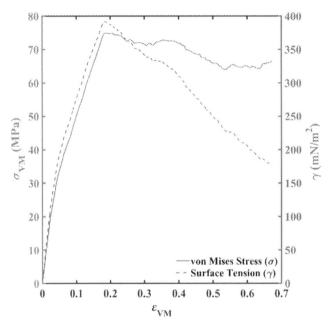

Fig. 4.4: Comparison of von Mises stress–strain behavior and surface tension behavior versus von Mises strain for a representative equibiaxial deformation $\dot{\varepsilon}_{VM}$ = 5.5 × 10^8 s^{-1}.

Furthermore, von Mises strain was chosen due to it allowing multiple strain states to be compared, which can be calculated as the following,

$$\varepsilon_{VM} = \left(\frac{2}{9} * \left[\left(\varepsilon_x - \varepsilon_y \right)^2 + \left(\varepsilon_y - \varepsilon_z \right)^2 + \left(\varepsilon_z - \varepsilon_x \right)^2 \right] + \frac{4}{3} * \left[\varepsilon_{xy} + \varepsilon_{yz} + \varepsilon_{xz} \right]^2 \right)^{0.5} \quad (4.4)$$

where ε_{VM} is von Mises strain, and ε_x, ε_y, and ε_z are the true strains for the x, y, and z dimensions of the simulation box. A few assumptions can be used to simplify these equations. First, because the membrane is planar, the primary focus is on the in-plane properties. Thus, the z dimension stress is negligible, and the z dimension's properties can be omitted. Second, the orthogonal simulation box has no shear strains and negligible shear stresses. Therefore, the shear components (xy, xz, and yz) can be assumed to be zero. These assumptions allow the von Mises stress and strain equations to be simplified to the following:

$$\sigma_{VM} = \left(\frac{1}{2} * \left[\left(\sigma_x - \sigma_y \right)^2 + \left(\sigma_y \right)^2 + \left(\sigma_x \right)^2 \right] \right)^{0} \quad (4.5)$$

$$\varepsilon_{VM} = \left(\frac{2}{9} * \left[\left(\varepsilon_x - \varepsilon_y \right)^2 + \left(\varepsilon_y \right)^2 + \left(\varepsilon_x \right)^2 \right] \right)^{0.5} \quad (4.6)$$

The von Mises strain equation was also used to determine the equivalent von Mises strain rate, $\dot{\varepsilon}_{VM}$, by replacing ε_x and ε_y with the corresponding strain rates in the x and y dimensions

($\dot{\varepsilon}_x$ and $\dot{\varepsilon}_y$). Here, $\dot{\varepsilon}_y$ was assumed to be zero when calculating the equivalent von Mises strain rate for the uniaxial strain state since it is not controlled, but ε_y was included in uniaxial strain calculations.

In addition to considering the method for combining the stress components, the type of stress going into these equations should also be considered. First, consider the commonly used equation for macroscale uniaxial stress

$$\sigma = \frac{F}{A} \tag{4.7}$$

where σ is the stress, F is the force, and A is the cross-sectional area. At the macroscale, this equation can be easily solved using simulated and experimental methods provided the force and the corresponding area can be determined. However, macroscale stress formulations assume that the material being examined is continuous (i.e., a continuum), which does not hold true when considering systems at the nanoscale and lower length scale simulation methods (Horstemeyer, 2012; Cranford and Buehler, 2012). Specifically, in MD, the volume area of individual atoms is not well defined and, therefore, cannot be considered continuous. Therefore, a different definition of stress is necessary.

This alternative stress definition is virial stress, which can be used for atomic stress calculations. Rather than determining the stress for each individual atom, the virial stress takes an averaged value of all atoms in the specified volume, which makes it inappropriate to use the virial stress as a point stress for individual atoms (Horstemeyer, 2012; Cranford and Buehler, 2012; Zimmerman et al., 2004). This formulation provides stress components for the entire system rather than across a cross-sectional area. This difference means that virial stress does not translate directly to macroscale stress formulations, such as the Piola–Kirchhoff (engineering) stress and the Cauchy (true) stress. These differences have led to much debate on how to best relate virial stress to macroscale stress (Zimmerman et al., 2004; Zhou, 2003; Liu and Qiu, 2009). Hence, care must be taken when considering the stress magnitudes observed on the nanoscale in relation to macroscale stress values for the larger structure.

4.2.6.2 Image analysis for stereological quantification of neuron membrane damage

For image analysis, an important point to consider is what visualization style and particle size will be used. These choices will be determined by the system and property being visualized, which may even affect what visualization software is chosen. For example, the program OVITO (Open Visualization Tool) (Stukowski, 2010) is often used for inorganic systems, and the program VMD (Visual Molecular Dynamics) (Humphrey et al., 1996) is typically used more for biological systems. This distinction often is due to the programs' originally designed purpose (i.e., what type of material or analysis was the program originally made for), which affects which visual settings and analysis tools are available.

68 Chapter 4

For the current simulations, the visualization software OVITO (Stukowski, 2010) was used due to its ability to natively read LAMMPS output files and easily output images of the system in a format that could be postprocessed. For creating the images, atoms were displayed as individual spheres using their van der Waals radii to represent their size (Bondi, 1964). Be aware that, depending on the software, this behavior may not be the default setting. While not normally a problem when only viewing the structure, this setting will affect any image analysis results and must be considered when performing this type of analysis. The size factor is also particularly important when considering simulations using coarse-grained particles. If a user leaves the default setting for particle size rather than defining the particle size based on the coarse-grained model's particle dimensions, it may even look like pores are present in the equilibrated structure, which would certainly make identifying and quantifying true pores difficult!

Another important consideration is how periodic boundary conditions are handled during image analysis, presuming they are used in the simulation. When measuring pores via image analysis and using periodic boundary conditions, it is possible for pores to cross from the unit cell to a neighboring cell. This effect can result in a single pore being registered as multiple pores with its full area being split accordingly. To address this problem, one can display the unit cell with the surrounding cells and identify and remove any duplicate pores. For example, in the current case of a planar phospholipid bilayer, displaying 3 by 3 by 1 periodic images (x by y by z) showing the perpendicular (top) view of the structure with water hidden shows a total of nine identical images of the bilayer structure. In this new visualization, all pores are displayed in their whole form in at least one location. To further increase the contrast, all visible atoms were also colored black while the background was colored white, as shown in Fig. 4.5. Then, any partial pores on the exterior periodic image boundaries and duplicate pores due to periodic images can be removed and unique pores can be identified and quantified using image analysis. A custom MATLAB (MATLAB, 2013) function was written to perform this analysis.

The results for these pores can then be used to determine pore density and mean pore area, which can be used to calculate damage using

$$\phi = \eta * \upsilon * c \qquad (4.8)$$

where ϕ is the damage, η is the nucleation, υ is the growth, and c is the coalescence (Horstemeyer et al., 2000). While pore density can be used directly for the nucleation (η) term, the growth and coalescence terms are not easily separately measured. Therefore, the product of the pore growth and coalescence ($\upsilon*c$) terms is represented by the mean pore area.

In addition to these pore properties, strains for separately the events of interest, such as when pores reach specific size thresholds, can be found. This portion of the image analysis requires that the pore dimensions be examined. For the current analysis, the minor axis length was

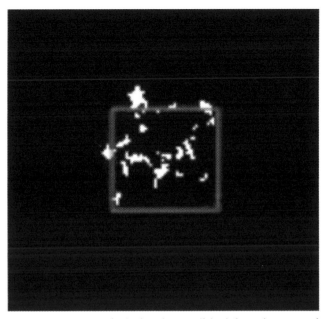

Fig. 4.5: Representative image analysis for the equibiaxial strain state when deformed at $\dot{\varepsilon}_{VM} = 5.5 \times 10^8$ s^{-1}. Duplicate pores from periodic images have been removed so that only unique pores are shown.

chosen as the size metric under the premise that molecules can only penetrate through the membrane when both dimensions are sufficiently large. Biologically relevant to TBI, a minor axis length greater than 3 Å was chosen to detect pores large enough to allow passage of water (Narten and Levy, 1971) and 10 Å to detect pores large enough to allow penetration of a calcium ion with two hydration layers (Schwenk and Rode, 2004). Additionally, to prevent the violation of periodic boundaries, the data cutoff for analyses can be determined for each simulation using this pore information. Specifically, some max pore dimension or area criteria, such as a pore having a major axis length that is greater than either half of or the whole simulation box, can be defined to create a consistent data cutoff criterion.

4.3 Results and discussion for the phospholipid bilayer (neuron membrane)

There are many potential ways to analyze structures when performing MD, and membrane deformations have been previously shown to be both stress state (Murphy et al., 2018) and strain (loading) rate (Tomasini et al., 2010; Koshiyama and Wada, 2011; Murphy et al., 2016) dependent. Therefore, the focus here will be on the stress–strain and image analysis (damage) results for the equibiaxial strain state as a demonstrative case. The concept of a membrane failure limit diagram (MFLD) is also introduced. A failure limit diagram illustrates the domains of a material's failure region (Horstemeyer, 2000). Readers are directed to previous

70 Chapter 4

publications for a more comprehensive examination of multiple strain states and strain rates and additional analyses.

4.3.1 Stress–strain and damage response

The stress–strain behavior of the membrane (and other structures) provides insights into how the membrane reacts to deformations, including how the resulting forces are affected by different strain (loading) rates and stress states. Importantly, both stress and strain are normalized against a reference frame (i.e., the current or initial system dimensions). We note that the strain state affects the stress magnitude and failure strain (Murphy et al., 2018), and strain rate has been shown to affect the stress magnitude and yield strain (Murphy et al., 2016).

Strain rate has been more thoroughly examined with respect to stress (Tomasini et al., 2010; Koshiyama and Wada, 2011; Murphy et al., 2016), which may be quantified as surface tension. The stress magnitude and yield strain have been shown to increase with strain rate at higher strain rates (Koshiyama and Wada, 2011; Murphy et al., 2016). Furthermore, the strain rate sometime affects the shape of the stress–strain behavior with lower strain rates not showing a pronounced yield (Murphy et al., 2016). Additionally, the stress magnitude has also been shown to match quasi-static stress results (i.e., to be strain-rate independent) at low loading rates (Koshiyama and Wada, 2011).

Of all the possible deformation paths, the equibiaxial strain state has been shown to be the most detrimental in-plane strain state (Murphy et al., 2018). Shown in Fig. 4.6(A), the equibiaxial strain state produced the greatest stresses and the lowest failure strains for an equivalent amount of von Mises strain (i.e., deformation) (Murphy et al., 2018).

For this particular applied strain rate, the stress quickly increases to a peak at yield and then decreases slowly, which is typical for this phospholipid bilayer membrane (Murphy et al., 2018). As mentioned, the equibiaxial strain state simulations exhibit the largest stresses, while the uniaxial and strip biaxial simulations exhibit the lowest stresses (Murphy et al., 2018).

Beyond the yield point, where weakening starts, the damage progression will finally end up in failure or rupture of the membrane. Damage from membrane mechanoporation allows assessing the combined effect of how many pores nucleated and grow. Fig. 4.6(B)shows the equibiaxial damage (pore volume fraction) behavior growing quickly after yield. The pore nucleation and growth (multiplicatively shown in Fig. 4.6(B) as f) are what weakens the membrane (shown as the negative slope in Fig. 4.6(A) after yield).

4.3.2 Membrane failure limit diagram

Looking to methods used to investigate other materials as inspiration can sometimes lead to interesting often yields interesting findings. For example, the idea behind MFLDs was originally taken from forming limit diagrams (Murphy et al., 2018), which are used for

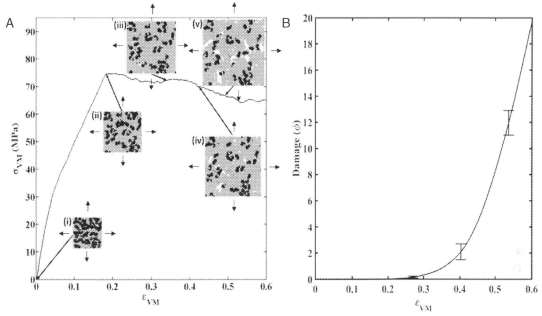

Fig. 4.6: (A) Representative von Mises stress–strain curve for the equibiaxial strain state at $\dot{\varepsilon}_{VM} = 5.5 \times 10^8$ s^{-1} and (B) Damage versus von Mises strain for the equibiaxial strain state. In (A), representative top view snapshots of the phospholipid bilayer (water hidden) show the phospholipid bilayer with red phospholipid headgroups and green phospholipid tails at the (i) initial structure, (ii) yield, (iii) first pore with diameter larger than 3 Å, (iv) initial water penetration (failure), and (v) the first pore with diameter larger than 10 Å.

investigating sheet metal failure for a given strain space (Horstemeyer, 2000; Keeler and Backofen, 1963; Safari et al., 2011). A simple MFLD is shown in Fig. 4.7.

The MFLD provides an easy-to-visualize way to determine where weakening and failure will occur depending on the principal in-plane strains. It also provides an easy way to identify strains at which a neuron cell will be weakened, which can lead to a loss of homeostasis and cellular death. As shown in Fig. 4.7, limit lines distinguish which principal strains should be considered "safe" versus unsafe ("failure"), which correlates to when the membrane integrity is known to be compromised. The safe zone can also be further broken into zones indicating into "safe" and "transition" zones, where the transition zone shows where failure at lower strains is suspected to be possible but unconfirmed.

In addition to being easy to visualize, MFLD limit lines also have the potential to allow simple mechanistic equations to be developed that could be implemented both on their own or in other models to help predict injury. For example, Murphy et al. (2018) showed that the relationship between pore size events (i.e., failure and pores large enough for water or calcium ions) exhibited a linear relationship across strain states. These equations can be used

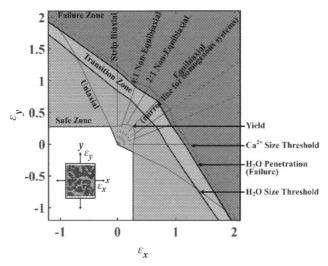

Fig. 4.7: Average ε_y-ε_x true strain membrane failure limit diagram (MFLD) for yield, water size threshold (d = 3 Å), failure, and calcium size threshold (d = 10 Å) for equibiaxial, nonequibiaxial, strip biaxial, and uniaxial tension deformations at $\dot{\varepsilon}_{VM}$ = 5.5 × 10^8 s^{-1}. The zones indicate expected damage with the failure zone (red) being dangerous strains, transition zone (blue) being potentially dangerous strains, and the safe zone (green) being safe strains.

to correlate membrane strain to potential for injury for a given system, which can to quickly identify if a set of strains would be deemed safe or unsafe (Murphy et al., 2018). Overall, the MFLD has the potential to be a useful tool for identifying membrane deformations that may lead to injury.

4.4 Summary

Investigations using MD to study membrane molecules on the outer layer of neuron can provide unique insights into the mechanisms affecting higher length scales related to weakening and failure. The MD analysis herein focused on the deformation of an equibiaxial strain state as a representative case, because it is typically the most deleterious to a structure. This example demonstrates how mechanisms affect the cellular response related to TBI can be modeled using nanoscale methods. These mechanical damage mechanisms and failure limits can now be upscaled into a macroscale MPISV material model and used to infer cellular and tissue death arising from mechanical impacts. Furthermore, a MFLD for a neuron cell is presented so that the macroscale failure strains can be defined.

What has yet to be done is to conduct this type of failure analysis for other structures of the neuron and also the axon. Also, the CHARMM (Russo and Van Duin, 2011) database, although fine for this study, is limited and would not be able to capture reactions prevalent in the brain. As such, the MEAMBO atomistic potential described in Chapter 3 is more likely to capture the rest of the entities in the brain. Hence, future studies are recommended to employ MEAMBO to achieve the full multiscale modeling of the brain.

Acknowledgments

This material is based upon work supported by the Department of Agricultural and Biological Engineering and the Center for Advanced Vehicular Systems at Mississippi State University. Further, material presented in this paper is a product of the CREATE-GV Element of the Computational Research and Engineering Acquisition Tools and Environments (CREATE) Program sponsored by the US Department of Defense HPC Modernization Program Office. This effort was sponsored under contract number W912HZ-13-C-0037.

References

Andersen, H.C., 1983. Rattle: a "velocity" version of the shake algorithm for molecular dynamics calculations. J. Comput. Phys. 52, 24–34. https://doi.org/10.1016/0021-9991(83)90014-1.

Arnarez, C., Uusitalo, J.J., Masman, M.F., Ingólfsson, H.I., de Jong, D.H., Melo, M.N., et al., 2015. Dry Martini, a coarse-grained force field for lipid membrane simulations with implicit solvent. J. Chem. Theory Comput. 11, 260–275. https://doi.org/10.1021/ct500477k.

Bondi, A., 1964. van der Waals volumes and radii. J. Phys. Chem. 68, 441–451. https://doi.org/10.1021/j100785a001.

Ciccotti, G., Ryckaert, J.P., 1986. Molecular dynamics simulation of rigid molecules. Comput. Phys. Rep. 4, 346–392. https://doi.org/10.1016/0167-7977(86)90022-5.

Cooper, S.T., McNeil, P.L., 2015. Membrane repair: mechanisms and pathophysiology. Physiol. Rev. 95, 1205–1240. https://doi.org/10.1152/physrev.00037.2014.

Cranford, S.W., Buehler, M.J., 2012. Biomateriomics. Springer, Dordrecht, Netherlands.

Cullen, D.K., Vernekar, V.N., LaPlaca, M.C., 2011. Trauma-induced plasmalemma disruptions in three-dimensional neural cultures are dependent on strain modality and rate. J. Neurotrauma 28, 2219–2233. https://doi.org/10.1089/neu.2011.1841.

Dannenhoffer-Lafage, T., Voth, G.A., 2020. Reactive coarse-grained molecular dynamics. J. Chem. Theory Comput. 16, 2020. https://doi.org/10.1021/acs.jctc.9b01140

Dickson, C.J., Madej, B.D., Skjevik, Å.A., Betz, R.M., Teigen, K., Gould, I.R., et al., 2014. Lipid14: the amber lipid force field. J. Chem. Theory Comput. 10, 865–879. https://doi.org/10.1021/ct4010307.

El Sayed, T., Mota, A., Fraternali, F., Ortiz, M., 2008. A variational constitutive model for soft biological tissues. J. Biomech. 41, 1458–1466. https://doi.org/10.1016/j.jbiomech.2008.02.023.

Evans, E., Heinrich, V., Ludwig, F., Rawicz, W., 2003. Dynamic tension spectroscopy and strength of biomembranes. Biophys. J. 85, 2342–2350. https://doi.org/10.1016/S0006-3495(03)74658-X.

Farkas, O., Povlishock, J.T., 2007. Cellular and subcellular change evoked by diffuse traumatic brain injury: a complex web of change extending far beyond focal damage. In: Weber, J.T., Andrew, I.R.M. (Eds.). Progress in Brain Research, 161. Elsevier, Amsterdam, Boston, pp. 43–59. https://doi.org/10.1016/S0079-6123(06)61004-2.

Garimella, H.T., Menghani, R.R., Gerber, J.I., Sridhar, S., Kraft, R.H., 2019. Embedded finite elements for modeling axonal injury. Ann. Biomed. Eng. 47, 1889–1907. https://doi.org/10.1007/s10439-018-02166-0.

Geddes, D.M., Cargill 2nd, R.S., LaPlaca, M.C., 2003. Mechanical stretch to neurons results in a strain rate and magnitude-dependent increase in plasma membrane permeability. J. Neurotrauma 20, 1039–1049. https://doi.org/10.1089/089771503770195885.

Gwag, B.J., Canzoniero, L.M.T., Sensi, S.L., DeMaro, J.A., Koh, J.Y., Goldberg, M.P., et al., 1999. Calcium ionophores can induce either apoptosis or necrosis in cultured cortical neurons. Neuroscience 90, 1339–1348. https://doi.org/10.1016/S0306-4522(98)00508-9.

Hochmuth, R.M., 2000. Micropipette aspiration of living cells. J. Biomech. 33, 15–22.

Horn, H.W., Swope, W.C., Pitera, J.W., Madura, J.D., Dick, T.J., Hura, G.L., et al., 2004. Development of an improved four-site water model for biomolecular simulations: TIP4P-Ew. J. Chem. Phys. 120, 9665–9678. https://doi.org/10.1063/1.1683075.

Horstemeyer, M.F., 2000. A numerical parametric investigation of localization and forming limits. Int. J. Damage Mech. 9, 255–285. https://doi.org/10.1177/105678950000900304.

Horstemeyer, M.F., 2012. Integrated Computational Materials Engineering (ICME) for Metals: Using Multiscale Modeling to Invigorate Engineering Design with Science. Wiley, Hoboken, New Jersy.

Horstemeyer, M.F., Lathrop, J., Gokhale, A.M., Dighe, M., 2000. Modeling stress state dependent damage evolution in a cast Al–Si–Mg aluminum alloy. Theor. Appl. Fract. Mech. 33, 31–47. https://doi.org/10.1016/S0167-8442(99)00049-X.

Hossain, D., Tschopp, M.A., Ward, D.K., Bouvard, J.L., Wang, P., Horstemeyer, M.F., 2010. Molecular dynamics simulations of deformation mechanisms of amorphous polyethylene. Polymer (Guildf) 51, 6071–6083. https://doi.org/10.1016/J.POLYMER.2010.10.009.

Humphrey, W., Dalke, A., Schulten, K., 1996. VMD: visual molecular dynamics. J. Mol. Graph 14, 33–38. https://doi.org/10.1016/0263-7855(96)00018-5.

Ikeguchi, M., 2004. Partial rigid-body dynamics in NPT, NPAT and NPγT ensembles for proteins and membranes. J. Comput. Chem. 25, 529–541. https://doi.org/10.1002/jcc.10402.

Izadi, S., Anandakrishnan, R., Onufriev, A.V., 2014. Building water models: a different approach. J. Phys. Chem. Lett. 5, 3863–3871. https://doi.org/10.1021/jz501780a.

Jorgensen, W.L., Chandrasekhar, J., Madura, J.D., Impey, R.W., Klein, M.L., 1983. Comparison of simple potential functions for simulating liquid water. J. Chem. Phys. 79, 926–935. https://doi.org/10.1063/1.445869.

Kampfl, A., Posmantur, R., Nixon, R., Grynspan, F., Zhao, X., Liu, S.J., et al., 1996. μ-Calpain activation and calpain-mediated cytoskeletal proteolysis following traumatic brain injury. J. Neurochem. 67, 1575–1583. https://doi.org/10.1046/j.1471-4159.1996.67041575.x.

Keeler, S.P., Backofen, W.A., 1963. Plastic instability and fracture in sheets stretched over rigid punches. Trans. Am. Soc. Met. Q. 56, 25–48.

Khalak, Y., Baumeier, B., Karttunen, M., 2018. Improved general-purpose five-point model for water: TIP5P/2018. J. Chem. Phys. 149, 224507. https://doi.org/10.1063/1.5070137.

Klauda, J.B., Venable, R.M., Freites, J.A., O'Connor, J.W., Tobias, D.J., Mondragon-Ramirez, C., et al., 2010. Update of the CHARMM all-atom additive force field for lipids: validation on six lipid types. J. Phys. Chem. B 114, 7830–7843. https://doi.org/10.1021/jp101759q.

Koshiyama, K., Wada, S., 2011. Molecular dynamics simulations of pore formation dynamics during the rupture process of a phospholipid bilayer caused by high-speed equibiaxial stretching. J. Biomech. 44, 2053–2058. https://doi.org/10.1016/j.jbiomech.2011.05.014.

LaPlaca, M., Thibault, L., 1997. An in vitro traumatic injury model to examine the response of neurons to a hydrodynamically-induced deformation. Ann. Biomed. Eng. 25, 665–677. https://doi.org/10.1007/BF02684844.

LaPlaca, M.C., Cullen, D.K., McLoughlin, J.J., Cargill II, R.S., 2005. High rate shear strain of three-dimensional neural cell cultures: a new in vitro traumatic brain injury model. J. Biomech. 38, 1093–1105. https://doi.org/10.1016/j.jbiomech.2004.05.032.

LaPlaca, M.C., Prado, G.R., 2010. Neural mechanobiology and neuronal vulnerability to traumatic loading. J. Biomech. 43, 71–78. https://doi.org/10.1016/j.jbiomech.2009.09.011.

Leach, A.A., 2001. Molecular Modelling: Principles and Applications, second ed. Prentice Hall, Harlow, England.

Lee, S., Tran, A., Allsopp, M., Lim, J.B., Hénin, J., Klauda, J.B., 2014. CHARMM36 united atom chain model for lipids and surfactants. J. Phys. Chem. B 118, 547–556. https://doi.org/10.1021/jp410344g.

Leontiadou, H., Mark, A.E., Marrink, S.J., 2004. Molecular dynamics simulations of hydrophilic pores in lipid bilayers. Biophys. J. 86, 2156–2164. https://doi.org/10.1016/S0006-3495(04)74275-7.

Li, C., Choi, P., Sundararajan, P.R., 2010. Simulation of chain folding in polyethylene: a comparison of united atom and explicit hydrogen atom models. Polymer (Guildf) 51, 2803–2808. https://doi.org/10.1016/j.polymer.2010.04.049.

Liu, B., Qiu, X., 2009. How to compute the atomic stress objectively? J. Comput. Theor. Nanosci. 6, 1081–1089.

Marrink, S.J., Risselada, H.J., Yefimov, S., Tieleman, D.P., de Vries, A.H., 2007. The MARTINI force field: coarse grained model for biomolecular simulations. J. Phys. Chem. B 111, 7812–7824. https://doi.org/10.1021/jp071097f.

MATLAB (9.9.0.1495850 (R2020b)), 2020. The MathWorks, Inc. Natick, MA.

Mendis, K.K., Stalnaker, R.L., Advani, S.H., 1995. A constitutive relationship for large deformation finite element modeling of brain tissue. J. Biomech. Eng. 117, 279–285. https://doi.org/10.1115/1.2794182.

Miller, K., 1999. Constitutive model of brain tissue suitable for finite element analysis of surgical procedures. J. Biomech. 32, 531–537. https://doi.org/10.1016/S0021-9290(99)00010-X.

Miyamoto, S., Kollman, P.A., 1992. Settle: an analytical version of the SHAKE and RATTLE algorithm for rigid water models. J. Comput. Chem. 13, 952–962. https://doi.org/10.1002/jcc.540130805.

Mun, S., Bowman, A.L., Nouranian, S., Gwaltney, S.R., Baskes, M.I., Horstemeyer, M.F., 2017. Interatomic potential for hydrocarbons on the basis of the modified embedded-atom method with bond order (MEAM-BO). J. Phys. Chem. A 121, 1502–1524. https://doi.org/10.1021/acs.jpca.6b11343.

Murphy, M.A., Horstemeyer, M.F.F., Gwaltney, S.R., Stone, T., Laplaca, M.C., Liao, J., et al., 2016. Nanomechanics of phospholipid bilayer failure under strip biaxial stretching using molecular dynamics. Model. Simul. Mater. Sci. Eng. 24, 055008. https://doi.org/10.1088/0965-0393/24/5/055008.

Murphy, M.A., Mun, S., Horstemeyer, M.F., Baskes, M.I., Bakhtiary, A., LaPlaca, M.C., et al., 2018. Molecular dynamics simulations showing 1-palmitoyl-2-oleoyl-phosphatidylcholine (POPC) membrane mechanoporation damage under different strain paths. J. Biomol. Struct. Dyn. 37, 1–14. https://doi.org/10.1080/07391102.2018.1453376.

Narten, A.H., Levy, H.A., 1971. Liquid water: molecular correlation functions from X-ray diffraction. J. Chem. Phys. 55, 2263–2269. https://doi.org/10.1063/1.1676403.

Needham, D., Nunn, R.S., 1990. Elastic deformation and failure of lipid bilayer membranes containing cholesterol. Biophys. J. 58, 997–1009. https://doi.org/10.1016/S0006-3495(90)82444-9.

Ovalle-García, E., Torres-Heredia, J.J.J., Antillón, A., Ortega-Blake, I.I., Ovalle-García, E., Torres-Heredia, J.J.J., et al., 2011. Simultaneous determination of the elastic properties of the lipid bilayer by atomic force microscopy: bending, tension, and adhesion. J. Phys. Chem. B 115, 4826–4833. https://doi.org/10.1021/jp111985z.

Picas, L., Milhiet, P.-E., Hernández-Borrell, J., 2012. Atomic force microscopy: a versatile tool to probe the physical and chemical properties of supported membranes at the nanoscale. Chem. Phys. Lipids 165, 845–860. https://doi.org/10.1016/j.chemphyslip.2012.10.005.

Plimpton, S., 1995. Fast parallel algorithms for short-range molecular dynamics. J. Comput. Phys. 117, 1–19. https://doi.org/10.1006/jcph.1995.1039.

Prabhu, R., Horstemeyer, M.F., Tucker, M.T., Marin, E.B., Bouvard, J.L., Sherburn, J.A., et al., 2011. Coupled experiment/finite element analysis on the mechanical response of porcine brain under high strain rates. J. Mech. Behav. Biomed. Mater. 4, 1067–1080. https://doi.org/10.1016/j.jmbbm.2011.03.015.

Price, D.J., Brooks, C.L., 2004. A modified TIP3P water potential for simulation with Ewald summation. J. Chem. Phys. 121, 10096–11103. https://doi.org/10.1063/1.1808117.

Rashid, B., Destrade, M., Gilchrist, M.D., 2014. Mechanical characterization of brain tissue in tension at dynamic strain rates. Forensic Biomech. 33, 43–54. https://doi.org/10.1016/j.jmbbm.2012.07.015.

Rawicz, W., Smith, B.A., McIntosh, T.J., Simon, S.A., Evans, E., 2008. Elasticity, strength, and water permeability of bilayers that contain raft microdomain-forming lipids. Biophys. J. 94, 4725–4736. https://doi.org/10.1529/biophysj.107.121731.

Russo, M.F., Van Duin, A.C.T., 2011. Atomistic-scale simulations of chemical reactions: bridging from quantum chemistry to engineering. Nucl. Instrum. Methods Phys. Res. Sect. B Beam Interact. Mater. Atoms. 269, 1549–1554. https://doi.org/10.1016/j.nimb.2010.12.053.

Safari, M., Hosseinipour, S.J., Azodi, H.D., 2011. Experimental and numerical analysis of forming limit diagram (FLD) and forming limit stress diagram (FLSD). Mater. Sci. Appl. 02, 496–502. https://doi.org/10.4236/msa.2011.25067.

Schwenk, C.F., Rode, B.M., 2004. Ab initio QM/MM MD simulations of the hydrated Ca^{2+} ion. Pure Appl. Chem. 76, 37–47. https://doi.org/10.1351/pac200476010037.

Senftle, T.P., Hong, S., Islam, M.M., Kylasa, S.B., Zheng, Y., Shin, Y.K., et al., 2016. The ReaxFF reactive force-field: development, applications and future directions. NPJ Comput. Mater. 2, 1–14. https://doi.org/10.1038/npjcompumats.2015.11.

Shigematsu, T., Koshiyama, K., Wada, S., 2014. Molecular dynamics simulations of pore formation in stretched phospholipid/cholesterol bilayers. Chem. Phys. Lipids 183, 43–49. https://doi.org/10.1016/j.chemphyslip.2014.05.005.

Shigematsu, T., Koshiyama, K., Wada, S., 2015. Effects of stretching speed on mechanical rupture of phospholipid/cholesterol bilayers: molecular dynamics simulation. Sci. Rep. 5, 15369. https://doi.org/10.1038/srep15369.

Stukowski, A., 2010. Visualization and analysis of atomistic simulation data with OVITO—the Open Visualization Tool. Model. Simul. Mater. Sci. Eng. 18, 15012.

Terasaki, M., Miyake, K., McNeil, P.L., 1997. Large plasma membrane disruptions are rapidly resealed by Ca^{2+}-dependent vesicle–vesicle fusion events. J. Cell Biol. 139, 63–74.

Tieleman, D.P., Leontiadou, H., Mark, A.E., Marrink, S.-J.J., 2003. Simulation of pore formation in lipid bilayers by mechanical stress and electric fields. J. Am. Chem. Soc. 125, 6382–6383. https://doi.org/10.1021/ja029504i.

Tolpekina, T.V., den Otter, W.K., Briels, W.J., 2004. Simulations of stable pores in membranes: system size dependence and line tension. J. Chem. Phys. 121, 8014–8020. https://doi.org/10.1063/1.1796254.

Tomasini, M.D., Rinaldi, C., Tomassone, M.S., 2010. Molecular dynamics simulations of rupture in lipid bilayers. Exp. Biol. Med. (Maywood) 235, 181–188. https://doi.org/10.1258/ebm.2009.009187.

Wang, J., Wolf, R.M., Caldwell, J.W., Kollman, P.A., Case, D.A., 2004. Development and testing of a general amber force field. J. Comput. Chem. 25, 1157–1174. https://doi.org/10.1002/jcc.20035.

Yoon, D.Y.Y., Smith, G.D.D., Matsuda, T., 1993. A comparison of a united atom and an explicit atom model in simulations of polymethylene. J. Chem. Phys. 98, 10037–10043. https://doi.org/10.1063/1.464436.

Zhang, Y., Feller, S.E., Brooks, B.R., Pastor, R.W., 1995. Computer simulation of liquid/liquid interfaces. I. Theory and application to octane/water. J. Chem. Phys. 103, 10252–10266. https://doi.org/10.1063/1.469927.

Zhou, M., 2003. A new look at the atomic level virial stress: on continuum-molecular system equivalence. Proc. R. Soc. Lond. Ser. A Math. Phys. Eng. Sci. 459, 2347–2392.

Zimmerman, J.A., WebbIII, E.B., Hoyt, J.J., Jones, R.E., Klein, P.A., Bammann, D.J., 2004. Calculation of stress in atomistic simulation. Model. Simul. Mater. Sci. Eng. 12, S319.

CHAPTER 5

Microscale mechanical modeling of brain neuron(s) and axon(s)

Mark F. Horstemeyer[a], A. Bakhtiarydavijani[b], Raj K. Prabhu[c]

[a]School of Engineering, Liberty University, Lynchburg, VA, United States [b]Center for Advanced Vehicular Systems (CAVS), Mississippi State University, Mississippi State, MS, United States [c]USRA, NASA HRP CCMP, NASA Glenn Research Center, Cleveland, OH, United States

5.1 Introduction

The human brain is made up of neurons, glial cells, astrocytes, oligodendrocytes, and microglia. While the neurons with approximately 86 billion cell count are not the most numerous, there are far more glial cells in the brain, is assumed to be the building block of the brain. During the development of the human brain, these neurons are pushed to the outer surface of the brain where their large numbers cause the surface of the brain to curve creating the outer convoluted surface (gray matter) of our brain. The axons then stretch through the interior of the brain (white matter) to build the circuitry of the brain. The simplest way to classify neurons is through their anatomy that is: One axon and the number of dendrites. While another method is to describe them based on their location in the nervous system and their distinct shape such as soma shape and size and dendrite characterizations. More recently a push has been made for a universal classification of neurons by considering the neuronal electrophysiology, morphology, and the individual transcriptome of the cells. All in all, this shows that our understanding of neurons and their structure is still limited and more so their injury biomechanics.

Injuries that occur in the brain are serious and need to be appropriately diagnosed. To understand the mechanisms of traumatic brain injury (TBI), different injury criteria such as head injury criterion (HIC) and brain injury criterion (BrIC) (Takhounts et al., 2013) have been introduced. However, these criteria are based on global kinematics, which deals with rigid body motions and not internal deformations. As such, one cannot measure actual local brain damage at the cellular or tissue size scales (Marjoux et al., 2008). However, using three-dimensional finite element analysis (FEA) of the head is another way to analyze both kinematics and deformations to assess TBI better and thus giving the ability to distinguish between different boundary conditions. In this context, most FEA of the head have not included finer morphological details nor heterogeneous geometries (Zhang et al., 2001; Horgan and Gilchrist, 2003; Takhounts et al., 2013; Yang et al., 2014). The simplifications

Multiscale Biomechanical Modeling of the Brain.
DOI: https://doi.org/10.1016/B978-0-12-818144-7.00016-5
Copyright © 2022 Elsevier Inc. All rights reserved.

78   Chapter 5

used in previous whole head FEA can save on computational costs as the complexity of the head model is reduced, but precision is lost. Furthermore, the global stress–strain behavior would give different values than the different stress–strain behaviors at the tissue level.

5.2 Modeling microscale neurons

Fig. 5.1 shows a multiscale hierarchical paradigm for modeling neuronal mechanical behavior specifically the mechanical deformation of the brain during a TBI. The microscale entity is the neuron, which has three parts as shown in Fig. 5.2: soma, dendrites, and an axon. With a diameter of up to 14 μm and a length of up to 1 m, the neuron is the largest cell in the human body, both in surface area and volume. As one of the most important cells in the body, the neuron transmits information through a chemo-electrical process throughout the brain. A neuron can be mechanically damaged thus requiring a coupled mechano-chemo-electrical model to capture the material behavior. The soma is the cell body of the neuron as shown in Fig. 5.2. Several short dendrites extend from the soma into the surroundings, one of which develops

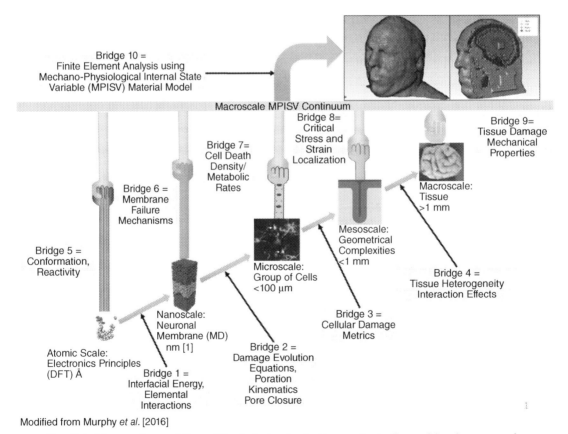

Modified from Murphy et al. [2016]

Fig. 5.1: Mesoscale modeling of brain behavior in the context of a multiscale approach.

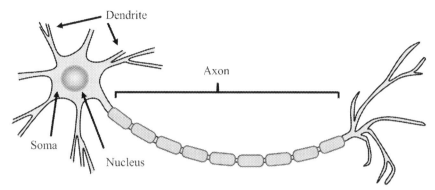

Fig. 5.2: Neuron picture showing the soma, dendrites, and axon.

into an axon. In contrast to dendrites, the axon is long with a high degree of anisotropy. The axonal cytoskeleton comprises longitudinally aligned microtubules and neurofilaments that connect from different crosslinks. An actin cortex held together by spectrin encapsulates the cytoskeleton. Extensive research has been devoted to neurons since their discovery in the 19th century with specific focus on axonal growth.

5.2.1 Modeling neurons

The neuron cell membrane is composed of a phospholipid bilayer. Neuronal membrane disruption has been shown as a pathway for neuronal death that can be achieved through mechanical loading on brain tissue and neurons (Cargill and Thibault, 1996; LaPlaca and Thibault, 1997; Geddes et al., 2003; Serbest et al., 2005). The disruption of the neuronal membrane can trigger cell death through different pathways (Raghupathi, 2004; BARBEE, 2006; Farkas et al., 2006). Due to the complexity of neuronal injury and the interdependence of different injury pathways (Prins et al., 2013), researchers often use the in vitro approach to isolate neuronal injury pathologies.

In vitro cell culture studies provide a means to reduce the number of variables in tissue injury and directly study cell-death mechanisms due to mechanical loads (Balentine et al., 1988; Cargill and Thibault, 1996; LaPlaca and Thibault, 1997; Geddes et al., 2003; Cullen and LaPlaca, 2006; LaPlaca et al., 2007; LaPlaca and Prado, 2010; Cullen et al., 2011). These studies highlight the dependence of neuronal death on strain rate (LaPlaca and Thibault, 1997; Geddes et al., 2003; LaPlaca et al., 2005; Bar-Kochba et al., 2016) and strain state (Geddes-Klein et al., 2006; Cullen et al., 2011) shown in Fig. 5.3. Neuronal cell death in in vitro studies includes significant standard deviations, some of which may be attributed to neuronal morphology and orientation variations (Cullen and LaPlaca, 2006). Defining injury criteria for neuronal membrane injury criteria can then help in defining the effects of neuronal morphology and orientation. As the neuronal membrane thickness is extremely small and exists at the nanoscale, such criteria need to be defined at the nanoscale. Nanoscale, in silico

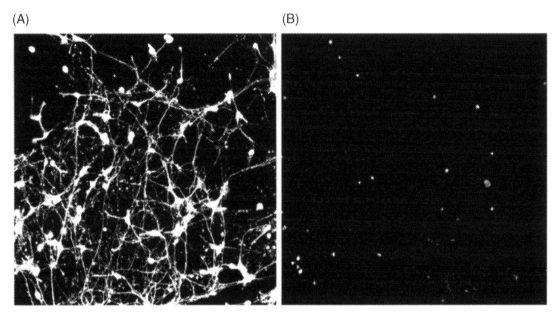

Fig. 5.3: Neuronal cell death in a 3D cell culture of Sprague-Dawley rat neurons after 0.50 shear strain at 30 s^{-1} strain rate showing (A) live and (B) dead neurons (Cullen and LaPlaca, 2006).

molecular dynamic (MD), simulations have looked at membrane deformation and resulting poration behavior under different stretching rates and states (Tieleman et al., 2003, 2006; Tomasini et al., 2010; Shigematsu et al., 2014; Murphy et al., 2016, 2018). At the same time, coarse-graining methods in MD studies have allowed for larger membranes to be modeled (Ingólfsson et al., 2017), but have yet to be extended to a cellular scale. At the neuron scale, Zubler and Douglas (2009) developed a diffusional–chemical model to capture the growth and interactions of neurons. In developing a possible mesh for mechanical analysis of neurons and axons, Bauer et al. (2014) developed a computational model that captures the patchy axonal connections in the brain. Because neurons and axons come in a wide variety of shapes and sizes, researchers have traced three-dimensionally thousands of neurons (see NeuroMorpho.Org; Ascoli et al., 2007 and ModelDB; Hines et al., 2004), so experimental like this can be used to calibrate and validate models.

Constitutive models of a neuron are almost absent from the literature. It is very difficult to conduct a tension or compression test on a neuron alone. Simon et al. (2016) used an atomic force microscope to garner load–displacement curves of individual neurons. Wu et al. (2018) tried to assess different methods of quantifying the mechanical behavior of neurons. Other than these two efforts to experimentally quantify the stress–strain behavior of a neuron, the models and data are missing.

5.2.2 Modeling mechanical behavior of axons

Similar to neurons, axon modeling has had limited progress, but there are a few that have made progress. Axons dynamically change their stiffness, viscosity, and internal stress, and we still do not know or understand the structure–property–function relationship today. Margulies and Thibault (1992) were probably the first researchers to model anything related to an axon although the analytical model was more of a phenomenological tool related to diffuse axonal injury and not the structure–property–function relationship. Anderson et al. (1999) later conducted a coupled experimental/numerical study of a head impact on a sheep in which FEA was conducted to study axon damage. Since Anderson et al. (1999) employed a continuum level FEA, the details of axons were not really identified. Abolfathi et al. (2009) modeled bulk axons to quantify the elastic modulus relationship with the volume of axons in a bundle. Garcia et al. (2012) developed a growth model for axons. Javid et al. (2014) used a genetic optimization algorithm to determine the mechanical viscoelastic properties of axons garnered from stress relaxation tests in tension. Zhang et al. (2017) stated that the structure–property–function relationship of the axonal structure is unclear to date. Zhang et al. (2017) tried to quantify some aspect of that relationship by using atomic force microscopy showing that the axon plasma membrane stiffness was greater than that of the dendrites and somas. Zhang et al. (2017) also ran coarse-grain MD to model the behavior of the axon membrane skeleton that was validated by the atomic force microscopy results. Probably the most detailed work to date of a microscale analysis of an axon is that of De Rooij et al. (2017) who conducted FEA of individual axons. They created an axon model of discrete microtubules that were connected to neighboring microtubules via discrete crosslinking mechanisms. They showed that using a viscoelastic constitutive model for each material and structure in the axon that two mechanisms passive dissipative crosslinking and active motor crosslinking were elucidated in quantifying axon mechanical behavior. Yousefsani et al. (2018) conducted transverse-plane hyperelastic mesoscale finite element simulations of axons in a continuum thus stating white matter using an embedded element technique. The simulations included a histology-informed probabilistic distribution of axonal fibers embedded within an extracellular matrix, both described using the generalized Ogden (1997) constitutive model. Although progress of modeling and simulating axons has been made, much more research is needed to fully appreciate axonal behavior.

5.3 Summary and future

Clearly, it is open season on modeling and simulating the mechanical behavior of a neuron or a set of neurons as very little mechanical–chemical–electrical modeling has been conducted directly on neurons. The next opportunities of research rely heavily in this area if progress on mechanical damage of neurons is realizable.

82 Chapter 5

References

Abolfathi, N., Naik, A., Sotudeh Chafi, M., Karami, G., Ziejewski, M., 2009. A micromechanical procedure for modelling the anisotropic mechanical properties of brain white matter. Comput. Methods Biomech. Biomed. Eng. 12, 249–262.

Anderson, R.W., Brown, C.J., Blumbergs, P.C., Scott, G., Finney, J.W., Jones, N.R., McLean, A.J., 1999. Mechanisms of axonal injury: an experimental and numerical study of a sheep model of head impact, Proceedings of International Conference on the Biomechanics of Impact (IRCOBI), 107–120.

Ascoli, G.A., Donohue, D.E., Halavi, M., 2007. NeuroMorpho.Org: a central resource for neuronal morphologies. J. Neurosci. 27, 9247–9251. doi:10.1523/JNEUROSCI.2055-07.2007.

Balentine, J.D., Greene, W.B., Bornstein, M., 1988. In vitro spinal cord trauma. Lab. Invest. [Internet] 58, 93–99. [cited 2019 May 11]Available at http://www.ncbi.nlm.nih.gov/pubmed/3336206.

Barbee, K.A., 2006. Mechanical cell injury. Ann. N. Y. Acad. Sci. [Internet] 1066, 67–84. Available at http://dx.doi.org/10.1196/annals.1363.006.

Bar-Kochba, E., Scimone, M.T., Estrada, J.B., Franck, C., 2016. Strain and rate-dependent neuronal injury in a 3D in vitro compression model of traumatic brain injury. Sci. Rep. [Internet] 6, 30550. [cited 2017 Aug 30] Available at http://www.ncbi.nlm.nih.gov/pubmed/27480807.

Bauer, R., Zubler, F., Hauri, A., Muir, D.R., Douglas, R.J., 2014. Developmental origin of patchy axonal connectivity in the neocortex: a computational model. Cerebral Cortex 24 (2), 487–500.

Cargill, R.S., Thibault, L.E., 1996. Acute alterations in $[Ca^{2+}]i$ in NG108-15 cells subjected to high strain rate deformation and chemical hypoxia: an in vitro model for neural trauma. J. Neurotrauma [Internet] 13, 395–407. Available at http://online.liebertpub.com/doi/abs/10.1089/neu.1996.13.395.

Cullen, D.K., LaPlaca, M.C., 2006. Neuronal response to high rate shear deformation depends on heterogeneity of the local strain field. J. Neurotrauma [Internet] 23, 1304–1319. Available at http://dx.doi.org/10.1089/neu.2006.23.1304.

Cullen, D.K., Vernekar, V.N., LaPlaca, M.C., 2011. Trauma-induced plasmalemma disruptions in three-dimensional neural cultures are dependent on strain modality and rate. J. Neurotrauma [Internet] 28, 2219–2233. [cited 2018 Apr 9] Available at http://www.ncbi.nlm.nih.gov/pubmed/22023556.

De Rooij, R., Miller, K.E., Kuhl, E., 2017. Modeling molecular mechanisms in the axon. Comput. Mech. 59 (3), 523–537.

Farkas, O., Lifshitz, J., Povlishock, J.T., 2006. Mechanoporation induced by diffuse traumatic brain injury: an irreversible or reversible response to injury? J. Neurosci. [Internet] 26, 3130–3140. Available at http://www.jneurosci.org/content/26/12/3130.

Garcia, J.A., Pena, J.M., McHugh, S., Jérusalem, A., 2012. A model of the spatially dependent mechanical properties of the axon during its growth. Comput. Model. Eng. Sci. (CMES) 87 (5), 411–432.

Geddes, D.M., Cargill, R.S., LaPlaca, M.C., 2003. Mechanical stretch to neurons results in a strain rate and magnitude-dependent increase in plasma membrane permeability. J. Neurotrauma [Internet] 20, 1039–1049. [cited 2017 Aug 25]Available at http://www.ncbi.nlm.nih.gov/pubmed/14588120.

Geddes-Klein, D.M., Schiffman, K.B., Meaney, D.F., 2006. Mechanisms and consequences of neuronal stretch injury in vitro differ with the model of trauma. J. Neurotrauma [Internet] 23, 193–204. [cited 2019 May 3] Available at http://www.liebertpub.com/doi/10.1089/neu.2006.23.193.

Hines, M.L., Morse, T., Migliore, M., Carnevale, N.T., Shepherd, G.M., 2004. ModelDB: a database to support computational neuroscience. J. Comput. Neurosci. 17, 7–11. doi:10.1023/B:JCNS.0000023869.22017.2e.

Horgan, T.J., Gilchrist, M.D., 2003. The creation of three-dimensional finite element models for simulating head impact biomechanics. Int. J. Crashworthiness 8 (4), 353–366.

Ingólfsson, HI, Carpenter, TS, Bhatia, H, Bremer, P-TT, Marrink, SJ, Lightstone, FC, Ingólfsson, HI, Carpenter, TS, Bhatia, H, Bremer, P-TT, et al., 2017. Computational lipidomics of the neuronal plasma membrane. Biophys. J 113, 2271–2280.

Javid, S., Rezaei, A., Karami, G., 2014. A micromechanical procedure for viscoelastic characterization of the axons and ECM of the brainstem. J. Mech. Behav. Biomed. Mater. 30, 290–299.

LaPlaca, M., Thibault, L., 1997. An in vitro traumatic injury model to examine the response of neurons to a hydrodynamically-induced deformation. Ann. Biomed. Eng. [Internet] 25, 665–677. Available at http://dx.doi.org/10.1007/BF02684844.

LaPlaca, M.C., Cullen, D.K., McLoughlin, J.J., Cargill II, R.S., 2005. High rate shear strain of three-dimensional neural cell cultures: a new in vitro traumatic brain injury model. J. Biomech. [Internet] 38, 1093–1105. Available at http://www.sciencedirect.com/science/article/pii/S0021929004002659.

LaPlaca, M.C., Prado, G.R., 2010. Neural mechanobiology and neuronal vulnerability to traumatic loading. J. Biomech. [Internet] 43, 71–78. [cited 2018 Aug 10]Available at https://www.sciencedirect.com/science/article/pii/S0021929009005028.

LaPlaca, M.C., Simon, C.M., Prado, G.R., Cullen, D.K., 2007. CNS injury biomechanics and experimental models. In: Weber, J.T., Andrew, I.R.M. (Eds.). Progress in Brain Research [Internet], 161. Elsevier, pp. 13–26. [place unknown]Available at http://www.sciencedirect.com/science/article/pii/S0079612306610029.

Margulies, S.S., Thibault, L.E., 1992. A proposed tolerance criterion for diffuse axonal injury in man. J. Biomech. 25 (8), 917–923.

Marjoux, D., Baumgartner, D., Deck, C., Willinger, R., 2008. Head injury prediction capability of the HIC, HIP, SIMon and ULP criteria. Accid. Anal. Prevent. 40 (3), 1135–1148.

Murphy, M.A., Horstemeyer, M.F.F., Gwaltney, S.R., Stone, T., LaPlaca, M.C., Liao, J., Williams, L.N., Prabhu, R., Stone, T., LaPlaca, M.C., et al., 2016. Nanomechanics of phospholipid bilayer failure under strip biaxial stretching using molecular dynamics. Model. Simul. Mater. Sci. Eng. [Internet] 24, 55008. Available at http://stacks.iop.org/0965-0393/24/i=5/a=055008.

Murphy, M.A., Mun, S., Horstemeyer, M.F., Baskes, M.I., Bakhtiary, A., LaPlaca, M.C., Gwaltney, S.R., Williams, L.N., Prabhu, R.K., 2018. Molecular dynamics simulations showing 1-palmitoyl-2-oleoyl-phosphatidylcholine (POPC) membrane mechanoporation damage under different strain paths. J. Biomol. Struct. Dyn. 37 (5), 1346–1359. Available at https://doi.org/10.1080/07391102.2018.1453376.

Ogden, R.W., 1997. Non-Linear Elastic Deformations, second ed. Dover, New York, NY.

Prins, M., Greco, T., Alexander, D., Giza, C.C., 2013. The pathophysiology of traumatic brain injury at a glance. Dis. Model. Mech. [Internet] 6, 1307–1315. [cited 2018 Jan 8]Available at http://www.ncbi.nlm.nih.gov/pubmed/24046353.

Raghupathi, R., 2004. Cell death mechanisms following traumatic brain injury. Brain Pathol. [Internet] 14, 215–222. Available at http://onlinelibrary.wiley.com/doi/10.1111/j.1750-3639.2004.tb00056.x/abstract.

Serbest, G., Horwitz, J., Barbee, K., 2005. The effect of poloxamer-188 on neuronal cell recovery from mechanical injury. J. Neurotrauma [Internet] 22, 119–132. [cited 2019 May 11]Available at http://www.liebertpub.com/doi/10.1089/neu.2005.22.119.

Shigematsu, T., Koshiyama, K., Wada, S., 2014. Molecular dynamics simulations of pore formation in stretched phospholipid/cholesterol bilayers. Chem. Phys. Lipids [Internet] 183, 43–49. [cited 2019 Jan 8]Available at https://www.sciencedirect.com/science/article/pii/S0009308414000681?via%3Dihub.

Simon, M., Dokukin, M., Kalaparthi, V., Spedden, E., Sokolov, I., Staii, C., 2016. Load rate and temperature dependent mechanical properties of the cortical neuron and its pericellular layer measured by atomic force microscopy. Langmuir 32 (4), 1111–1119.

Takhounts, E.G., Craig, M.J., Moorhouse, K., McFadden, J. and Hasija, V., 2013. Development of Brain Injury Criteria (BrIC) (No. 2013-22-0010). SAE Technical Paper.

Tieleman, D.P., Leontiadou, H., Mark, A.E., Marrink, S.-J.J., 2003. Simulation of pore formation in lipid bilayers by mechanical stress and electric fields. J. Am. Chem. Soc. [Internet] 125, 6382–6383. [cited 2019 Jan 8] Available at http://dx.doi.org/10.1021/ja029504i.

Tieleman, D.P., MacCallum, J.L., Ash, W.L., Kandt, C., Xu, Z., Monticelli, L., 2006. Membrane protein simulations with a united-atom lipid and all-atom protein model: lipid–protein interactions, side chain transfer free energies and model proteins. J. Phys. Condens Matter [Internet] 18, S1221–S1234. [cited 2018 Oct 4]Available at http://stacks.iop.org/0953-8984/18/i=28/a=S07?key=crossref.dd5b8fe65bb3fe7f37871d7ef29ee46e.

Tomasini, M.D., Rinaldi, C., Tomassone, M.S., 2010. Molecular dynamics simulations of rupture in lipid bilayers. Exp. Biol. Med. [Internet] 235, 181–188. [cited 2019 Jan 10]Available at http://journals.sagepub.com/doi/10.1258/ebm.2009.009187.

Wu, P.-H., Aroush, D.R.-B., Asnacios, A., Chen, W.-C., Dokukin, M.E., Doss, B.L., Durand-Smet, P., Ekpenyong, A., Guck, J., Guz, N.V., Janmey, P.A., Lee, J.S.H., Moore, N.M., Ott, A., Poh, Y.-C., Ros, R., Sander, M., Sokolov, I., Staunton, J.R., Wang, N., Whyte, G., Wirtz, D., 2018. A comparison of methods to assess cell mechanical properties. Nat. Methods 15 (7), 491–498 https://doi.org/10.1038/s41592-018-0015-1.

Yang, B., Tse, K.M., Chen, N., Tan, L.B., Zheng, Q.Q., Yang, H.M., Hu, M., Pan, G., Lee, H.P., 2014. Development of a finite element head model for the study of impact head injury. BioMed Res. Int., Article ID 408278. doi:https://doi.org/10.1155/2014/408278.

Yousefsani, S.A., Farahmand, F., Shamloo, A., 2018. A three-dimensional micromechanical model of brain white matter with histology-informed probabilistic distribution of axonal fibers. J. Mech. Behav. Biomed. Mater. 88, 288–295.

Zhang, L., Yang, K.H., Dwarampudi, R., Omori, K., Li, T., Chang, K., Hardy, W.N., Khalil, T.B., King, A.I., 2001. Recent advances in brain injury research: a new human head model development and validation. Stapp Car Crash J. 45, 369–394. https://doi.org/10.4271/2001-22-0017.

Zhang, Y., Abiraman, K., Li, H., Pierce, D.M., Tzingounis, A.V., Lykotrafitis, G., 2017. Modeling of the axon membrane skeleton structure and implications for its mechanical properties. PLoS Comput. Biol. 13 (2), e1005407.

Zubler, F., Douglas, R., 2009. A framework for modeling the growth and development of neurons and networks. Front. Comput. Neurosci. 3, 25.

CHAPTER 6

Mesoscale finite element modeling of brain structural heterogeneities and geometrical complexities

A. Bakhtiarydavijani[a], R. Miralami[a], A. Dobbins[b], Mark F. Horstemeyer[c], Raj K. Prabhu[d]

[a]*Center for Advanced Vehicular Systems (CAVS), Mississippi State University, Mississippi State, MS, United States* [b]*Biomedical Engineering Department, University of Alabama-Birmingham, Birmingham, AL, United States* [c]*School of Engineering, Liberty University, Lynchburg, VA, United States* [d]*USRA, NASA HRP CCMP, NASA Glenn Research Center, Cleveland, OH, United States*

6.1 Introduction

The human brain consists of different multifunctional parts arising from different structure–function relationships. Literally, the brain is a network with multiple length scales of connectivity that are mainly in forms of anatomical connectivity (structural connectivity), functional connectivity, or effective connectivity. Besides different and separated anatomical parts, including left and right hemisphere and parietal, temporal, occipital, and frontal lobes, the brain, like other tissue of the central nervous system, can be divided into gray and white matters. Gray matter, which has a pinkish-light gray color, is located on the outer surface of the brain, and contains nerve cell bodies and neuronal somas as well as all nerve synapses. The white matter is surrounded by the gray matter and contains myelinated axon tracts. The folding of the gray matter increases the brain's surface area and allows for large cortical sheets within a limited volume to be able to admit greater functions. Gyri and sulci are the names of the convex and concave folds of the cerebral cortex.

Injuries that occur in the brain are serious and need to be appropriately diagnosed. To understand the mechanisms of traumatic brain injuries (TBI), different injury criteria such as head injury criterion and brain injury criterion (Takhounts et al., 2013) have been introduced. However, these criteria are based on global kinematics, which deals with rigid body motions and not internal deformations. As such, one cannot measure actual local brain damage at the cellular or tissue size scales (Marjoux et al., 2008). However, using three-dimensional finite element analysis (FEA) of the head is another way to analyze both kinematics and deformations to assess TBI better and thus giving the ability to distinguish between different boundary

Multiscale Biomechanical Modeling of the Brain.
DOI: https://doi.org/10.1016/B978-0-12-818144-7.00013-X
Copyright © 2022 Elsevier Inc. All rights reserved.

86 Chapter 6

conditions. In this context, most FEA of the head have not included finer morphological details nor heterogeneous geometries (Zhang et al., 2001; Horgan and Gilchrist, 2003; Takhounts et al., 2013; Yang et al., 2014). The simplifications used in previous whole head FEA can save on computational costs as the complexity of the head model is reduced, but precision is lost. Furthermore, the global stress–strain behavior would give different values than the different stress–strain behaviors at the tissue level.

The brain is complicated as multiple length scales and materials interconnect; hence, to employ FEA is computationally expensive. As such, high performance computing is required to capture the finer details of the brain. As an example of modeling a localized region where experimental data is available in the literature, we conducted FEA to examine the local response of sulci on chronic traumatic encephalopathy (CTE), which is a subset of common disease to a range of contact sports. CTE is a progressive neurodegenerative disease, and there is a correlation of CTE localization with the negative curvature sulci during the early stages of neurodegeneration (McKee et al., 2009). Observing neurofibrillary tangles accumulate below the sulci depth in the frontal lobe is an early diagnosis of CTE (McKee et al., 2016; Mez et al., 2017). Understanding the effects of complex brain geometries and heterogeneities to further studying CTE and its underlying causes can help identify preventive measures. As most whole head FE models do not include cortical folding, a complimentary means of accounting for the effects of said convolutions is required. This chapter highlights the potential of mesoscale FEA in capturing geometrical complexities that are generally unattainable with macroscale FEA.

6.1.1 Modeling length scale

Fig. 6.1 shows a multiscale hierarchical paradigm for modeling neuronal mechanical behavior specifically the mechanical deformation of the brain during a TBI at the mesoscale, with the mesoscale length scale being isolated.

The mesh size of macroscale FEA, which characterizes the whole head in our simulations, directly correlates with the computational cost. The cost increases noticeably by using a finer mesh size with smaller elements. Introducing the complex brain geometry also produces complex interface interactions that exacerbate errors in the simulations. A coarser mesh size, on the other hand, limits the details such that the pertinent geometries might be missed. A very coarse mesh with large elements, especially at the places where significant stresses take place, can decrease accuracy and lead to large errors in the results. The tension between simple and detailed geometries is always present when converging on a mesh size.

In the context of a multiscale modeling approach (Fig. 6.1), mesoscale FEA allows us to quantify the effects of detailed geometrical variations arising from sulci that can be validated from experimental measurements. We, by using mesoscale simulations, can add a level of granularity that is missing in higher scale FEA with a reasonable computational cost. Specifically, mesoscale two-dimensional (2D) FEA can provide the anatomical prescription of

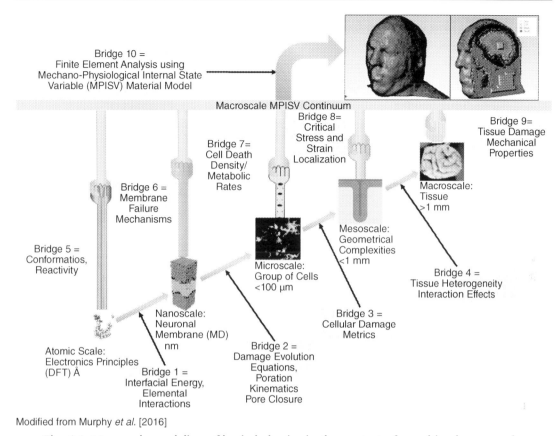

Modified from Murphy et al. [2016]

Fig. 6.1: Mesoscale modeling of brain behavior in the context of a multiscale approach.

the brain's sulci and gyri, gray–white matter interface, and pia mater. This greater resolution provides a more detailed description of the stress wave propagation and the associated stress concentrations around the sulci that could not be captured in macroscale simulations (Ho and Kleiven, 2009; Song et al., 2015). Therefore, in mesoscale 2D FE simulations, we are able to capture the following:

- structural and geometrical complexities including gray matter, white matter, pia mater, sulci, and brain–cerebrospinal fluid (CSF) interface properties, and
- local geometrical variations arising from sulci.

6.2 Methods

6.2.1 Computational methods for properties

The mesoscale 2D FEA represents mechanical impacts to the frontal, occipital, temporal, and parietal lobes along with other structural and geometrical details, including gray matter, white matter, and sulci. The effects of various material heterogeneities such as sulci and gray matter

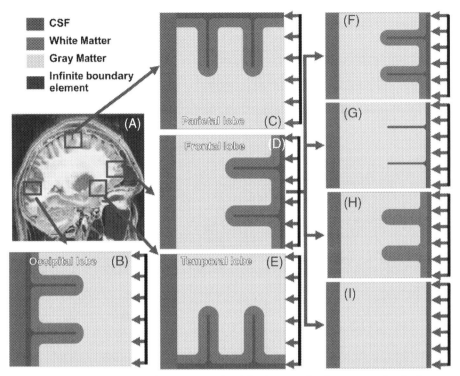

Fig. 6.2: (A) Magnetic resonance image (MRI) of the human brain parasagittal plane to show the correlated mesoscale 2D finite element (FE) models (B–I). These models represent the (B) occipital lobe, (C) parietal lobe, (D) frontal lobe, and (E) the temporal lobe. Further, four mesoscale models are used to study the inclusion of (F) gray matter, white matter, and sulci, (G) sulci and homogeneous brain, (H) gray matter, white matter, and (I) homogeneous brain.

in different brain lobes during frontal head impacts were studied via three different sets of 2D FEA (Fig. 6.2): (1) football blunt impact in the frontal lobe with different structural complexities (sulci, gray–white matter differentiation, pia mater); (2) football blunt impact in the parietal, temporal, occipital, and frontal lobes with two sulci; and (3) football blunt impact in the frontal lobe with only one sulcus.

Table 6.1 summarizes the different simulations that were conducted to examine various parametric effects: (1) homogeneous brain with no sulci versus the heterogeneous brain with sulci, (2) gray and white matter, (3) brain–CSF boundary, and (4) pia matter. The sulci geometry included two 1 mm wide sulci that were 25 mm deep with a 16 mm distance between each other. To differentiate gray matter from white matter, we included a 3 mm thick layer for the gray matter.

For our FEA, the brain tissue and gray and white matter material properties were defined using the MSU TP 1.1. material model that is an elasto-viscoplastic internal state variable (ISV) material model (Bouvard et al., 2010), calibrated to experimental brain data (Prabhu

Table 6.1: Two-dimensional (2D) finite element analysis (FEA) built considering the anatomical features, locations, and impact scenario.

Input load	Gray–white matter differentiation	Sulcus	Pia mater	Brain–CSF interface	Lobe
Football blunt impact—two sulci	No	No	No	Tied	Frontal
	No	Yes	No	Tied	Frontal
	Yes	No	No	Tied	Frontal
	Yes	Yes	No	Tied	Frontal
	Yes	Yes	Yes	Tied	Frontal
	Yes	Yes	No	Frictionless	Frontal
	Yes	Yes	No	7.5%	Frontal
	Yes	Yes	No	15%	Frontal
	Yes	Yes	No	30%	Frontal
	Yes	Yes	No	Tied	Temporal
	Yes	Yes	No	Tied	Occipital
	Yes	Yes	No	Tied	Parietal
Football blunt impact—one sulcus	Yes	Sulcus length (mm) 7.5 15 24.5	No	Tied	Frontal

et al., 2011) and validated to whole head cadaver tests (Johnson et al., 2016). In this material model, the deformation gradient is defined as the following,

$$\mathbf{F} = \mathbf{F}^e \ \mathbf{F}^P$$

Herein, $\mathbf{F}^e$ denotes the elastic deformation gradient, and $\mathbf{F}^P$ denotes the plastic deformation gradient. We can derive velocity gradients ($\mathbf{L}^e$ and $\mathbf{L}^P$) from deformation gradients by applying the conventional continuum mechanics approach

$$\mathbf{L}^e = \dot{\mathbf{F}}^e \mathbf{F}^{e-1} , \ \overline{\mathbf{L}}^P = \dot{\mathbf{F}}^P \mathbf{F}^{p-1}$$

The Cauchy stress can be calculated as follows,

$$\sigma = \mathbf{J}^{e-1} \ \mathbf{F}^e \overline{\mathbf{S}} \mathbf{F}^{eT}$$

where $\overline{\mathbf{S}}$ indicates the second Piola–Kirchhoff stress,

$$\overline{\mathbf{S}} = 2 \frac{\partial \hat{\overline{\psi}}}{\partial \overline{\mathbf{C}}^e}$$

Then, we define the strain energy as a function of the Cauchy–Green strain, $\overline{\mathbf{C}}^e$, and three ISVs $\overline{\xi}_1$, $\overline{\xi}_2$, and $\overline{\mathbf{E}}^\beta$ that were first founded from atomistic simulations (Tschopp et al., 2011)

$$\overline{\psi} = \overline{\psi} \left(\overline{\mathbf{C}}^e, \overline{\xi}_1, \overline{\xi}_2, \overline{\mathbf{E}}^\beta \right)$$

90 Chapter 6

Furthermore, the Bouvard et al. (2010) model was able to capture the nonlinear behavior of soft tissue at large strains by the following equation,

$$\hat{\psi}_{E^{\beta}} = -\mu_R \lambda_L \ln\left[1 - \frac{\lambda_1^{\beta} + \lambda_2^{\beta} + \lambda_3^{\beta} - 3}{\lambda_L}\right].$$

Here, we calculate the derivative of the strain energy function with respect to time as the following,

$$\dot{\hat{\psi}} = \frac{\partial\hat{\psi}}{\partial\bar{\mathbf{C}}^e} : \dot{\bar{\mathbf{C}}}^e + \frac{\partial\hat{\psi}}{\partial\bar{\xi}_1} : \dot{\bar{\xi}}_1 + \frac{\partial\hat{\psi}}{\partial\bar{\xi}_2} : \dot{\bar{\xi}}_2 + \frac{\partial\hat{\psi}}{\partial\bar{\mathbf{E}}^{\beta}} : \dot{\bar{\mathbf{E}}}^{\beta}$$

In this equation, we have three thermodynamic conjugates of the strain, similar to ISVs as follows,

$$\bar{\kappa}_1 = \frac{\partial\hat{\psi}}{\partial\bar{\xi}_1}, \quad \bar{\kappa}_2 = \frac{\partial\hat{\psi}}{\partial\bar{\xi}_2}, \quad \bar{\alpha} = \frac{\partial\hat{\psi}}{\partial\bar{\mathbf{E}}^{\beta}}$$

If we assume that there is no plastic spin, then we can write the plastic flow rule as the following,

$$\dot{\mathbf{F}}^p = \bar{\mathbf{D}}^p \mathbf{F}^p, \quad \bar{\mathbf{D}}^p = \frac{1}{\sqrt{2}}\dot{\bar{\gamma}}^p \bar{\mathbf{N}}^p.$$

$\bar{\mathbf{N}}^p$ denotes the plastic flow direction, and $\dot{\bar{\gamma}}^p$ denotes the viscous shear strain rate

$$\dot{\bar{\gamma}}^p = \dot{\bar{\gamma}}_0^p\left[\sinh\left(\frac{\bar{\tau} - \left(\bar{\kappa}_1 + \bar{\kappa}_2 + \alpha_p\bar{\pi}\right)}{Y}\right)\right]^m$$

In the formula above, $\dot{\bar{\gamma}}_0^p$ refers to a reference strain rate, $\bar{\tau}$ refers to the equivalent shear stress, and $\bar{\pi}$ refers to an effective pressure.

Now, the stress-like conjugates of the ISVs mentioned above can be defined as the following (Bouvard et al., 2010),

$$\dot{\bar{\xi}}_1 = h_0\left(1 - \frac{\bar{\xi}_1}{\bar{\xi}^*}\right)\dot{\bar{\gamma}}^p, \quad \dot{\bar{\xi}}^* = g_0\left(1 - \frac{\bar{\xi}^*}{\bar{\xi}_{sat}^*}\right)\dot{\bar{\gamma}}^p, \quad \bar{\xi}_1(\mathbf{X},0) = 0, \quad \bar{\xi}^*(\mathbf{X},0) = \bar{\xi}_0^*$$

$$\dot{\bar{\xi}}_2 = h_1\left(\bar{\lambda}^p - 1\right)\left(1 - \frac{\bar{\xi}_2}{\bar{\xi}_{2sat}}\right)\dot{\bar{\gamma}}^p, \quad \bar{\lambda}^p = \frac{1}{\sqrt{3}}\sqrt{\mathrm{Tr}\left(\bar{\mathbf{B}}^p\right)}, \quad \bar{\mathbf{B}}^p = \mathbf{F}^p\mathbf{F}^{p^{\mathrm{T}}}$$

$$\dot{\bar{\beta}} = R_{s_1}\left(\bar{D}^p\bar{\beta} + \bar{\beta}\bar{D}^p\right), \quad \bar{\beta}(\mathbf{X},0) = I, \quad \bar{\mathbf{E}}^{\beta} = \ln\left(\beta\right)$$

Table 6.2 lists the constants for this material model described herein. The constants relate to the macroscale mechanical properties are $\mu, K, \dot{\lambda}_0^p, m, Y,$ and r. The constants that define ξ_1 behavior are $\bar{\xi}_0^*, \bar{\xi}_{sat}^*, h_0, g_0,$ and C_{κ_1}. The constants that define ξ_2 are $h_1, \bar{\xi}_{2sat},$ and $C_{\kappa2}$. The constants that define $\bar{\mathbf{E}}^{\beta}$ are $R_{s1}, \lambda_L,$ and μ_R. Initial conditions are defined via $\alpha_p, \bar{\xi}_{10},$ and $\bar{\xi}_{20}$.

Table 6.2 contains calibrated constants from Johnson et al. (2016). Along with implementing these constant, the CSF was treated as an incompressible elastic material with a very low shear modulus, and the pia mater was employed as a second-order viscoplastic material (Aimedieu and Grebe, 2004). The viscoelastic behavior is given by the following,

Mesoscale finite element modeling of brain structural heterogeneities 91

Table 6.2: MSU TP 1.1 elastic–viscoplastic brain model material constants (Johnson et al., 2016).

Constant	Gray matter	White matter
ρ	1.04e−9	1.04e−9
μ (MPa)	4.5	5.5
K (MPa)	2100	3000
$\dot{\bar{\gamma}}_0^p\left(\mathrm{s}^{-1}\right)$	95	95
α_p	0	0
m	0.1	0.1
Y (MPa)	0.001	0.001
$C_{\kappa 1}$ (MPa)	0.07	0.07
$C_{\kappa 2}$ (MPa)	0.001	0.001
h_0	85	85
$\bar{\xi}_{10}$	0.04	0.04
$\bar{\xi}_0^*$	0.23	0.23
$\bar{\xi}_0^*$	0.001	0.001
g_0	0.13	0.13
h_1	5	5
$\bar{\xi}_{20}$	0	0
$\bar{\xi}_{2\,sat}$	0.1	0.1
R_{s_1}	0.6	0.6
μ_R	0.11	0.11
λ_L	10	10

$$G(t)=G_0\left(1+\sum_{i=1}^{n} G_R \exp\left(-\frac{t}{\tau_i}\right)\right).$$

where $G(t)$ is the shear modulus at time (t), G_0 is the shear modulus at $t = 0$, G_∞ is the shear modulus at $t = \infty$, and G_R is the ratio of G_∞ to G_0. The elastic equation for the CSF is incompressible Hooke's Law (1678),

$$\sigma = E\varepsilon$$

The pia mater was employed as a second-order viscoplastic material (Aimedieu and Grebe, 2004) with the material constants shown in Table 6.3. The brain–CSF interface behavior was defined by the tangential friction coefficient. To employ pia mater, we implemented a 0.1 mm one element thick viscoelastic membrane into the base simulation between the CSF and the brain. Also, the effect of the sulcus length was examined through simulations with one sulcus at different lengths.

92 Chapter 6

Table 6.3: Pia mater (Aimedieu and Grebe, 2004) and CSF (Johnson et al., 2016) material properties.

	Pia mater	CSF
Density ρ (g/cm^3)	1.040	1.004
Poisson's ratio	0.450	0.495
Elastic modulus (MPa)	20.00	0.15
Relaxation (s)	0.919	-
Decay	2	-

The elements used for simulations were reduced quadratic plane strain elements (Abaqus CPE4R) for the brain and infinite boundary elements (Abaqus CINPE4) for the infinite boundaries. Our model had 9000 elements for the mesh convergence study. To pick the number of elements, we performed simulations with a different number of elements and observed that the generated result (average pressure) with 9000 elements was within 5% standard error of the 280,000 element simulation. For a more detailed field output gradients at a depth of the sulcus, the mesh was refined to 0.1 mm element size to capture the strain gradients related to the sulcus, while the mesh with larger elements captured the average and maximum stress values.

The 2D FEA simulations were conducted in the Abaqus/ExplicitTM software for two reasons:

1. To validate the multiscale load transfer between the macroscale and mesoscale FEA simulations. Stress wave propagation through the cerebral cortex was examined by applying time–pressure histories in the mesoscale FEA. The resulting stress waves were comparable to clinical studies of CTE.
2. To quantify the geometric and material influences on the local stress state of the sulci/gyri. Parametric analysis of variations was performed in cortical morphometry, gray–white matter differentiation, pia mater presence, brain–CSF interfacial properties, and brain lobes to identify their influence on the maximum von Mises stresses and pressures below the sulci.

6.2.2 Model validation and boundary conditions

The pressure–time histories that informed the boundary conditions for the mesoscale simulations were garnered from the macroscale helmeted head FE simulations of American football helmet-to-helmet hits (Fig. 6.3). In these simulations, 6 m/s head impacts were performed where the Frankfort (eye–ear) (Lack, 1930) plane was rotated forward (Fig. 6.3B), which resulted in a mid-sagittal head impact (Johnson et al., 2016).

The macroscale simulation included the facemask, helmet shell, helmet liner, flesh, cortical bone, cancellous bone, CSF, and the brain. The macroscale model was meshed with nearly 2.6 million elements and validated to intracranial pressure (Nahum et al., 1977) by Johnson et al. (2016). These pressures were then applied to mesoscale 2D FE simulations that

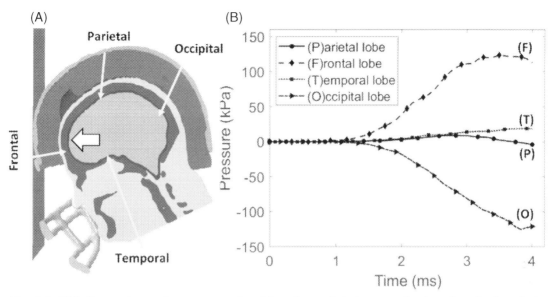

Fig. 6.3: This figure shows the pressure–time histories and their respective locations in the whole head model. (A) The macroscale human head finite element simulation of a helmeted American football player (cut along the sagittal plane) with the white arrow showing the direction of the head movement. (B) The pressure–time histories were extracted from the macroscale head model at the frontal, temporal, parietal, and occipital lobes.

produced a detailed time evolution of the von Mises stresses and pressures inside the mesoscale brain models. The average von Mises stresses, maximum von Mises stresses, and the maximum measured pressures for all simulations were generated as presented in Table 6.4.

In all our mesoscale 2D FE simulations, we used symmetric boundary conditions for boundaries adjacent to the loading surface. For boundaries opposite of the loading surface, we used a single infinite boundary to prevent pressure wave reflections at that surface.

6.3 Results and discussion

The mesoscale 2D FE simulations were performed to better understand the effects of structural and geometrical complexities of brain tissue on brain injury biomechanics. These simulations showed how the structural heterogeneities of the brain, along with the difference in brain and CSF mechanical properties, generate stress gradients, and stress localization that can dictate the location of CTE initiation.

To achieve this goal, we loaded all simulations with the pressure–time histories obtained from the macroscale FEA of a frontal head impact of a helmeted football player. Considering these boundary conditions, the evolution of the Von Mises stress and pressure in the frontal lobe at the time of peak loading, along with the direction of the pressure waves, the direction of the

94   Chapter 6

Table 6.4: Average and peak von Mises stress ($\overline{\sigma}_{vm}$), maximum von Mises stress (σ_{vm}^{max}), maximum hydrostatic pressure (P) derived from the mesoscale finite element simulations (Figs. 6.4–6.8).

	$\overline{\sigma}_{vm}$ (kPa)	σ_{vm}^{max} (kPa)	P (kPa)
Geometric and structural complexities in the frontal lobe			
Differentiated brain, sulci, pia mater	5.578	18.06	126.3
Sulci, differentiated brain	5.675	17.99	131.0
Differentiated brain	0.457	0.585	122.7
Homogeneous brain, sulci	5.792	17.72	127.0
Homogeneous brain	0.433	0.455	122.5
Brain lobes			
Frontal	5.675	17.99	131.0
Parietal	0.969	6.26	11.29
Temporal	2.224	11.08	23.83
Occipital	5.730	18.05	-126.0
Sulcus length			
7.5 mm sulcus	4.222	16.60	132.1
15 mm sulcus	5.410	17.80	130.9
24.5 mm sulcus	6.764	18.07	123.3
Element size below the sulcus			
0.1 mm*	5.675	17.99	131.0
All 0.25 mm	5.361	17.53	127.9
All 0.5 mm	5.691	17.68	127.3
All 1 mm	5.719	17.05	126.5
All 2 mm	6.373	15.94	126.5

*Localized mesh refinement below the sulcus.

largest principal stress, and the localization of pressure and von Mises shear stress around a sulcus were generated. The pressure profiles showed a maximum stress localization below the sulci for all simulations.

6.3.1 Geometrical complexities

As previously mentioned, the mesoscale FE simulations used to study effects of sulci and brain structure on stress wave propagation included:

- sulci and differentiated gray–white matter
- sulci and homogeneous brain matter
- differentiated gray–white matter without sulci
- homogenous brain with no sulci.

These differences of material properties introduce multiple complexities in the local stresses and strains. First, the difference between the mechanical properties of the three sections (gray matter versus white matter versus the CSF) affects the stress wave speed, since it is dependent on the elastic modulus and density, meaning that a time difference exists between the stress wave front in each material. Second, the differences in the strain gradients at the biomaterial interfaces are exacerbated by the CSF's extremely low shear modulus, which effectively removes the accumulation and transfer of shear stresses. The effect of these differences would be expected to be most significant at the brain–CSF and gray–white matter interface. Hence, the biomaterial interfaces becomes a location of interest where the biomechanics of injury are important. To analyze this, we study the pressure (Fig. 6.4) and Von Mises stress evolutions (Fig. 6.6) in the frontal lobe for the helmeted frontal head impact scenario.

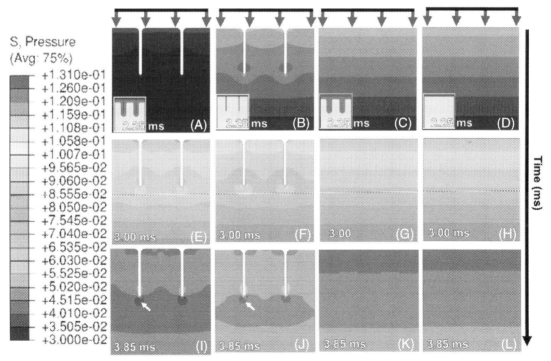

Fig. 6.4: Pressure (MPa) evolution dependence on structural model complexities in mesoscale two-dimensional finite element simulations. All simulations were compressively loaded from the top with the pressure–time histories observed in the frontal lobe during the football player head impact (Fig. 5.3). Each column corresponds to a unique model that includes (A, E, I) sulci and differentiated gray–white matter; (B, F, J) sulci and homogeneous brain; (C, G, K) differentiated gray–white matter with no sulci; and (D, H, L) homogenous brain with no sulci. From top to bottom, each row corresponds to the same time at 2.25, 3.00, and 3.85 milliseconds, respectively. The cerebral spinal fluid (CSF) and the infinite boundary are hidden for visual clarity.

96   Chapter 6

The sulci introduce an internal stress concentration because of the local radius of curvature. The third column in Fig. 6.4 shows the slight difference in the pressure wave propagation for the gray and white matter. The mismatch between the pressure waves does, however, introduce some von Mises stress (shearing). The peak pressure in the brain tissue was approximately 585 Pa in the white matter. On the other hand, the difference between the CSF and brain tissue (for example, Column 1 versus Column 3 in Fig. 6.4) not only significantly distorts the pressure wave transmission from straight contours to curved contours but also generates stress concentrations (localized increased pressure) as designated by the white arrows in Fig. 6.4I,J. Fig. 6.6I,J illustrates the same phenomenon with the von Mises stress, with white arrows in the brain tissue at the sulcus base. These sulci generate an additional 18 kPa von Mises stress when introduced (Fig. 6.6C and F), which is much more significant than those introduced at the gray white matter interface.

Although the sulci introduce a fairly strong stress concentration in the brain, the pia mater had much less influence on the progression of the pressure and stress waves. The inclusion of the pia mater increased the maximum von Mises stress by less than 1% and reduced the maximum pressure by only 3%.

6.3.1.1 The effects of the sulci

The simulations clearly show that sulci and gyri morphology generated stress concentrations in localized regions thus increasing the pressure and von Mises stress just below the sulci (Fig. 6.5C). The reason behind these localized pressure concentrations is that the geometrical notch root radii of sulci introduces two pressure waves, with magnitudes greater than those in the CSF (Fig. 6.5A), which move along the sulcus sides and interpose at its end (Fig. 6.5B). The effects of this stress wave interaction increased the pressure (Fig. 6.4K, L) and shear stresses (Fig. 6.6I, L). The differences in the gray and white matter through different elasto-viscoplastic constitutive properties led to the generation of minor von Mises stress magnitudes at the gray matter below the depths of gray–white matter interface. This arose because the stress wave propagation was faster in the brain tissue when compared to the CSF.

The pressure waves traveling through the brain tissue and on each side the sulcus transmitted the stresses locally at the sulcus end and at a short distance beyond the radius through their interaction with one another. The local pressure and von Mises stress concentrations generated by a sulcus just below it at its depth was 39 times greater when compared to simulations without the sulcus. This local pressure and stress concentration arose, because of the impedance difference between the CSF and the brain tissue thus causing stress waves to travel at different speeds (they are slower in the CSF compared to the brain tissue). The location of these stress concentrations below the sulcus agrees with the location of tau protein accumulation (Mez et al., 2017), which is a damage indicator at the early stages of CTE.

Mesoscale finite element modeling of brain structural heterogeneities   97

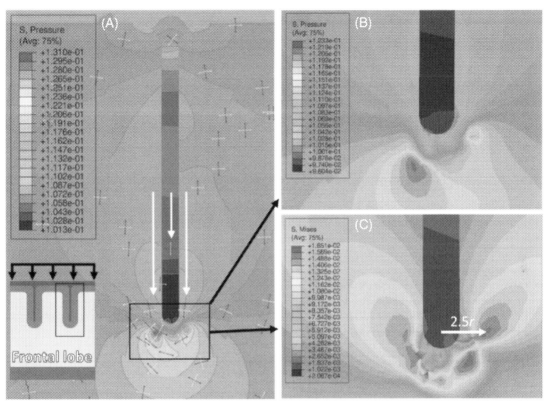

Fig. 6.5: (A) Mesoscale two-dimensional finite element simulation results of the (B) pressure and (C) von Mises stress (MPa) illustrating the stress concentration below the sulcus in the frontal lobe at 3.85 milliseconds. Here, the pressure wave acts in parallel with the sulcus axis.

Von Mises shear bands also extend from the sulcus base and introduce the second site for stress localization where they interact with one another. When only one sulcus is considered, the von Mises stress bands originated at the sulcus end and extended into the brain tissue (Fig. 6.8A–C). When two sulci of the same length were considered, the shear bands overlapped to add a new area of stress localization, which was 23% greater (black arrow in Fig. 6.6C, F) compared to the von Mises stress band in the single sulcus simulations (white circle in Fig. 6.8C). However, the location and existence of this secondary stress localization were dependent on the sulcal length and orientation relative to the incoming pressure wave. Also, the distance between sulci plays a role in the location, magnitude, and extent of this stress concentration because of the stress field interaction.

6.3.1.2 Sulcus orientation

During a frontal head impact, as the pressure wave traverses through the brain, it interacts with sulci at different angles. Here we look at four different sulci orientations as set up in

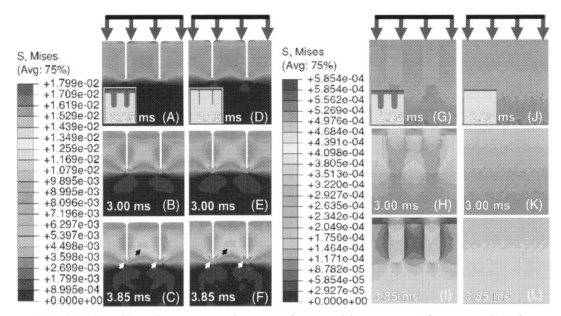

Fig. 6.6: The effect of geometry and structural compositions on von Mises stress (MPa) evolution. Different models that include (A, B, C) sulci and differentiated gray–white matter; (D, E, F) sulci and homogeneous brain; (G, H, I) differentiated gray–white matter without sulci; and (J, K, L) homogenous brain with no sulci are used to study the von Mises stress (MPa) evolution under pressure–time histories in the frontal lobe (Fig. 6.3). The three rows, from top to bottom present the resulting von Mises stress contours at 2.25, 3.00, and 3.85 milliseconds, respectively. The white arrows point to the location of the von Mises stress concentrations at the sulcus depth. The black arrows point to the interaction of shear waves produced from the sulcus base. The cerebral spinal fluid (CSF) and the infinite boundary are hidden for visual clarity.

Fig. 6.2B, E that correspond to the occipital, parietal, frontal, and temporal lobes, respectively. In these models the sulci orientation can be parallel or perpendicular to the incoming pressure wave. This orientation affects the mechanical responses and the evolution of the stresses arising from the sulcus end. When the sulci are parallel to the incoming pressure wave, the von Mises stresses concentrate at the sulcus end, initially appear in the brain tissue at the CSF/brain tissue interface (Fig. 6.7A, J), and then move further into the brain tissue and to the depth of the sulci ends (Fig. 6.7C, L). On the other hand, when sulci are perpendicular to the incoming pressure wave, the von Mises stresses, initiate and stay at the CSF/ brain tissue interface at the sulcus end (Fig. 6.7D, I).

6.3.1.3 Sulcus length

We can then assess the effect of sulcus length using single sulcus models. Overall, we found that by increasing the sulcus length, the maximum von Mises stress increased as the applied pressure (Fig. 6.8G, H) peaked. When the sulcus length increased from 7.5 mm to 24.5 mm (three times), the maximum von Mises stress below the sulcus also increased from 16.6 kPa

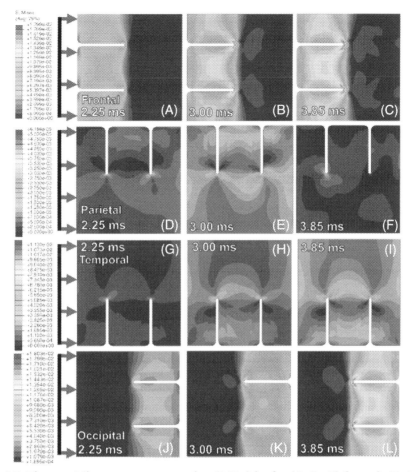

Fig. 6.7: The von Mises stress progression (MPa) in the (A, B, C) frontal; (D, E, F) occipital; (G, H, I) parietal; and (J, K, L) temporal lobes during the frontal head impact of a helmeted football player. The three columns from left to right represent von Mises stress contours at 2.25, 3.00, and 3.85 milliseconds, respectively. All models include gray matter, white matter, cerebral spinal fluid (CSF) and infinite boundary elements with the CSF and the infinite boundary being hidden for visual clarity.

to 18.1 kPa (approximately 9%). This occurs based on how the pressure wave propagates through the geometric complexities. The increased sulcus length results in an increased mismatch between the two wave fronts in the CSF and gray matter. This means that at the sulcus end, the pressure difference between the CSF and brain tissue increases with the sulcus length. This increased pressure difference then created the localized shearing that produced greater von Mises stresses at the sulcus base (Fig. 6.8G, I). At the same time, the shear bands that extended from the sulcus base were greater. While the magnitude of these shear bands did not significantly change, the region that they covered was greater, allowing for more overlap in multisulci models that then created a larger area susceptible to injury.

100   Chapter 6

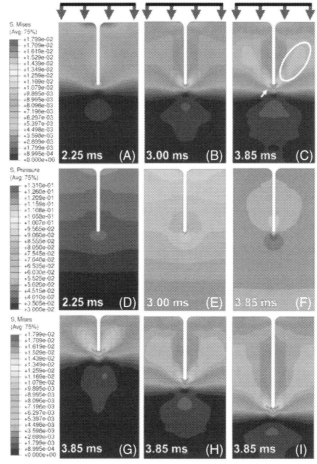

Fig. 6.8: (A, B, C) von Mises stress (MPa) and (D, E, F) pressure (MPa) time evolution in a two-dimensional mesoscale model that includes a 14.5 mm sulcus and the von Mises stress contours considering a sulcus length of (G) 7.5 mm, (H) 14.5 mm, and (I) 24.5 mm in the frontal lobe. All models include gray matter, white matter, cerebral spinal fluid (CSF) and infinite boundary elements with the CSF and the infinite boundary being hidden for visual clarity. The white arrow shows the von Mises stress localization arising from the inclusion of a sulcus that is the von Mises stress band is circled white. The CSF and the infinite boundary are not shown for visual clarity.

6.4 Summary

As has been repeatedly shown throughout this book, TBI is a multiscale problem with injury biomechanics being transferred from one length scale to another. In this chapter, we examined the injury biomechanics introduced by brain geometrical complexities using mesoscale models illustrating the stress concentration at the sulci ends. Since gray matter grows faster than white matter, folds of gyri and sulci are formed in time that introduce stress concentrations, which are locations of potential brain damage upon mechanical impacts. Specifically, the injury mechanism whereby the interactions of the initial pressure wave with brain folds produce shear stress localizations and pressure concentrations just below the sulci. These findings were matched with clinical studies of CTE as shown in Fig. 6.9 (Horstemeyer et al., 2019) and its pathology in postmortem studies. The one-way coupled multiscale modeling methodology presented in this chapter is a cost-effective means of assessing such local phenomena.

References

Aimedieu, P., Grebe, R., 2004. Tensile strength of cranial pia mater: preliminary results. J. Neurosurg. 100 (1), 111–114. http://thejns.org/doi/abs/10.3171/jns.2004.100.1.0111.

Bouvard, J.L., Ward, D.K., Hossain, D., Marin, E.B., Bammann, D.J., Horstemeyer, MF., 2010. A general inelastic internal state variable model for amorphous glassy polymers. Acta Mech. 213 (1–2), 71–96. http://link.springer.com/article/10.1007/s00707-010-0349-y.

Ho, J., Kleiven, S., 2009. Can sulci protect the brain from traumatic injury? J. Biomech. 42 (13), 2074–2080. http://www.sciencedirect.com/science/article/pii/S0021929009003364.

Hooke, R., De Potentia Restitutiva, or of Spring. Explaining the Power of Springing Bodies, London, 1678.

Horgan, T.J., Gilchrist, MD., 2003. The creation of three-dimensional finite element models for simulating head impact biomechanics. Int. J. Crashworthiness 8 (4), 353–366. http://link.springer.com/article/10.1533/ijcr.2003.0243.

Horstemeyer, M.F., Berthelson, P.R., Moore, J., Persons, A.K., Dobbins, A., Prabhu, R.K., 2019. A mechanical brain damage framework used to model abnormal brain tau protein accumulations of National Football League players. Ann. Biomed. Eng. 47 (9), 1873–1888.

Johnson, K.L.L., Chowdhury, S., Lawrimore, W.B.B., Mao, Y., Mehmani, A., Prabhu, R., Rush, G.A.A., Horstemeyer, MFF., 2016. Constrained topological optimization of a football helmet facemask based on brain response. Mater. Des. 111, 108–118. [accessed 2018 May 16]. https://www.sciencedirect.com/science/article/pii/S0264127516311273.

Lack, L.H., 1930. The endocranial equivalents of the Frankfurt plane and the exocranial position of the internal auditory meatus. J. Anat. 65 (Pt 1), 96.

Marjoux, D., Baumgartner, D., Deck, C., Willinger, R., 2008. Head injury prediction capability of the HIC, HIP, SIMon and ULP criteria. Accid. Anal. Prevent. 40 (3), 1135–1148. https://www.sciencedirect.com/science/article/pii/S0001457507002175?via%3Dihub.

McKee, A.C., Cairns, N.J., Dickson, D.W., Folkerth, R.D., Keene, C.D., Litvan, I., Perl, D.P., Stein, T.D., Vonsattel, J.-.P., Stewart, W., et al., 2016. The first NINDS/NIBIB consensus meeting to define neuropathological criteria for the diagnosis of chronic traumatic encephalopathy. Acta Neuropathol. 131 (1), 75–86. http://www.ncbi.nlm.nih.gov/pubmed/26667418.

McKee, A.C., Cantu, R.C., Nowinski, C.J., Hedley-Whyte, E.T., Gavett, B.E., Budson, A.E., Santini, V.E., Lee, H.-.S., Kubilus, C.A., Stern, RA., 2009. Chronic traumatic encephalopathy in athletes: progressive tauopathy after repetitive head injury. J. Neuropathol. Exp. Neurol. 68 (7), 709–735. http://www.ncbi.nlm.nih.gov/pubmed/19535999.

Mez, J., Daneshvar, D.H., Kiernan, P.T., Abdolmohammadi, B., Alvarez, V.E., Huber, B.R., Alosco, M.L., Solomon, T.M., Nowinski, C.J., McHale, L., Cormier, K.A., 2017. Clinicopathological evaluation of chronic traumatic encephalopathy in players of American football. J. Am. Med. Assoc. 318 (4), 360–370.

Nahum, A.M., Smith, R., Ward, CC., 1977. Intracranial pressure dynamics during head impact, Proceedings of the 21st Stapp Car Crash Conference, 339–366. http://papers.sae.org/770922/.

Prabhu, R., Horstemeyer, M.F.F., Tucker, M.T.T., Marin, E.B.B., Bouvard, J.L.L., Sherburn, J.A.A., Liao, J., Williams, LN., 2011. Coupled experiment/finite element analysis on the mechanical response of porcine brain under high strain rates. J. Mech. Behav. Biomed. Mater. 4 (7), 1067–1080. https://www.sciencedirect.com/science/article/pii/S1751616111000609.

Song, X., Wang, C., Hu, H., Huang, T., Jin, J., 2015. A finite element study of the dynamic response of brain based on two parasagittal slice models. Comput. Math. Methods Med. 2015, 1–14. http://www.hindawi.com/journals/cmmm/2015/816405/.

Takhounts, E.G., Craig, M.J., Moorhouse, K., McFadden, J., Hasija, V., 2013. Development of brain injury criteria (BrIC). Stapp Car Crash J. 57, 243–266.

Tschopp, M.A., Bouvard, J.L., Ward, D.K., Horstemeyer, M.F., 2011. Atomic scale deformation mechanisms of amorphous polyethylene under tensile loading, The Minerals, Metals and Materials Society (TMS) 2011 Conference Proceedings, Supplemental Proceedings: Materials Fabrication, Properties, Characterization, and Modeling, 789–794.

Yang, B., Tse, K.M., Chen, N., bin, T.L., Zheng, Q.Q., Yang, H.M., Hu, M., Pan, G., Lee, H.P., 2014. Development of a finite element head model for the study of impact head injury. BioMed Res. Int. 2014, 408278.

Zhang, L., Yang, K.H., Dwarampudi, R., Omori, K., Li, T., Chang, K., Hardy, W.N., Khalil, T.B., King, AI., 2001. Recent Advances in Brain Injury Research: A New Human Head Model Development and Validation, SAE Technical Papers. SAE International, United States.

CHAPTER 7

Modeling mesoscale anatomical structures in macroscale brain finite element models

T. Wu, J.S. Giudice, A. Alshareef, M.B. Panzer

School of Engineering, University of Virginia, Charlottesville, VA, United States

7.1 Introduction

Computational human brain models have been crucial for understanding the mechanisms of traumatic brain injuries (TBIs) at multiple scales because of their ability to link external head impact conditions to the mesoscopic (~ 1 mm, e.g., axonal tracts) and even microscopic (~ 1 μm, e.g., axons) responses of brain tissue that leads to injury. While the brain gray matter (GM) consists primarily of neuronal cell bodies, white matter (WM) mainly consist of axons, extended from neuronal cell bodies and distributed into bundles called tracts. The microstructure of the brain WM is heterogeneous and anisotropic, but most current computational brain models (Mao et al., 2013; Miller et al., 2016; Takhounts et al., 2008) have adopted an isotropic representation of the material. More importantly, the lack of the heterogeneous, and anisotropic WM structures may limit their capability in predicting TBI, because the damage to axons is believed to be one of the critical mechanisms of TBI (Meaney and Smith, 2011).

While micromechanical modeling of the WM (axons) is currently only applicable on a representative volume element of brain tissue instead of the whole brain (Abolfathi et al., 2009; Javid et al., 2014; Yousefsani et al., 2018), modeling WM anisotropy considering the mesoscopic axonal tracts (bundles of axons) in macroscale brain simulation has become an increasing focus of research in the last few years. This chapter will critically review current methods to incorporate the mesoscopic brain anatomical structures into macroscale brain computational models, starting with an overview of typical macroscale computational brain models and mesoscopic brain anatomical structures. Finally, the chapter will conclude by identifying the limitations of the current state of knowledge and future perspectives to facilitate investigations along this line of research.

7.2 Macroscale brain finite element model

Current state-of-the-art brain computational models generally include a detailed representation of the macroscale neuroanatomy obtained from medical imaging (computed tomography or magnetic resonance imaging [MRI]) and sophisticated constitutive models derived from

Multiscale Biomechanical Modeling of the Brain.
DOI: https://doi.org/10.1016/B978-0-12-818144-7.00008-6
Copyright © 2022 Elsevier Inc. All rights reserved.

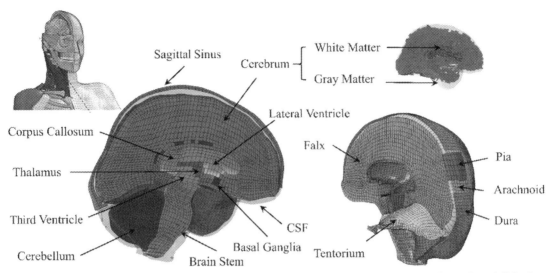

Fig. 7.1: Typical human brain finite element model modeled with a feature-based multiblock technique, Global Human Body Models Consortium (GHBMC) model developed initially in Mao et al. (2013).

experimental material testing of the brain parenchyma (Dixit and Liu, 2017; Giudice et al., 2019; Madhukar and Ostoja-Starzewski, 2019). Many computational models of the human brain have been developed using a wide array of numerical methods (e.g., element type, mesh size, element formulation, hourglass control methods) and material models. In general, the macroscale brain was typically modeled with solid three-dimensional (3D) elements using a feature-based multiblock technique, in which the brain is divided into different parts based on macroscale neuroanatomy (e.g., cerebral WM, cerebral GM, cerebellum, brain stem) (Fig. 7.1). Although different shear modulus/stiffness have been assigned to different parts, many models used in TBI research are still based on isotropic material.

The structural scale solid brain parts are constrained in the skull through meninges modeled with two-dimensional shell elements. Fig. 7.1 shows that the meninges include the dura mater, the arachnoid mater, and the pia mater from the outermost layer inward and primary dural extensions: the falx cerebri between the two cerebral hemispheres, and the tentorium cerebelli between the cerebral hemispheres and the cerebellum (note that some models do not include all the layers). The spaces between meninges and the brain are filled with a cerebrospinal fluid generally modeled as a fluid-like 3D solid. In brain biomechanics modeling, the meninges and cerebrospinal fluid provide complex boundary conditions for the brain tissues and influence the mechanical responses of the brain in the skull (Chafi et al., 2009; Hernandez et al., 2019; Ho et al., 2017; Lu et al., 2019; Panzer et al., 2012).

7.3 Mesoscale anatomical structures and imaging techniques

Besides the macroscale structures of the brain discussed in the previous section, the brain's structure is not always evident at the mesoscale level. At this length scale, different brain regions are connected to each other through axonal tracts (bundles of axons) to form an enormously complicated network system (Fig. 7.2). This structural connection (connectome) serves as a critical constraint on brain functionality and provides fundamental insight into the understanding of the pathophysiology underlying different brain injuries. The organization of the connectome in the brain can only be observed through advanced imaging techniques.

Diffusion-weighted MRI (DW-MRI), which reflects the amount of restriction experienced during water molecule diffusion (Brownian motion), is currently the only method capable of mapping the fiber architecture of nervous tissue in vivo. However, since the DW-MRI approach is simply to characterize the water diffusion, deriving specific microstructural traits from DW-MRI data is a challenging inverse problem with nonunique solutions that require strong modeling assumptions. Various algorithms have been designed to obtain local (voxel-wise) fiber orientations and reconstruct continuous fiber pathways by combining voxel-based models of fiber orientation. These techniques are collectively referred to as tractography analysis. Using different predetermined assumptions on water diffusion distribution pattern, different model-based tractography methods have been developed and used to image the human

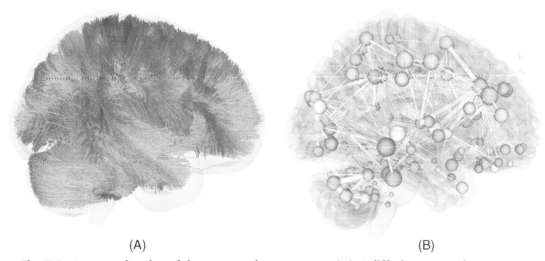

Fig. 7.2: An example atlas of the structural connectome. (A) A diffusion magnetic resonance image (MRI) revealed the underlying structural characteristics of axonal fiber bundles in a color-coded surface (red–blue–green indicates the orientation at the x–y–z axis, respectively). (B) The atlas was then used to build the connectome graph showing the connections between brain regions using a AAL2 (Rolls et al., 2015) cortical parcellation. The atlas was created and processed in DSI Studio (http://dsi-studio.labsolver.org/).

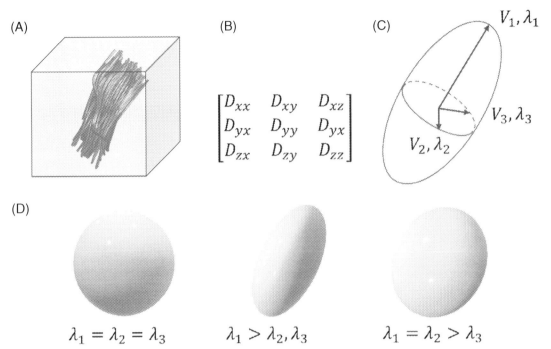

Fig. 7.3: Illustration of diffusion tensor imaging technique. (A) Fiber tracts have an arbitrary orientation to the voxel in scanner coordinates (x, y, z axes). (B) The 3 × 3 tensor model of diffusion derived from diffusivity (the velocity of diffusion) measurements in at least six noncollinear directions. (C) The diffusion tensor is characterized by an ellipsoid whose orientation is represented by three eigenvectors (V_1, V_2, V_3) and whose shape is determined by three eigenvalues (λ_1, λ_2, λ_3). The major principal axes of the ellipsoid represent the direction of maximum diffusivities. (D) Spherical, prolate, oblate tensors. The diffusion tensor model captures this orientated diffusion process with the diffusion ellipsoid. The tensor can also be oblate if the first two eigenvalues are equal or spherical if all three are equal.

connectome: diffusion tensor imaging (DTI; Basser et al., 1994), ball-and-sticks model (Behrens et al., 2003), neurite orientation dispersion and density imaging (Zhang et al., 2012), and more complicated models, such as composite hindered and restricted model of diffusion (Assaf and Basser, 2005) and AxCaliber (Assaf et al., 2008). Model-free methods, which make no underlying distribution assumption and obtain inference using empirical distribution, are also popular in tractography analysis: diffusion spectrum imaging (Wedeen et al., 2005), q-ball imaging (Tuch, 2004), and generalized Q-sampling imaging (Yeh et al., 2010).

One of the most popular tractography methods is DTI, which assumes that the velocity of water diffusion follows a 3D Gaussian distribution, whose covariance matrix is diffusion tensor. The diffusion tensor is a 3 × 3 symmetric, positive-definite matrix, which can be represented with three orthogonal eigenvectors and three positive eigenvalues through matrix diagonalization (Fig. 7.3). The principal eigenvector of the diffusion tensor (the direction of the fastest diffusion) is assumed to be the orientation of the axonal fiber tracts. This assumption is only strictly correct in regions where fiber tracts do not cross, fan, or branch, which is an essential limitation of the DTI method as between 66% and 90% of WM voxels were estimated to have multiple fiber crossing configurations (Jeurissen et al., 2013).

7.4 The importance of structural anisotropy in macroscale models of TBI

One important anatomic phenomenon of TBI was the findings of disrupted WM tracts and normal GM in autopsy (Strich, 1956). This type of injury was later called diffuse axonal injury. Diffuse axonal injury is thought to be caused by mechanical disruption of axonal cytoskeletons resulting in axonal swelling, retraction bulb, axonal degeneration, and downstream deafferentation. Histopathology studies on animal models (Gennarelli et al., 1982; Ibrahim et al., 2010) also found proteolysis, swelling, and other microscopic changes to the neuronal structure in injured subjects. Similar abnormalities of axons were recreated in vitro through tissue deformation in experimental conditions (Nakadate et al., 2017; Tang-Schomer et al., 2010), indicating a possible correlation between axonal strain (strain sustained by axons) and injury. Thus, axonal strain information is of great interest in computational modeling, and axonal strain is gaining more prominence as a reliable injury criterion because of its physiological relevance (Carlsen and Daphalapurkar, 2015; Garimella and Kraft, 2017a; Wright and Ramesh, 2012).

Although macroscale models have had a profound impact on advancing the understanding of TBI and have informed the development of protective equipment and restraint systems, most computational head models for predicting brain injury lack a detailed representation of the mesoscale structures, which may limit their capability in studying TBI and assessing tissue-based injury metrics such as axonal strain. As illustrated in Fig. 7.4, axonal strains cannot be easily correlated to the tissue strain without considering the axonal orientations (Zhou et al., 2021). Cellular level

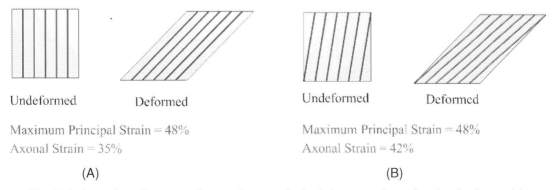

Fig. 7.4: Axonal strain versus the maximum principal tissue strain under simple shear with different fiber orientations. (A) Undeformed fiber direction is vertical to tissue shear direction. (B) Undeformed fiber direction is 80 degree to tissue shear direction.

108 Chapter 7

heterogeneities have an important influence on the axonal strain, leading to an orientation and location-dependent sensitivity of the tissue to mechanical loads.

The fact that the axonal strain was significantly different from the principal strain in values and distribution revealed the potential importance of incorporating anisotropic axonal fibers into brain finite element (FE) models (Wu et al., 2019). The significance of WM anisotropy on brain tissue responses has been recently studied (Sahoo et al., 2014; Wright et al., 2013; Zhao and Ji, 2019) and it is believed that the incorporation of WM anisotropy is necessary for the development of more precise injury metrics (Hajiaghamemar et al., 2020; Sahoo et al., 2016; Wu et al., 2021; Giordano and Kleiven, 2014a).

7.5 Material-based method

Recent attempts to incorporate WM anisotropy in brain FE models have been made based on tractography information. The majority of these studies (Table 7.1) have implicitly incorporated fiber tractography to inform anisotropic, fiber-reinforced constitutive models (Colgan et al., 2010; Ganpule et al., 2017; Giordano and Kleiven, 2014b; Kraft et al., 2012; Sahoo et al., 2014; Wright et al., 2013; Zhao and Ji, 2019). These anisotropic constitutive models were adapted from formulations once used for characterizing collagen fibers in arterial walls (Gasser et al., 2006) or ligaments (Quapp and Weiss, 1998). To derive the anisotropic hyperelastic strain energy potential W for the brain, it was assumed that it can be separated into an isotropic strain energy potential of the ground substance, W_{GS} and anisotropic strain energy potential of axonal fibers, W_f. Therefore, the anisotropic strain energy potential function is Eq. (7.1).

$$\mathrm{W}\left(\tilde{C},\right) = W_{GS}\left(\tilde{C}\right) + \sum_{i=1}^{N} W_{fi}\left(\tilde{C}, \tilde{E}_a\left(n_{0i}, k_i\right)\right) \tag{7.1}$$

in which $\tilde{C}$ is the isochoric part of the right Cauchy–Green strain tensor C. N is the number of fiber families. Currently, all the TBI studies that implemented the material-based method only model one fiber orientation per element ($N = 1$). n_{0i} is the weighted-average orientation of the fibers informed from DTI, and k_i is a dimensionless structure parameter related to the degree of anisotropy (or fiber dispersion). In modeling brain tissue, k_i was normally calculated from a scalar called fractional anisotropy (FA), which is computed from the eigenvalues of the diffusion tensor. For example, the dimensionless structure parameter in the Holzapfel–Gasser–Ogden model has been related with FA through Eq. (7.2) by assuming similarity between mechanical and diffusion anisotropy (Giordano and Kleiven, 2014b). Note that this derivation is only strictly true when $\lambda_1 \geq \lambda_2 = \lambda_3$ in the DTI because the Holzapfel–Gasser–Ogden model assume fibers are distributed with rotational symmetry about a mean preferred direction. At the lower limit, $k_i = 0$ (FA = 1), axons are perfectly aligned and at the upper limit, $k_i = 1/3$ (FA = 0), axons are randomly oriented and isotopically distributed.

$$k_i = \frac{1}{2}\frac{-6 + 4\mathrm{FA}^2 + 2\sqrt{3\mathrm{FA}^2 - 2\mathrm{FA}^4}}{-9 + 6\mathrm{FA}^2} \tag{7.2}$$

Modeling Mesoscale Anatomical Structures in Macroscale Brain Finite Element Models   109

Table 7.1: A list of computational studies in which white matter anisotropy is modeled with an anisotropic constitutive model.

Main reference	Research group**	Material model (reference)	Tractography source
Colgan et al., 2010	UCD	Holzapfel–Gasser–Ogden model (Gasser et al., 2006)	One DTI sample
Giordano and Kleiven, 2014b	KTH		One DTI sample
Wright et al., 2013*	JHU		Subject-specific DTI
Ganpule et al., 2017	JHU		Subject-specific DTI
Zhao and Ji, 2019	WPI		Subject-specific DTI
Kraft et al., 2012	ARL	Puso–Weiss model (Puso and Weiss, 1998)	Subject-specific DTI
Sahoo et al., 2014	UDS		DTI template of 12 volunteers

* Two-dimensional finite element model.
** *ARL*, U.S. Army Research Laboratory; *JHU*, Johns Hopkins University; *KTH*, KTH Royal Institute of Technology; *UCD*, University College Dublin; *UDS*, University of Strasbourg; *WPI*, Worcester Polytechnic Institute.

7.6 Structure-based method

Another approach to include WM anisotropy based on the fiber tractography is by incorporating them explicitly as 1-D cable elements into the isotropic computational brain model. The fibers are mathematically coupled into solid meshes using the embedded element technique. This technique also called the "s-element method" or the "superimposed mesh method" ensures that the axonal fibers and volumetric ground substance obey momentum balance and continuity and have the same accelerations and velocities. Initially developed for modeling rebar-reinforced concrete composites, this method has been successfully implemented in modeling brain tissue by Garimella and Kraft (2017) and Wu et al. (2019) in the computational software ABAQUS (Dassault Systèmes Simulia Corp., Providence, RI, USA) and LS-Dyna (LSTC, Livermore, CA, USA), respectively. Although some differences exist in those studies, the general schematic of the process (Fig. 7.5) involves three steps:

1. A FE mesh of fiber networks was created based on the tractography. The fiber elements (1-D elements) can be categorized based on the FA values of the tracts.
2. The fiber elements were aligned with the isotropic brain model in the same space. Procedures like morphing or registration would be performed to assure the geometry of the tractography template match with the geometry of the baseline brain FE model if subject-specific tractography data were not available.
3. Finally, the fiber mesh was mathematically embedded into solid elements of the baseline model. Mechanical properties of both the axon tract elements and the isotropic solid element should be assigned based on multimodal tissue data in the literature. Both the ground substance and fiber materials might be modeled as hyperviscoelastic material and contribute to the heterogeneous and anisotropic mechanical responses of the brain tissue.

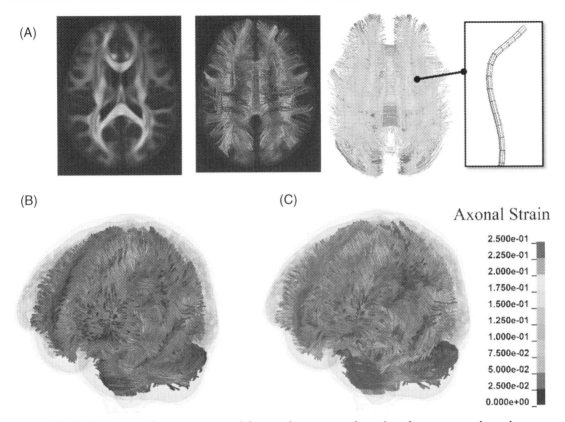

Fig. 7.5: A general process to model axonal tractography using the structure-based method. (A) DWI (right), axonal tractography (middle), 1-D cable finite elements of axonal tracts (left). (B) Brain model embedded with axonal FE tracts. (C) Example of maximum axonal strain distribution under a sagittal rotation of the head with mechanical properties described in Wu et al. (2019).

7.7 Summary and future perspectives

Advanced brain FE models are fundamental for investigating and understanding TBI. With the increasing interest in understanding injury mechanisms at the mesoscale level, the biofidelity of brain models needs to be improved in both anatomical representations and predicting biomechanical responses. In this chapter, we presented the recent efforts to incorporate the mesoscopic brain anatomical structures into macroscale brain computational models. However, other techniques for calculating strains during postprocessing have also been utilized (Chatelin et al., 2011; Ji et al., 2015; Zhao et al., 2016). In these studies, axonal strains are calculated by projecting the strains from the bulk volume elements onto the fiber directions derived from DTI after the FE simulation. While these techniques are able to differentiate between the maximum principal strain in solid elements and axonal strain in the tracts, they do not consider the mechanical interaction between macro- and mesoscale structures.

The material-based method is generally more accessible to the research society since some fiber-reinforced constitutive models are well-established and available in commercial FE simulation software. But it oversimplifies the brain parenchyma heterogeneity and requires the use of weighted-average fiber orientation for each element, which may not align with the actual direction of the axonal fiber bundles (Zhao and Ji, 2019). On the other hand, the structure-based method allows for the incorporation of multiple fiber orientations for a single element and takes advantage of the full axonal fiber tractography. Unlike the material-based method that relies on the discretization of information from WM fiber tracts, the structure-based method maintains the continuity of fiber tracts in the model.

However, the embedding method introduces new challenges associated with, but not limited to, mechanical characterization of axonal fibers and interaction between fibers and the ground substance. In the structure-based method, the ground substance occupies the full volume of the brain, including the volume under the fiber reinforcement. The addition of reinforcing axonal fibers that have finite cross-sectional area results in mass and stiffness redundancies, which often needs special consideration (Garimella et al., 2019), for example, artificially decreasing the density of the fibers to be negligible and subtracting the constitutive contribution of the ground substance material from the constitutive contribution of the axonal fibers.

One challenge inherent to explicitly modeling axonal fiber tracts is isolating the material properties of axonal fibers since the mechanical properties of axons are ambiguous in the literature (Bernal et al., 2007; Ning et al., 2006; Ouyang et al., 2013). Even if reliable axon material property data are available, it would be challenging to incorporate it into the model because the stiffness contribution from the axonal fibers depends on structural features (e.g., cross-sectional area of the fibers), as well as their constitutive model. The axonal fibers have not only different orientations in different regions of the brain but also have various volume fractions in different parts of the WM (Javid et al., 2014; Labus and Puttlitz, 2016). Currently, there is insufficient histological data to model the axonal volume fractions accurately. The fiber properties can only be estimated based on the stiffness differences between the WM and the GM.

Model mesoscale anatomical structures in macroscale brain FE models require the assumption that the mechanical properties of nervous tissue are correlated to its underlying microstructure. This highly heterogeneous microstructure is thought to be responsible for the interregional and intraregional heterogeneity of the mechanical behavior of WM, reported by previous studies (Budday et al., 2020; MacManus et al., 2017). The regional dependence or mechanical heterogeneity was typically found in compression, tension, and shear tests (Jin et al., 2013), indentation tests (Budday et al., 2015), and in vivo magnetic resonance elastography (Johnson et al., 2013). Fiber-rich regions like the brainstem and corona radiata were generally stiffer than fiber-deficient regions such as the cortex and thalamus. However,

112   Chapter 7

the authors acknowledge the contrary evidence in the literature on the mechanical anisotropy of WM. Velardi et al. (2006) found a significantly stiffer response in the fiber direction than perpendicular to it under uniaxial tests. Prange and Margulies (2002), Arbogast and Margulies (1997), Feng et al. (2013), and Jin et al. (2013) found significant anisotropy in shear, but their conclusions were contradictory, and the directional dependence did not correlate well with expected fiber orientation. Pervin and Chen (2009), Nicolle et al. (2004), and Budday et al. (2017) revealed no statistically significant dependencies on fiber orientation. In part, the contradictory results of the experimental observations could be a result of the complexity of the microstructure in the brain; over 80% of the WM voxels in the HCP-842 brain template had more than one fiber orientation even at 2.5 mm^3 resolution (Yeh et al., 2018). Consequently, extracting specimens that exhibit distinct fiber direction would be difficult. Understanding the directional and regional dependence of brain mechanical properties both in vitro and in vivo is inconclusive and is still a topic of intense ongoing research.

Another limitation is the fundamental ambiguities inherent in tract reconstruction techniques. The whole-brain fiber tractography was obtained from DWI and subsequent deterministic tractography analyses (which means extrapolating beyond the DWI data). While most studies get WM tracts from a single subject, a population-based atlas is preferred for increasing the validity of the fiber tractography and modeling representative topological interconnectivity in the general population (Yeh et al., 2018). Potentially, group-averaged tractography would reduce random errors associated with the individual fiber tracking process, but there would still be errors between the tractography and the true fiber architecture. Recently, it was shown that invalid bundles occur systematically across different research groups using different tract reconstruction methods when evaluated with ground truth bundles (Maier-Hein et al., 2017).

Multiscale computational models incorporating mesoscale anatomical structures captured through neuroimaging techniques, such as DTI, have the potential to improve our understanding of injury by providing a framework to integrate different physical models for a variety of injury mechanisms. It would give a possible linchpin for bridging the gap between the research performed in biomechanics and radiology and would have numerous clinical and research implications. Following this line of research, the following pieces of research are considered critical to push the frontiers in understanding the mechanism of human brain injury.

Recently, the significance of WM anisotropy on predicting TBI was studied (Hajiaghamemar et al., 2020; Sahoo et al., 2016; Wu et al., 2021; Giordano and Kleiven, 2014a). However, whether the inclusion of mesoscale connectional neuroanatomies (WM tracts) is sufficient for improving the predictive capabilities of a model has yet to be validated. With improvements in medical imaging techniques and the availability of real-time measurements of kinematic data during injury events (e.g., impacts in sports), researchers will have more resources available

Modeling Mesoscale Anatomical Structures in Macroscale Brain Finite Element Models 113

to evaluate the predictability of an injury metric/model. Computational modeling of the brain would also benefit from more strict validation using high-fidelity experimental measurements of deformation in the brain (Alshareef et al., 2018; Chan et al., 2018; Knutsen et al., 2014).

Current studies are particularly interested in predicting TBI using global metrics (e.g., maximum axonal strain, maximum principal strain), which do not account for the functional connectivity disruption due to the pattern of deformation (Kraft et al., 2012), and intersubject variability in connectional brain architectures (Giordano et al., 2017). Graph theory, the mathematical investigation of networks, provides a powerful and comprehensive approach to study the global and local topological network properties of sophisticated structural and functional brain connectivity. Combining computational modeling with connectome-based analysis would provide interdisciplinary insight (both biomechanical and neurophysiological) for understanding brain injury.

In addition to the connectional neuroanatomies (WM tracts) described in this chapter, the brain is also perfused with a highly complex vasculature system. The complete cerebrovasculature is rarely modeled in the macroscale computational models, although the influence of the large vasculature on the dynamic response of the brain has been investigated in some studies (Ho and Kleiven, 2007; Zhao and Ji, 2020; Unnikrishnan et al., 2019; Zhang et al., 2002). The knowledge of cerebrovasculature is crucial in understanding vascular brain injury, yet it has not been systematically mapped (imaged). One of the challenges is that the intersubjective variability of human cerebrovasculature is high compared to the axonal tractography. Recently, interest in venous and arterial physiology resulted in the publication of vasculature atlases (Bernier et al., 2018; Huck et al., 2019; Nowinski et al., 2009; Ward et al., 2018). These atlases will provide a basis for incorporating the complex cerebrovasculature into future computational brain models.

With these latest advancements and continuing efforts to include mesoscale neuroanatomies, computational brain models are expected to be a useful tool for understanding the mechanisms of TBI, evaluating tissue-based injury metrics, and developing injury mitigation systems.

References

Abolfathi, N., Naik, A., Sotudeh Chafi, M., Karami, G., Ziejewski, M., 2009. A micromechanical procedure for modelling the anisotropic mechanical properties of brain white matter. Comput. Methods Biomech. Biomed. Eng. 12, 249–262.

Alshareef, A., Giudice, J.S., Forman, J., Salzar, R.S., Panzer, M.B., 2018. A novel method for quantifying human in situ whole brain deformation under rotational loading using sonomicrometry. J. Neurotrauma 35, 780–789. https://doi.org/10.1089/neu.2017.5362.

Arbogast, K.B., Margulies, S.S., 1997. Regional differences in mechanical properties of the porcine central nervous system. SAE Technical Paper.

Assaf, Y., Basser, P.J., 2005. Composite hindered and restricted model of diffusion (CHARMED) MR imaging of the human brain. Neuroimage 27, 48–58.

Assaf, Y., Blumenfeld-Katzir, T., Yovel, Y., Basser, P.J., 2008. AxCaliber: a method for measuring axon diameter distribution from diffusion MRI. Magn. Resonance Med. 59, 1347–1354.

114 Chapter 7

Basser, P.J., Mattiello, J., LeBihan, D., 1994. Estimation of the effective self-diffusion tensor from the NMR spin echo. J. Magn. Resonance Ser. B 103, 247–254.

Behrens, T.E., Woolrich, M.W., Jenkinson, M., Johansen-Berg, H., Nunes, R.G., Clare, S., Matthews, P.M., Brady, J.M., Smith, S.M., 2003. Characterization and propagation of uncertainty in diffusion-weighted MR imaging. Magn. Resonance Med. 50, 1077–1088.

Bernal, R., Pullarkat, P.A., Melo, F., 2007. Mechanical properties of axons. Phys. Rev. Lett. 99, 018301.

Bernier, M., Cunnane, S.C., Whittingstall, K., 2018. The morphology of the human cerebrovascular system. Hum. Brain Map. 39, 4962–4975.

Budday, S., Nay, R., de Rooij, R., Steinmann, P., Wyrobek, T., Ovaert, T.C., Kuhl, E., 2015. Mechanical properties of gray and white matter brain tissue by indentation. J. Mech. Behav. Biomed. Mater. 46, 318–330.

Budday, S., Ovaert, T.C., Holzapfel, G.A., Steinmann, P., Kuhl, E., 2020. Fifty shades of brain: a review on the mechanical testing and modeling of brain tissue. Arch. Comput. Methods Eng. 27 (4), 1187–1230.

Budday, S., Sommer, G., Birkl, C., Langkammer, C., Haybaeck, J., Kohnert, J., Bauer, M., Paulsen, F., Steinmann, P., Kuhl, E., et al., 2017. Mechanical characterization of human brain tissue. Acta Biomater. 48, 319–340.

Carlsen, R.W., Daphalapurkar, N.P., 2015. The importance of structural anisotropy in computational models of traumatic brain injury. Front. Neurol. 6, 28.

Chafi, M.S., Dirisala, V., Karami, G., Ziejewski, M., 2009. A finite element method parametric study of the dynamic response of the human brain with different cerebrospinal fluid constitutive properties. Proc. Inst. Mech. Eng. Part H: J. Eng. Med. 223, 1003–1019.

Chan, D.D., Knutsen, A.K., Lu, Y.-C., Yang, S.H., Magrath, E., Wang, W.-T., Bayly, P.V., Butman, J.A., Pham, D.L., 2018. Statistical characterization of human brain deformation during mild angular acceleration measured in vivo by tagged magnetic resonance imaging. J. Biomech. Eng. 140, 101005.

Chatelin, S., Deck, C., Renard, F., Kremer, S., Heinrich, C., Armspach, J.-P., Willinger, R., 2011. Computation of axonal elongation in head trauma finite element simulation. J. Mech. Behav. Biomed. Mater. 4, 1905–1919.

Colgan, N.C., Gilchrist, M.D., Curran, K.M., 2010. Applying DTI white matter orientations to finite element head models to examine diffuse TBI under high rotational accelerations. Prog. Biophys. Mol. Biol. 103, 304–309.

Dixit, P., Liu, G., 2017. A review on recent development of finite element models for head injury simulations. Arch. Comput. Methods Eng. 24, 979–1031.

Feng, Y., Okamoto, R.J., Namani, R., Genin, G.M., Bayly, P.V., 2013. Measurements of mechanical anisotropy in brain tissue and implications for transversely isotropic material models of white matter. J. Mech. Behav. Biomed. Mater. 23, 117–132.

Ganpule, S., Daphalapurkar, N.P., Ramesh, K.T., Knutsen, A.K., Pham, D.L., Bayly, P.V., Prince, J.L., 2017. A three-dimensional computational human head model that captures live human brain dynamics. J. Neurotrauma 34, 2154–2166.

Garimella, H.T., Kraft, R.H., 2017a. A new computational approach for modeling diffusion tractography in the brain. Neural Regener. Res. 12, 23.

Garimella, H.T., Kraft, R.H., 2017b. Modeling the mechanics of axonal fiber tracts using the embedded finite element method. Int. J. Numer. Methods Biomed. Eng. 33, e2823.

Garimella, H.T., Menghani, R.R., Gerber, J.I., Sridhar, S., Kraft, R.H., 2019. Embedded finite elements for modeling axonal injury. Ann. Biomed. Eng. 47 (9), 1889–1907.

Gasser, T.C., Ogden, R.W., Holzapfel, G.A., 2006. Hyperelastic modelling of arterial layers with distributed collagen fibre orientations. J. R. Soc. Interf. 3, 15–35.

Gennarelli, T.A., Thibault, L.E., Adams, J.H., Graham, D.I., Thompson, C.J., Marcincin, R.P., 1982. Diffuse axonal injury and traumatic coma in the primate. Ann. Neurol. 12, 564–574.

Giordano, C., Kleiven, S., 2014a. Evaluation of axonal strain as a predictor for mild traumatic brain injuries using finite element modeling. SAE Technical Paper.

Giordano, C., Kleiven, S., 2014b. Connecting fractional anisotropy from medical images with mechanical anisotropy of a hyperviscoelastic fibre-reinforced constitutive model for brain tissue. J. R. Soc. Interf. 11, 20130914.

Giordano, C., Zappalà, S., Kleiven, S., 2017. Anisotropic finite element models for brain injury prediction: the sensitivity of axonal strain to white matter tract inter-subject variability. Biomec. Model. Mechanobiol. 16, 1269–1293.

Giudice, J.S., Zeng, W., Wu, T., Alshareef, A., Shedd, D.F., Panzer, M.B., 2019. An analytical review of the numerical methods used for finite element modeling of traumatic brain injury. Ann. Biomed. Eng. 47 (9), 1855–1872.

Hajiaghamemar, M., Wu, T., Panzer, M.B., Margulies, S.S., 2020. Embedded axonal fiber tracts improve finite element model predictions of traumatic brain injury. Biomech. Model. Mechanobiol. 19 (3), 1109–1130.

Hernandez, F., Giordano, C., Goubran, M., Parivash, S., Grant, G., Zeineh, M., Camarillo, D., 2019. Lateral impacts correlate with falx cerebri displacement and corpus callosum trauma in sports-related concussions. Biomech. Model. Mechanobiol. 18 (3), 631–649.

Ho, J., Kleiven, S., 2007. Dynamic response of the brain with vasculature: a three-dimensional computational study. J. Biomech. 40, 3006–3012.

Ho, J., Zhou, Z., Li, X., Kleiven, S., 2017. The peculiar properties of the falx and tentorium in brain injury biomechanics. J. Biomech. 60, 243–247.

Huck, J., Wanner, Y., Fan, A.P., Jäger, A.-T., Grahl, S., Schneider, U., Villringer, A., Steele, C.J., Tardif, C.L., Bazin, P.-L., et al., 2019. High resolution atlas of the venous brain vasculature from 7 T quantitative susceptibility maps. bioRxiv 444349.

Ibrahim, N.G., Ralston, J., Smith, C., Margulies, S.S., 2010. Physiological and pathological responses to head rotations in toddler piglets. J. Neurotrauma 27, 1021–1035.

Javid, S., Rezaei, A., Karami, G., 2014. A micromechanical procedure for viscoelastic characterization of the axons and ECM of the brainstem. J. Mech. Behav. Biomed. Mater. 30, 290–299.

Jeurissen, B., Leemans, A., Tournier, J.-D., Jones, D.K., Sijbers, J., 2013. Investigating the prevalence of complex fiber configurations in white matter tissue with diffusion magnetic resonance imaging. Hum. Brain Map. 34, 2747–2766.

Ji, S., Zhao, W., Ford, J.C., Beckwith, J.G., Bolander, R.P., Greenwald, R.M., Flashman, L.A., Paulsen, K.D., McAllister, T.W., 2015. Group-wise evaluation and comparison of white matter fiber strain and maximum principal strain in sports-related concussion. J. Neurotrauma 32, 441–454.

Jin, X., Zhu, F., Mao, H., Shen, M., Yang, K.H., 2013. A comprehensive experimental study on material properties of human brain tissue. J. Biomech. 46, 2795–2801.

Johnson, C.L., McGarry, M.D., Gharibans, A.A., Weaver, J.B., Paulsen, K.D., Wang, H., Olivero, W.C., Sutton, B.P., Georgiadis, J.G., 2013. Local mechanical properties of white matter structures in the human brain Neuroimage 79, 145–152.

Knutsen, A.K., Magrath, E., McEntee, J.E., Xing, F., Prince, J.L., Bayly, P.V., Butman, J.A., Pham, D.L., 2014. Improved measurement of brain deformation during mild head acceleration using a novel tagged MRI sequence. J. Biomech. 47, 3475–3481.

Kraft, R.H., Mckee, P.J., Dagro, A.M., Grafton, S.T., 2012. Combining the finite element method with structural connectome-based analysis for modeling neurotrauma: connectome neurotrauma mechanics. PLoS Comput. Biol. 8, e1002619.

Labus, K.M., Puttlitz, C.M., 2016. An anisotropic hyperelastic constitutive model of brain white matter in biaxial tension and structural–mechanical relationships. J. Mech. Behav. Biomed. Mater. 62, 195–208.

Lu, Y.-C., Daphalapurkar, N.P., Knutsen, A.K., Glaister, J., Pham, D.L., Butman, J.A., Prince, J.L., Bayly, P.V., Ramesh, K.T., 2019. A 3D computational head model under dynamic head rotation and head extension validated using live human brain data, including the falx and the tentorium. Ann. Biomed. Eng. 47 (9), 1923–1940.

MacManus, D.B., Pierrat, B., Murphy, J.G., Gilchrist, M.D., 2017. Region and species dependent mechanical properties of adolescent and young adult brain tissue. Sci. Rep. 7, 13729.

Madhukar, A., Ostoja-Starzewski, M., 2019. Finite element methods in human head impact simulations: a review. Ann. Biomed. Eng. 47 (9), 1832–1854.

Maier-Hein, K.H., Neher, P.F., Houde, J.-C., Côté, M.-A., Garyfallidis, E., Zhong, J., Chamberland, M., Yeh, F.-C., Lin, Y.-C., Ji, Q., 2017. The challenge of mapping the human connectome based on diffusion tractography. Nat. Commun. 8, 1349.

116   Chapter 7

Mao, H., Zhang, L., Jiang, B., Genthikatti, V.V., Jin, X., Zhu, F., Makwana, R., Gill, A., Jandir, G., Singh, A., et al., 2013. Development of a finite element human head model partially validated with thirty five experimental cases. J. Biomech. Eng. 135, 111002.

Meaney, D.F., Smith, D.H., 2011. Biomechanics of concussion. Clin. Sports Med. 30, 19–31.

Miller, L.E., Urban, J.E., Stitzel, J.D., 2016. Development and validation of an atlas-based finite element brain model. Biomech. Model. Mechanobiol. 15, 1201–1214.

Nakadate, H., Kurtoglu, E., Furukawa, H., Oikawa, S., Aomura, S., Kakuta, A., Matsui, Y., 2017. Strain-rate dependency of axonal tolerance for uniaxial stretching. Stapp Car Crash J. 61, 53–65.

Nicolle, S., Lounis, M., Willinger, R., 2004. Shear properties of brain tissue over a frequency range relevant for automotive impact situations: new experimental results. SAE Technical Paper.

Ning, X., Zhu, Q., Lanir, Y., Margulies, S.S., 2006. A transversely isotropic viscoelastic constitutive equation for brainstem undergoing finite deformation. J. Biomech. Eng. 128, 925–933.

Nowinski, W.L., Volkau, I., Marchenko, Y., Thirunavuukarasuu, A., Ng, T.T., Runge, V.M., 2009. A 3D model of human cerebrovasculature derived from 3T magnetic resonance angiography. Neuroinformatics 7, 23–36.

Ouyang, H., Nauman, E., Shi, R., 2013. Contribution of cytoskeletal elements to the axonal mechanical properties. J. Biol. Eng. 7, 21.

Panzer, M.B., Myers, B.S., Capehart, B.P., Bass, C.R., 2012. Development of a finite element model for blast brain injury and the effects of CSF cavitation. Ann. Biomed. Eng. 40, 1530–1544.

Pervin, F., Chen, W.W., 2009. Dynamic mechanical response of bovine gray matter and white matter brain tissues under compression. J. Biomech. 42, 731–735.

Prange, M.T., Margulies, S.S., 2002. Regional, directional, and age-dependent properties of the brain undergoing large deformation. J. Biomech. Eng. 124, 244–252.

Puso, M., Weiss, J., 1998. Finite element implementation of anisotropic quasi-linear viscoelasticity using a discrete spectrum approximation. J. Biomech. Eng. 120, 62–70.

Quapp, K., Weiss, J., 1998. Material characterization of human medial collateral ligament. J. Biomech. Eng. 120, 757–763.

Rolls, E.T., Joliot, M., Tzourio-Mazoyer, N., 2015. Implementation of a new parcellation of the orbitofrontal cortex in the automated anatomical labeling atlas. Neuroimage 122, 1–5.

Sahoo, D., Deck, C., Willinger, R., 2014. Development and validation of an advanced anisotropic visco-hyperelastic human brain FE model. J. Mech. Behav. Biomed. Mater. 33, 24–42.

Sahoo, D., Deck, C., Willinger, R., 2016. Brain injury tolerance limit based on computation of axonal strain. Accid. Anal. Prevent. 92, 53–70.

Strich, S.J., 1956. Diffuse degeneration of the cerebral white matter in severe dementia following head injury. J. Neurol. Neurosurg. Psychiatry 19, 163.

Takhounts, E.G., Ridella, S.A., Hasija, V., Tannous, R.E., Campbell, J.Q., Malone, D., Danelson, K., Stitzel, J., Rowson, S., Duma, S., 2008. Investigation of traumatic brain injuries using the next generation of simulated injury monitor (SIMon) finite element head model. SAE Technical Paper.

Tang-Schomer, M.D., Patel, A.R., Baas, P.W., Smith, D.H., 2010. Mechanical breaking of microtubules in axons during dynamic stretch injury underlies delayed elasticity, microtubule disassembly, and axon degeneration. FASEB J. 24, 1401–1410.

Tuch, D.S., 2004. Q-ball imaging. Magn. Resonance Med. 52, 1358–1372.

Unnikrishnan, G., Mao, H., Sundaramurthy, A., Bell, E.D., Yeoh, S., Monson, K., Reifman, J., 2019. A 3-D rat brain model for blast-wave exposure: effects of brain vasculature and material properties. Ann. Biomed. Eng. 47 (9), 2033–2044.

Velardi, F., Fraternali, F., Angelillo, M., 2006. Anisotropic constitutive equations and experimental tensile behavior of brain tissue. Biomech. Model. Mechanobiol. 5, 53–61.

Ward, P.G., Ferris, N.J., Raniga, P., Dowe, D.L., Ng, A.C., Barnes, D.G., Egan, G.F., 2018. Combining images and anatomical knowledge to improve automated vein segmentation in MRI. NeuroImage 165, 294–305.

Wedeen, V.J., Hagmann, P., Tseng, W.-Y.I., Reese, T.G., Weisskoff, R.M., 2005. Mapping complex tissue architecture with diffusion spectrum magnetic resonance imaging. Magn. Resonance Med. 54, 1377–1386.

Wright, R.M., Post, A., Hoshizaki, B., Ramesh, K.T., 2013. A multiscale computational approach to estimating axonal damage under inertial loading of the head. J. Neurotrauma 30, 102–118.

Wright, R.M., Ramesh, K., 2012. An axonal strain injury criterion for traumatic brain injury. Biomech. Model. Mechanobiol. 11, 245–260.

Wu, T., Alshareef, A., Giudice, J.S., Panzer, M.B., 2019. Explicit modeling of white matter axonal fiber tracts in a finite element brain model. Ann. Biomed. Eng. 47 (9), 1908–1922.

Wu, T., Hajiaghamemar, M., Giudice, J.S., Alshareef, A., Margulies, S.S., Panzer, M.B., 2021. Evaluation of tissue-level brain injury metrics using species-specific simulations. J. Neurotrauma 38 (13), 1879–1888.

Yeh, F.-C., Panesar, S., Fernandes, D., Meola, A., Yoshino, M., Fernandez-Miranda, J.C., Vettel, J.M., Verstynen, T., 2018. Population-averaged atlas of the macroscale human structural connectome and its network topology. NeuroImage 178, 57–68.

Yeh, F.-C., Wedeen, V.J., Tseng, W.-Y.I., 2010. Generalized q-sampling imaging. IEEE Trans. Med. Imaging 29, 1626–1635.

Yousefsani, S.A., Farahmand, F., Shamloo, A., 2018. A three-dimensional micromechanical model of brain white matter with histology-informed probabilistic distribution of axonal fibers. J. Mech. Behav. Biomed. Mater. 88, 288–295.

Zhang, H., Schneider, T., Wheeler-Kingshott, C.A., Alexander, D.C., 2012. NODDI: practical in vivo neurite orientation dispersion and density imaging of the human brain. Neuroimage 61, 1000–1016.

Zhang, L., Bae, J., Hardy, W.N., Monson, K.L., Manley, G.T., Goldsmith, W., Yang, K.H., King, A.I., 2002. Computational study of the contribution of the vasculature on the dynamic response of the brain. SAE Technical Paper.

Zhao, W., Ford, J.C., Flashman, L.A., McAllister, T.W., Ji, S., 2016. White matter injury susceptibility via fiber strain evaluation using whole-brain tractography. J. Neurotrauma 33, 1834–1847.

Zhao, W., Ji, S., 2019. White matter anisotropy for impact simulation and response sampling in traumatic brain injury. J. Neurotrauma 36 (2), 250–263.

Zhao, W., Ji, S., 2020. Incorporation of vasculature in a head injury model lowers local mechanical strains in dynamic impact. J. Biomech. 104, 109732.

Zhou, Z., Domel, A.G., Li, X., Grant, G., Kleiven, K., Camarillo, D., Zeineh, M., 2021. White matter tract-oriented deformation is dependent on real-time axonal fiber orientation. J. Neurotrauma 38 (12), 1730–1745.

CHAPTER 8

A macroscale mechano-physiological internal state variable (MPISV) model for neuronal membrane damage with subscale microstructural effects

A. Bakhtiarydavijani[a], M.A. Murphy[a], Raj K. Prabhu[b], T.R. Fonville[c], Mark F. Horstemeyer[c]

[a]Center for Advanced Vehicular Systems (CAVS), Mississippi State University, Mississippi State, MS, United States [b]USRA, NASA HRP CCMP, NASA Glenn Research Center, Cleveland, OH, United States [c]School of Engineering, Liberty University, Lynchburg, VA, United States

8.1 Introduction

In Chapter 4, molecular dynamics (MD) simulations of a representative lipid bilayer were proposed in which membrane mechanoporation was defined based on criteria for water diffusion, water bridging, and calcium ion diffusion. In this chapter, we bring that information into a macroscale model in terms of "damage" in a continuum model.

The importance of neuronal membrane integrity in neuronal survivability and cell death as a result of mechanical disruption of the membrane integrity has been well documented (Farkas et al., 2006; Raghupathi, 2004). Significant neuron deformation and membrane strain disrupt the membrane's ability to regulate the neuronal ion homeostasis (LaPlaca and Prado, 2010; LaPlaca et al., 2005, 2019; Cullen et al., 2011; Geddes et al., 2003). This ion homeostasis is central in regulating cellular functions and is closely controlled inside the neuron (Siesjö, 1993; LaPlaca et al., 2019; Stoica and Faden, 2010; Yuan et al., 2003). Hence, when looking at mechanoporation related cell death, injury thresholds can be determined based on the extent of the neuronal ion homeostasis disruption.

The current chapter addresses the development of damage evolution equations for neuronal mechanoporation that will bridge nanoscale mechanoporation behavior (Bridge 6 in Fig. 8.1) to the macroscale mechano-physiological internal state variable (MPISV) model. To achieve this, mechanoporation, or damage, is first multiplicatively decoupled into the pore number density (the number of pores in a unit area) and pore area per pore Horstemeyer et al. (1999, 2019). These two independent internal state variables (ISVs) are then phenomenologically

Multiscale Biomechanical Modeling of the Brain.
DOI: https://doi.org/10.1016/B978-0-12-818144-7.00003-7
Copyright © 2022 Elsevier Inc. All rights reserved.

120   Chapter 8

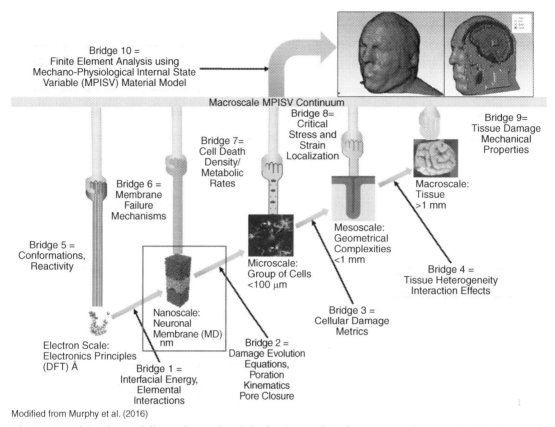

Modified from Murphy et al. (2016)

Fig. 8.1: Multiscale modeling schematic of the brain. In this chapter, we focus on Bridge 6, which connects the atomistic scale with the macroscale mechano-physiological internal state variable (MPISV) model.

defined as a function of the observable state variables and material properties as per Coleman and Gurtin (1967). Defining damage in this way allows for mechanoporation behavior to be scaled from the nanoscale to the microscale, mesoscale, or macroscale in a thermodynamically consistent manner. Pore number density and growth rate equations also need to conform to the ISV standards previously defined by Coleman and Gurtin (1967). The damage evolution equations are then calibrated to the MD simulation results (Murphy et al., 2016, 2019). A sensitivity analysis is performed related to the constants and material properties to visualize the underlying uncertainties and material property dependencies. As an example, the damage evolution equations are then extended to a spherical neuronal soma membrane and coupled with Nernst–Planck diffusion equations to compare with experimental cell culture studies.

8.1.1 Definitions

Because the MD simulations that output the mechanoporation data are in the current configuration, the stress and strain definitions that describe mechanoporation behavior need to be defined in the same configuration. Also, considering the large strains associated with mechanoporation, true stress and strain definitions are used. The true principal strains, ε_i, are defined as the following,

$$\varepsilon_i = \ln\left(l_i / L_i\right) \tag{8.1}$$

where, l_i and L_i are the current and initial box dimensions in the direction "i." The von Mises strain, ε_{vM}, for the simulation box is given by,

$$\varepsilon_{vm} = \frac{\sqrt{2}}{3}\left[\left(\varepsilon_1 - \varepsilon_2\right)^2 + \left(\varepsilon_2 - \varepsilon_3\right)^2 + \left(\varepsilon_1 - \varepsilon_3\right)^2\right]^{0.5} \tag{8.2}$$

The true strain rates, $\dot{\varepsilon}_i(t)$, are expressed as the rate of change of the strain:

$$\dot{\varepsilon}_i(t) = \frac{\partial \varepsilon_i(t)}{\partial t}. \tag{8.3}$$

This form is used as the simulation box is deformed under constant velocities. The equivalent von Mises strain rate, $\dot{\varepsilon}_{vM}(t)$, is then defined as the following,

$$\dot{\varepsilon}_{vM}(t) = \frac{\sqrt{2}}{3}\left[\left(\dot{\varepsilon}_1(t) - \dot{\varepsilon}_2(t)\right)^2 + \left(\dot{\varepsilon}_2(t) - \dot{\varepsilon}_3(t)\right)^2 + \left(\dot{\varepsilon}_1(t) - \dot{\varepsilon}_3(t)\right)^2\right]^{0.5} \tag{8.4}$$

Using the von Mises form for strain provides a means of comparing different strain states and allows for a meaningful combination when developing the damage evolution equations and its integration into the ISV constitutive material model.

The stresses used in the current work, σ_1 and σ_2, refer to the force over membrane cross section in their respective box dimension and are the principal stresses. As the box of atoms is relaxed in the z-direction given a plane stress condition, $\sigma_3 = 0$. The stress invariants of the box stresses that are used in the damage evolution equations are given as follows,

$$\begin{aligned} I_1 &= \sigma_{xx} + \sigma_{yy} + \sigma_{zz} = \sigma_1 + \sigma_2 \\ J_2 &= \frac{1}{2}\left(s_1^2 + s_2^2 + s_3^2\right) = \frac{1}{6}\left[\left(\sigma_1 - \sigma_2\right)^2 + \sigma_1^2 + \sigma_2^2\right] \\ J_3 &= \frac{1}{3}\left(s_1^3 + s_2^3 + s_3^3\right) \end{aligned} \tag{8.5}$$

where I_1 is the first invariant of the stress tensor, J_2 and J_3 are the second and third invariants of the deviatoric stress tensor, respectively, and s_i are the principal stresses of the deviatoric stress tensor,

$$s_i = \sigma_i - \frac{I_1}{3} \tag{8.6}$$

8.2 Membrane disruption

Cell membrane disruption under mechanical loading has long been of interest (Zeldovich, 1943; Deryagin and Gutop, 1962). Direct visualization of cell membrane rupture has been difficult considering light microscopy captures rates compared to the rupture rate. This obstacle, however, was overcome using more viscous mediums that slow the failure process (Sandre et al., 1999; Brochard-Wyart et al., 2000) where pores nucleate, grow, coalesce, and even close in a stretched vesicle. Here, the leakage of internal fluid allows for enough relaxation that the pore becomes energetically unfavorable. Significant attention has also been given to studying the vesicle failure stresses (Needham and Nunn, 1990; Rawicz et al., 2000; Bloom and Evans, 1991; Evans and Smith, 2011b; Evans et al., 2003). Further, pore nucleation in lipid bilayer membranes has been shown to be rate dependent (Evans and Smith 2011b; Evans et al., 2003). Considering the difficulty of capturing membrane poration behavior using experimental methods due to size and time frame of failure, computational methods can be used to capture the kinetics and kinematics of membrane failure (Murphy et al., 2019; Tieleman et al., 2003, 2006; Tomasini et al., 2010; Shigematsu et al., 2014, 2015; Leontiadou and Marrink, 2004; Murphy et al., 2016; Bakhtiarydavijani et al., 2019).

When attempting to extrapolate the poration behavior from membrane vesicles to capture neuronal rupture, some complexities exist:

- The neuronal membrane is significantly more complex than lipid bilayer vesicles.
- The neuronal membrane is connected to the cytoskeleton that stabilizes the neuronal anatomy. This means that pore nucleation in the neuronal membrane is not always accompanied by its failure (Cooper and McNeil, 2015).
- The neuronal membrane can implement repair mechanisms that effectively closes the pores (Cooper and McNeil, 2015).

Considering these general outlines, we develop a set of damage rate equations that captures the nucleation of pores in the membrane, the pores' growth due to stretching, and pores' resealing considering the repair mechanisms.

8.3 Development of damage evolution equation

Since the representative neuronal membrane modeled in the MD simulations were performed under periodic boundary conditions, the membrane can be assumed to be two-dimensional. Hence, damage in the membrane is defined as pore number density and pore growth in two dimensions. First, we express damage (φ), that is dimensionless, as the current total pore area, A_p, to the current membrane area, A_1:

$$\phi = \frac{A_p}{A_1} \tag{8.7}$$

The total pore area can then be written as:

$$A_p = N a_p \tag{8.8}$$

where N is the number of pores and a_p is the average pore area. Assuming pore number density, η as a function of the number of pores and the current membrane area:

$$\eta = \frac{N}{A_1} \tag{8.9}$$

The number of pores in Eq. (8.8) can be replaced by Eq. (8.9):

$$A_p = \eta A_1 a_p \tag{8.10}$$

Replacing the total pore area in Eq. (8.8) with Eq. (8.7) then gives:

$$\phi = \frac{\eta A_1 a_p}{A_1} = \eta a_p \tag{8.11}$$

Hence, damage is defined as a function of pore number density, or pore nucleation. The rate of damage, $\dot{\phi}$, is then the time derivative of damage:

$$\dot{\phi} = \dot{\eta} a_p + \eta \dot{a}_p \tag{8.12}$$

The rate form of damage used here allows for history-dependent property characterization. The next step is to define the two variables, pore number density and pore growth rate as a function of observable variables (stress, strain rate, and itself). Further, these equations should be dependent on membrane microstructures representative of the length scale.

Nucleation of pores on a membrane has been investigated using experimental approaches on simplified bilayers (Bloom and Evans, 1991; Evans and Smith, 2011b, 2011a; Rawicz et al., 2000; Zhelev and Needham, 1993) and MD simulations of representative membranes (Tieleman et al., 2003; 2006; Tomasini et al., 2010; Shigematsu et al., 2014, 2015; Leontiadou et al., 2004; Murphy et al., 2016, 2019; Bakhtiarydavijani et al., 2019). Experimental studies show that surface tension versus the edge tension energies play an important role in defining the pore nucleation on a lipid bilayer (Needham and Nunn, 1990; Evans et al., 2003). These experiments are generally accompanied by catastrophic failure due to the ramped forces. MD simulations are then used to look at the pore nucleation and growth behavior under different stress states and strain rates (Murphy et al., 2019; Koshiyama and Wada, 2011; Shigematsu et al., 2014; Koshiyama et al., 2010). Here we define the pore nucleation, or pore number density, and pore growth equations considering poration thermodynamics identified through experimental and theoretical work (Deryagin and Gutop, 1962; Evans et al., 2003). The pore nucleation and pore growth ISVs are informed by results from the MD simulations (Bakhtiarydavijani et al., 2019; Murphy et al., 2019) while ensuring thermodynamic consistency (Coleman and Gurtin, 1967).

8.3.1 Pore number density rate

The pore number density equation should be defined as a function of the strain rate and stress state as well as the membrane mechanical properties. Furthermore, the pore number density rate needs to be a function of itself to be consistent with the ISV theory.

$$\dot{\eta} = h\left(\eta, \dot{\varepsilon}_{vm}, \tau_{eq}, f(\sigma)\right) \tag{8.13}$$

124   Chapter 8

Significant effort was made to study the thermodynamics of lipid bilayer rupture through the micropipette aspiration technique (Sandre et al., 1999; Martínez-Balbuena et al., 2015; Evans and Smith, 2011b; Evans et al., 2003). These studies show that an energy landscape exists for poration in the lipid bilayer vesicles where the critical energy for membrane rupture directly correlates with the edge energy and the applied stress or surface tension. From here, pore nucleation is assumed to correlate with the equivalent tension (τ_{eq}) on the membrane.

$$\dot{\eta} \propto \tau_{eq} \tag{8.14}$$

where

$$\tau_{eq} = \frac{\tau_{mod}\varepsilon_{vm}}{\tau_{peak}}\left(1 - \phi\right) \tag{8.15}$$

$\tau_{mod}\varepsilon_{vm}$ is the surface tension that is normalized by the peak surface tension τ_{peak}. However, as pores grow on the membrane, this surface tension is relaxed. To account for this relaxation, we correct the membrane strain with $(1 - \varphi)$. This can be better understood by assuming that the effective membrane area, A_{eff}, is the current box area minus the sum of the pore areas. Hence, we correct the surface tension using damage. This ratio can further be simplified considering Eq. (8.7):

$$\frac{A_{eff}}{A_1} = \frac{A_1 - A_p}{A_1} = 1 - \phi \tag{8.16}$$

MD simulations show the stress state dependence of pore nucleation in a representative bilayer membrane (Murphy et al., 2016). To capture the stress state dependence while conforming to limitations set on these variables by the thermodynamics of continua (Coleman and Gurtin, 1967), the stress state dependence is described using stress invariants as previously defined for other materials (Horstemeyer and Gokhale, 1999):

$$f\left(\sigma\right) = b_1\left[\frac{4}{27} - \frac{J_3^2}{J_2^3}\right] + b_2\frac{J_3}{J_2^{3/2}} + b_3\frac{I_1}{\sqrt{J_2}} \tag{8.17}$$

I_1, J_2, and J_3 were given in Eq. (8.5). b_1, b_2, and b_3 are the material constants that represent the stress state sensitivity. In this equation, each nondimensional term is only a function of stress invariants. Under torsion or simple shear, the second and third terms are zero, while under compression and tension, the first term is zero. Table 8.1 gives the value of each nondimensional term under the applied stress states considering in-plane ε_1 and ε_2 ratio, where σ_{zz} is zero due to the relaxation of the MD model in the z-direction. This approach follows previous works that look at stress state dependence (Horstemeyer and Gokhale, 1999; Miller and McDowell, 1992) who captured the stress state dependence of damage by introducing an invariants-based nondimensional function into their void density equations. Finally, combining the formulas into one equation gives the following pore nucleation rate equation,

$$\dot{\eta} = d_1\dot{\varepsilon}_{vm}\eta\tau_{eq}\left(b_1\left[\frac{4}{27} - \frac{J_3^2}{J_2^3}\right] + b_2\frac{J_3}{J_2^{3/2}} + b_3\frac{I_1}{\sqrt{J_2}}\right) \tag{8.18}$$

Table 8.1: The nondimensional stress state dependent terms give unique values for each deformation/stress state.

Deformation state	Strain state ratio ($\varepsilon_1/\varepsilon_2$)	$\frac{4}{27}-\frac{J_3^2}{J_2^3}$	$\frac{J_3}{J_2^{3/2}}$	$\frac{I_1}{\sqrt{J_2}}$
Equibiaxial deformation	1	0	$-2/(3\sqrt{3})$	$2\sqrt{3}$
Nonequibiaxial deformation	0.5	0.030	−0.344	3.422
Strip biaxial deformation	0	$\sqrt{2}$	0.08	2.872

here d_1 is a nondimensional constant dependent on the membrane heterogeneity. Finally, η_0 is the initial pore number density when Eq. (8.18) is integrated.

8.3.2 Pore growth rate

Pore growth, similar to pore nucleation, is represented by a function of strain rate and stress state from MD simulations, surface tension from theoretical work, and pore area considering the ISV theory,

$$\dot{a}_p = h\left(a_p, \dot{\varepsilon}_{vm}, \tau_{eq}, f(\sigma)\right) \tag{8.19}$$

Two energies govern pore evolution in the membrane: strain energy relaxed as a result of pore growth, $a_p\tau_{eq}$, and the energy increase as a result of the increase in pore perimeter ($2\sqrt{\pi a_p}$) considering the pore edge energy (γ), $-2\sqrt{\pi a}\gamma$ (Fig. 8.2). Since pore growth correlates with the excess force in the membrane, the following proportion emerges,

$$\dot{a}_p \propto \left(\tau_{eq}a_p - 2\sqrt{\pi a_p}\frac{\gamma}{\gamma_{ref}}\right) \tag{8.20}$$

γ_{ref} is a material characteristic that is constant, while the equivalent surface tension depends on the strain in the membrane A_{mem}. The stress state dependence of pore growth can be

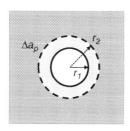

Fig. 8.2: Schematic of pore size change where r_1 is the original radius and r_2 is the final radius and Δa_p is the change in area of the pore.

126 Chapter 8

captured using Eq. (8.20). Additionally, MD simulations (Murphy et al., 2016) have shown that pore growth to be dependent on the strain rate:

$$g = \ln\left(\dot{\varepsilon}_{vm} / \dot{\varepsilon}_{vm0}\right)\exp\left(-r_1 \ln\left(\frac{\dot{\varepsilon}_{vm}}{\dot{\varepsilon}_{vm0}}\right)^{r_2}\right).$$ (8.21)

Here, r_1 and r_2 are material constants. Finally, pore growth can be written as the following,

$$\dot{a}_p = d_2(a_p\tau_{eq} - 2\sqrt{\pi a_p}\gamma)g\dot{\varepsilon}_{vm}\left(c_1\left[\frac{4}{27} - \frac{J_3^2}{J_2^3}\right] + c_2\frac{J_3}{J_2^{\frac{3}{2}}} + c_3\frac{I_1}{\sqrt{J_2}}\right)$$ (8.22)

here d_2 is a nondimensional constant. Further, c_i must be calibrated to stress state data. Combining the pore number density rate (Eq. (8.18)) and pore growth (Eq. (8.22)) will give the damage of the representative membrane.

8.3.3 Pore resealing

After the deformation process, boundary conditions return to zero, which allows for the membrane stresses and strains to relax. Following this relaxation pores on the representative membrane can either grow bigger or smaller. Researchers (Sandre et al., 1999; Brochard-Wyart et al., 2000) have shown that small pores on the cell membrane that are too small (nanometer range) are unstable and automatically will close. On the other hand, the sealing of larger pores requires active-repair that is implemented through calcium-triggered exocytosis where cells can implement different mechanisms to close such pores (Cooper and McNeil, 2015; Terasaki et al., 1997). For instance, cells transfer small vesicles to the membrane that fuse with the membrane, increasing the cell surface area and reducing surface tension that seal the pore (Cooper and McNeil, 2015). The finer details of these mechanisms, however, are not yet fully understood. Traditionally, exponential decay equations have been used to capture pore resealing on the membrane (Bier et al., 1999),

$$a_{p_1} = a_{p_f} \exp\left(-\frac{t}{t_h}\right)$$ (8.23)

where t is the time and t_h is the half-life of the pore, derived from experimental results. a_{p_f} is then the average pore area after the mechanical loads have been removed and a_{p_1} is the current average pore area as the pore reseals with time. The decay function considering a half-life of 0.5 is shown in Fig. 8.3.

8.4 Garnering data from molecular dynamics simulations

The constants for the continuum ISV pore number density and pore area equations are garnered from the MD simulation results (Bakhtiarydavijani et al., 2019; Murphy et al., 2016, 2019).

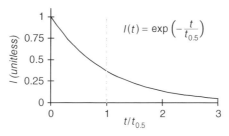

Fig. 8.3: Nondimensionalized pore resealing rate.

Computational measurements of poration behavior is performed by counting all pores and their diameter at the current time. Fig. 8.4 presents the derivation of number of pores and average pore area. Furthermore, multiple initial structures need to be modeled to ensure the independence of the resulting mechanoporation from the loosely controlled system properties such as the initially equilibrated structure. These MD simulations included three equibiaxial, three nonequibiaxial, and three strip biaxial membrane stretching at the same von Mises engineering strain rate (5.5×10^8 s^{-1}) along with a nonequibiaxial and a strip biaxial stretching at lower von Mises engineering strain rates Table 8.2. A two-sample Kolmogorov-Smirnov test for continuous distributions is used to look at the dependence/ or independence of damage between the three initial configurations for different strain states as shown in Table 8.2.

8.5 Calibration of the mechano-physiological internal state variable damage rate equations

In order to calibrate the constitutive model to experimental data, one must minimize the error difference between the model and experiment via an iterate solver because the set of equations are nonlinear. Here, mechanoporation is defined through the pore number density and growth rate equations that need to be calibrated to the data captured from MD simulations. These equations are defined as a function of time, strain rate, and stress state. Furthermore, the strain rate and stress state dependencies of this MPISV damage model are independent of one another. On the other hand, both the pore number density (Eq. (8.18)) and pore growth rate equations (Eq. (8.22)) interact with other as per the total damage (Eq. (8.11)). Hence, while the strain rate and stress state dependent constants can be independently determined, the pore number density and pore growth rate equations can only be calibrated together. The initial material constant values need to be determined for both the pore number density and pore growth rate equations, under a constant strain rate and stress state. Hence, calibration of the MPISV equations was performed through a point simulator code in MATLAB that incrementally calculated strains under a stress state and strain rate. Table 8.3 gives the calibrated constants for the strain rate and stress states given in Table 8.2. The calibrated pore number density, and pore area, along with the predicted damage are then presented in Fig. 8.5.

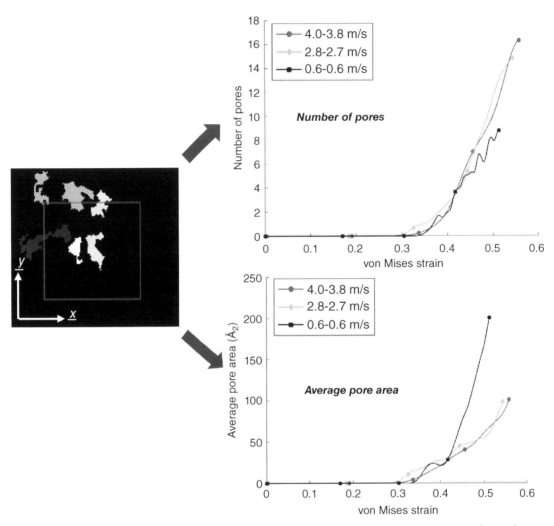

Fig. 8.4: Molecular dynamics (MD) picture of a damaged state showing the neuronal membrane mechanoporation and associated quantified number of pores and average pore area.

8.6 Sensitivity analysis of damage model at this length scale

Next, we show the sensitivity of the current damage model to the applied boundary conditions and initial material property settings. These include the downscaled boundary conditions, that is, strain, strain rate, and stress state; material properties: neuron size, pore half-life, surface tension, line tension, and initial damage values (initial pore number density and initial pore area). The base value for these parameters is those used to calibrate the MPISV as shown in Table 8.4. The sensitivity of membrane damage to each of these parameters is then studied

A macroscale mechano-physiological internal state variable (MPISV) model for neuronal

Table 8.2: Summary of molecular dynamics (MD) simulation results, including the applied velocity boundary conditions, the resulting stress ratio for the two deformed dimensions, and the number of initial structures (simulation runs) that were used for each deformation rate. Additionally, P values calculated through the Kolmogorov–Smirnov test are presented (statistical significance of $P < 0.01$).

	Applied velocities			
	Direction 1	Direction 2	Stress component	Kolmogorov–Smirnov
Strain state	(m/s)	(m/s)	ratio $\dfrac{\sigma_x}{\sigma_y}$	test (P value)
Equibiaxial	0.2	0.2	1	-
Equibiaxial	0.6	0.6	1	-
Equibiaxial	2.8	2.7	1	-
Equibiaxial	4.0	3.8	1	A versus B 0.011
				B versus C 0.017
				C versus A 0.326
Nonequibiaxial	4.6	2.2	1.2	A versus B 0.991
				B versus C 0.068
				C versus A 0.222
Strip biaxial	4.0	0.0	2.28	A versus B 0
				B versus C 0
				C versus A 0.075

by independently varying each parameter by a factor of two, as shown in Fig. 8.6. The results show that the current damage constitutive model is very dependent on the initial neuron size and pore number density, while not very dependent on the surface tension.

8.7 Comparison of model with cell culture studies

To compare the model dependencies with cell cultures, we assume a 4 μm diameter spherical neuron. Global strains from previously performed cell culture studies can then be mapped onto the elements on the spherical neuron membrane considering their orientation to the

Table 8.3: Calibrated internal state variable (ISV) damage constants for pore number density, growth, and pore closure rates.

Pore number density rate (#/nm^2/s)				Pore growth rate (nm^2/s)			Pore closure rate (s)	
η_0 (#/nm^2)	0.0012	a_{p0} (nm^2)	1432.96	r_1	0.016		τ (s)	6.54
d_1 (#/nm^2)	0.005	d_2 (nm^2)	37.15	r_2	1.607		Length scale	1000
b_1	−0.639	c_1	−0.464				parameter	
b_2	0.063	c_2	0.082					
b_3	0.051	c_3	0.052					

130 Chapter 8

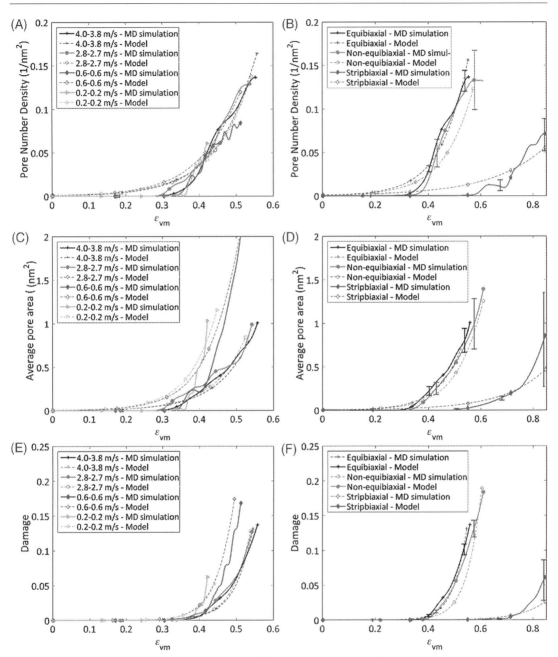

Fig. 8.5: The internal state variable (ISV) damage model calibrated with molecular dynamics (MD) results for the pore number density (A, B), average area (C, D), and damage (E, F) with their strain rate (A, C, E) and stress state (B, D, F) dependence. Data from (Bakhtiarydavijani et al., 2019).

A macroscale mechano-physiological internal state variable (MPISV) model for neuronal   131

Table 8.4: Base material properties for sensitivity analysis.

Parameter	Value
Strain	0.12
Strain rate	54.7 s^{-1}
Stress state	Strip biaxial
Surface tension modulus	1.46 N/m
Line tension	0.5 pN
Pore half-life	6.54 s
Neuron radius	1 μm
Initial pore number density	0.04 #/nm^2
Initial pore area	5 × 10^{-5} nm^2

reference configuration. This produces spatially dependent strain and strain gradients on the neuronal membrane. These strains are then coupled with the damage evolution equations to capture membrane mechanoporation. Coupling these equation sets with the Nernst–Planck diffusion equations (Kirby, 2010) can then give a representation of the intracellular ion concentration change with the considered ion concentrations inside and outside the neuron.

The Nernst–Planck diffusion equation set can determine the flux of ions across a narrow chamber considering the concentration and the electric potential difference between the two sides (Kirby, 2010) while assuming Brownian motion (Cárdenas et al., 2000). The flux of each ion type and the direction of flux is determined by the following,

$$j_B = -D_B \left(\frac{d[B]}{dz} - \frac{n_B F}{RT} \frac{E_m}{L} [B] \right) \tag{8.24}$$

where L is the thickness of the damaged hole in the phospholipid layer, which is the outer cell of the neuron, D_B is the diffusion coefficient of B in the cytosol, n_B is the electron valance of the B, and $[B]_{\text{ext}}$ and $[B]_{\text{int}}$ are the extracellular and intracellular concentrations of B, respectively. F is the Faraday's constant, R is the ideal gas constant, T is the temperature, and E_m is the electric potential occurring across the membrane as a result of the sum of the ion imbalance calculated using the difference in the current concentration of the intracellular and extracellular ions,

$$E_m = \frac{RT}{F} \ln \left(\frac{\sum_i^N D_{A_i} \left[A_i^+ \right]_{\text{out}} + \sum_i^K D_{M_i} \left[M_i^- \right]_{\text{in}}}{\sum_i^K D_{M_i} \left[M_i^i \right]_{\text{in}} + \sum_i^N D_{A_i} \left[A_i^- \right]_{\text{out}}} \right) \tag{8.25}$$

Eq. (8.24), for each ion in the system (Table 8.5), is informed by Eq. (8.25) and solved for the unit area and then combined with the mechanical damage evolution equations (Eq. (8.12)), that is, pore number density rate (Eq. (8.18)), pore growth rate (Eq. (8.22)), and pore closure (Eq. (8.23)) to capture the change in intracellular ion concentrations and to produce a mechano-physiological damage state.

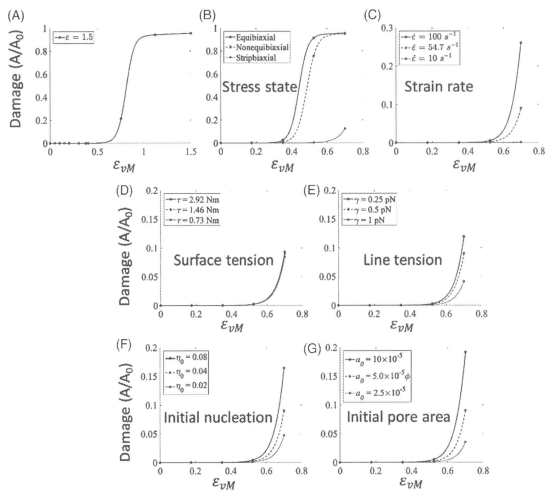

Fig. 8.6: Sensitivity analysis of the internal state variable (ISV) damage model with base parameters defined in Table 8.3. (A) Finite strain and considering a 50% variation of (B) von Mises strain rate, (C) von Mises stress state, (D) surface tension, (E) edge tension, (F) initial number of pores, and (G) initial area of pores are presented.

In order to study the sensitivity of intracellular ion homeostasis, the same base parameters used to study damage (Table 8.4) were used considering the initial ion concentrations (Table 8.5). We then studied the changes of intracellular calcium ion [Ca^{2+}] and its dependence on the boundary conditions, material properties, and neuron size (8.7). [Ca^{2+}] showed the most significant difference between intracellular and extracellular concentration and was also measured in some cell culture studies (LaPlaca and Thibault, 1997).

Table 8.5: Initial intracellular and extracellular ion concentrations for diffusion equation input (Bear et al., 2006).

Ion	Intracellular concentration (mM)	Extracellular concentration (mM)
$[Ca^{2+}]$	0.59×10^{-4}	1.8
$[Na^+]$	15	150
$[Cl^-]$	13	150
$[K^+]$	100	5
$[Bal^-]$	9	102

8.8 Discussion

Mechanoporation related neuronal injuries during TBI involves multiple length scales (Bakhtiarydavijani et al., 2019). Neuronal membrane mechanoporation initiates at the nanoscale with pores nucleating under mechanical loads. These pores then grow to reduce the system energy in the deformed membrane. Once the pore size is large enough (Murphy et al., 2019) ions and proteins diffuse into and out of the neuron, the extent of which can disrupt the neuronal ion homeostasis. Significant disruption of intracellular ion homeostasis will then cause neuronal cell death, possibly also affecting surrounding cells in severe injury conditions (Stoica and Faden, 2010; Yuan et al., 2003). At the macroscale, the cumulative effects of these injured neurons will then define TBI thresholds.

The current work develops damage evolution equations for the representation of membrane disruption in microscale domains. Here, we developed MPISV damage rate equations to capture the strain rate, stress state, and history of neuronal membrane mechanoporation behavior informed from MD simulations (8.6). The strain rate and stress state dependence of injury has been experimentally shown in cell culture studies (Bar-Kochba et al., 2016; Cullen et al., 2011). Further, cell culture studies show that neuronal injury is diffuse with healthy neurons in the vicinity of those with disrupted membranes (Cullen and LaPlaca, 2006). This implies that microstructural features need to be included into injury metrics to capture the local differences. The analytical study of the MPISV on a spherical neuron (Fig. 8.7) in this chapter shows that the extent of intracellular calcium ion concentration disruption is dependent on the boundary conditions, neuron size, and material properties. Interestingly, the current MPISV predicts that the neuron size is an important factor on the extent of ion homeostasis disruption (Fig. 8.7G), while surface tension does not seem to play as an important role (Fig. 8.7D). Because there is a distribution of neuron sizes in a brain, one can easily see the causal relationship to individual cell injury thresholds. Furthermore, the strain rate and stress state dependence of the MPISV damage equations can also apply to fully three-dimensional structures as opposed to the plane stress membrane studied herein. These strain rate and stress state dependencies then need to be introduced into microscale neuron models to accurately capture TBI's microstructural dependence.

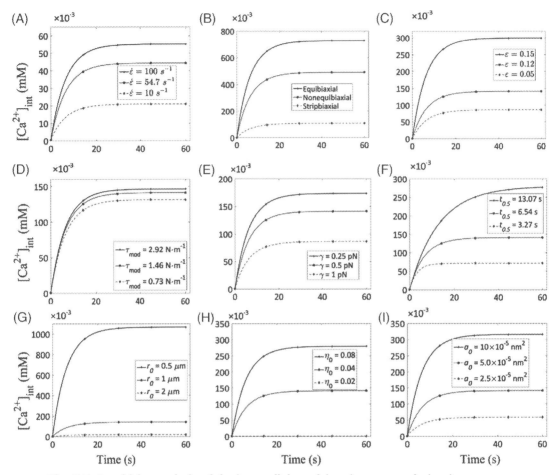

Fig. 8.7: Sensitivity analysis of the intracellular calcium ion accumulation in a neuron with base parameters and an initial ion concentration defined in Table 8.4. Here the (A) von Mises strain rate, (B) von Mises stress state, (C) finite strain, (D) surface tension, (E) edge tension, (F) pore half-life, (G) neuronal radius, (H) initial number of pores, and (I) initial area of pores are presented.

The current Nernst–Planck model gives insight into the disruption of intracellular ion concentration through mechanoporation considering a spherical neuron. The next step is to generate different neuron geometries since they are more complex with the extension of neurites. Furthermore, finite element simulations need to be implemented to study the effects of the extracellular matrix on the translating boundary conditions onto the membrane. Implementing the MPISV damage evolution equations into these finite element simulations can then help more accurately capture the effects of mechanoporation related neuronal injuries.

As computational methods advance and new modeling approaches are verified, researchers have been able to produce larger membranes that include a wider range of phospholipids, cholesterol, and proteins. These membranes are experimentally deformed at slower rates. These larger membranes would allow for boundary conditions that are more similar to real-world scenarios. The calibration of the MPISV model to said simulations can help to study of other membrane injury mechanisms such as loss of membrane integrity without poration.

The presented damage evolution model can be further enhanced through rigorous experimental validation at the microscale. At that length scale, force displacements on cell culture neurons accompanied by larger biomarkers can be used to further validate the mechanoporation behavior on neurons that benefit from a complex membrane with its embedded proteins and cholesterol.

8.9 Summary

In this chapter we presented a set of MPISV damage rate equations denoted by pore nucleation, pore growth, and pore resealing that are calibrated from lower length scale MD simulation results. This MPISV model is (1) strain rate dependent, (2) stress state dependent, (3) history dependent, and (4) incorporates lower length scale information that informs microscale neuron deformation. Coupled with Nernst–Planck diffusion equations, the model not only predicts holes in membranes but also the chemistry concentrations locally. Finally, the MPISV model coupled with the Nernst–Planck equation can be a powerful tool in introducing microstructural dependence to quantify TBI at the macroscale.

References

Bakhtiarydavijani, A., Murphy, M.A., Mun, S., Jones, M.D., Bammann, D.J., Laplaca, M.C., Horstemeyer, M.F., Prabhu, R.K., 2019. Damage biomechanics for neuronal membrane mechanoporation. Model. Simul. Mater. Sci. Eng. 27 (6), 065004. https://doi.org/10.1088/1361-651X/ab1efe.

Bakhtiarydavijani, A.H., Murphy, M.A., Mun, S., Jones, M.D., Horstemeyer, M.F., Prabhu, R.K., 2019. Multiscale modeling of the damage biomechanics of traumatic brain injury. Biophys. J. 116 (3), 322a. https://doi.org/10.1016/j.bpj.2018.11.1748.

Bar-Kochba, E., Scimone, M.T., Estrada, J.B., Franck, C., 2016. Strain and rate-dependent neuronal injury in a 3D in vitro compression model of traumatic brain injury. Sci. Rep. 6 (August), 30550. https://doi.org/10.1038/srep30550.

Bear, M.F., Connors, B.W., Paradiso, M.A., 2006. In: Purves, D., Augustine, G.J., Fitzpatric, D., Hall, W.C., White, L.D. (Eds.), Neurosciencethird ed. Lippincott Williams & Wilkins, Baltimore, MD.

Bier, M., Hammer, S.M., Canaday, D.J., Lee, R.C., 1999. Kinetics of sealing for transient electropores in isolated mammalian skeletal muscle cells. Bioelectromagnetics 20 (3), 194–201. http://files/563/597694.pdf.

Bloom, M., Evans, E., 1991. Physical properties of the fluid lipid-bilayer component of cell membranes: a perspective. Q. Rev. Biophys. 24 (3), 293–397. https://doi.org/10.1017/S0033583500003735.

Brochard-Wyart, F., de Gennes, P.G., Sandre, O., 2000. Transient pores in stretched vesicles: role of leak-out. Phys. A: Stat. Mech. Its Applic. 278 (1–2), 32–51. https://doi.org/10.1016/S0378-4371(99)00559-2.

Cárdenas, A.E., Coalson, R.D., Kurnikova, M.G., 2000. Three-dimensional Poisson-Nernst-Planck theory studies: influence of membrane electrostatics on gramicidin a channel conductance. Biophys. J. 79 (1), 80–93. https://doi.org/10.1016/S0006-3495(00)76275-8.

136 Chapter 8

Coleman, B.D., Gurtin, M.E., 1967. Thermodynamics with internal state variables. J. Chem. Phys. 47 (2), 597–613. https://doi.org/10.1063/1.1711937.

Cooper, S.T., McNeil, P.L., 2015. Membrane repair: mechanisms and pathophysiology. Physiol. Rev. 95 (4), 1205–1240. https://doi.org/10.1152/physrev.00037.2014.

Cullen, D.K., LaPlaca, M.C., 2006. Neuronal response to high rate shear deformation depends on heterogeneity of the local strain field. J. Neurotrauma 23 (9), 1304–1319. https://doi.org/10.1089/neu.2006.23.1304.

Cullen, D.K., Vernekar, V.N., LaPlaca, M.C., 2011. Trauma-induced plasmalemma disruptions in three-dimensional neural cultures are dependent on strain modality and rate. J. Neurotrauma 28 (11), 2219–2233. https://doi.org/10.1089/neu.2011.1841.

Deryagin, B.V., Gutop, Y.V., 1962. Theory of the breakdown (rupture) of free films. Kolloidn. Z. 24, 370.

Evans, E., Heinrich, V., Ludwig, F., Rawicz, W., 2003. Dynamic tension spectroscopy and strength of biomembranes. Biophys. J. 85 (4), 2342–2350 https://doi.org/10.1016/S0006-3495(03)74658-X.

Evans, E., Smith, B.A., 2011a. Kinetics of hole nucleation in biomembrane rupture. New J. Phys., 13. April. https://doi.org/10.1088/1367-2630/13/9/095010.

Evans, E., Smith, B.A., 2011b. Kinetics of hole nucleation in biomembrane rupture. New J. Phys. 13 (9), 095010. https://doi.org/10.1088/1367-2630/13/9/095010.

Farkas, O., Lifshitz, J., Povlishock, J.T., 2006. Mechanoporation induced by diffuse traumatic brain injury: an irreversible or reversible response to injury? J. Neurosci. 26 (12), 3130–3140. https://doi.org/10.1523/JNEUROSCI.5119-05.2006.

Geddes, D.M., Cargill, R.S., LaPlaca, M.C., 2003. Mechanical stretch to neurons results in a strain rate and magnitude-dependent increase in plasma membrane permeability. J. Neurotrauma 20 (10), 1039–1049. https://doi.org/10.1089/089771503770195885.

Horstemeyer, M.F., Gokhale, A.M., 1999. A void-crack nucleation model for ductile metals. Int. J. Solids Struct. 36 (33), 5029–5055. https://doi.org/10.1016/S0020-7683(98)00239-X.

Horstemeyer, M.F., Berthelson, P.R., Moore, J., Persons, A.K., Dobbins, A., Prabhu, R.K., 2019. A mechanical brain damage framework used to model abnormal brain tau protein accumulations of National Football League players. Ann. Biomed. Eng. *47* (9), 1873–1888.

Kirby, B.J., 2010. Micro- and Nanoscale FluidMechanics: Transport in Microfluidic Devices: Chapter 11. Cambridge University Press, Cambridge, United Kingdom. https://books.google.com/books?id=y7PB9f5zmU4C.

Koshiyama, K., Wada, S., 2011. Molecular dynamics simulations of pore formation dynamics during the rupture process of a phospholipid bilayer caused by high-speed equibiaxial stretching. J. Biomech. 44 (11), 2053–2058. https://doi.org/10.1016/j.jbiomech.2011.05.014.

Koshiyama, K., Yano, T., Kodama, T., 2010. Self-organization of a stable pore structure in a phospholipid bilayer. Phys. Rev. Lett. 105 (1), 18105. https://doi.org/10.1103/PhysRevLett.105.018105.

LaPlaca, M.C., Cullen, D.K., McLoughlin, J.J., Cargill II, R.S., 2005. High rate shear strain of three-dimensional neural cell cultures: a new in vitro traumatic brain injury model. J. Biomech. 38 (5), 1093–1105. https://doi.org/10.1016/j.jbiomech.2004.05.032.

LaPlaca, M.C., Lessing, M.C., Prado, G.R., Zhou, R., Tate, C.C., Geddes-Klein, D., Meaney, D.F., Zhang, L., 2019. Mechanoporation is a potential indicator of tissue strain and subsequent degeneration following experimental traumatic brain injury. Clin. Biomech. 64, 2–13. April. https://doi.org/10.1016/J.CLINBIOMECH.2018.05.016.

LaPlaca, M.C., Prado, G.R., 2010. Neural mechanobiology and neuronal vulnerability to traumatic loading. J. Biomech. 43 (1), 71–78. https://doi.org/10.1016/J.JBIOMECH.2009.09.011.

LaPlaca, M.C., Thibault, L.E., 1997. An in vitro traumatic injury model to examine the response of neurons to a hydrodynamically-induced deformation. Ann. Biomed. Eng. 25 (4), 665–677. https://doi.org/10.1007/BF02684844.

Leontiadou, H., Mark, A.E., Marrink, S.J., 2004. Molecular dynamics simulations of hydrophilic pores in lipid bilayers. Biophys. J. 86 (4), 2156–2164. https://doi.org/10.1016/S0006-3495(04)74275-7.

Martínez-Balbuena, L., Hernández-Zapata, E., Santamaría-Holek, I., 2015. Onsager's irreversible thermodynamics of the dynamics of transient pores in spherical lipid vesicles. Eur. Biophys. J. 44 (6), 473–481. https://doi.org/10.1007/s00249-015-1051-8.

Miller, M.P., McDowell, D.L., 1992. Stress state dependence of finite strain inelasticity, ASME Summer Mechanics and Materials Conferences, 32, 27–44. https://www.scopus.com/record/display.uri?eid=2-s2.0-0026487503&origin=resultslist&sort=plf-f&src=s&st1=stress+state+dependence+of+finite+strain+inelasticity&st2=&sid=3DCBECDA557112396B8373F6187942B9.wsnAw8kcdt7IPYLO0V48gA%3A10&sot=b&sdt=b&sl=68&s=TITLE-A.

Murphy, M.A., Horstemeyer, M.F.F., Gwaltney, S.R., Stone, T., LaPlaca, M.C., Liao, J., Williams, L.N., et al., 2016. Nanomechanics of phospholipid bilayer failure under strip biaxial stretching using molecular dynamics. Model. Simul. Mater. Sci. Eng. 24 (5), 55008. https://doi.org/10.1088/0965-0393/24/5/055008.

Murphy, M.A., Mun, S., Horstemeyer, M.F., Baskes, M.I., Bakhtiary, A., LaPlaca, M.C., Gwaltney, S.R., Williams, L.N., Prabhu, R.K., 2019. Molecular dynamics simulations showing 1-palmitoyl-2-oleoyl-phosphatidylcholine (POPC) membrane mechanoporation damage under different strain paths. J. Biomol. Struct. Dyn. 37 (5), 1346–1359. https://doi.org/10.1080/07391102.2018.1453376.

Needham, D., Nunn, R.S., 1990. Elastic deformation and failure of lipid bilayer membranes containing cholesterol. Biophys. J. 58 (4), 997–1009. http://www.ncbi.nlm.nih.gov/pmc/articles/PMC1281045/.

Raghupathi, R., 2004. Cell death mechanisms following traumatic brain injury. Brain Pathol. 14 (2), 215–222. https://doi.org/10.1111/j.1750-3639.2004.tb00056.x.

Rawicz, W., Olbrich, K.C., McIntosh, T., Needham, D., Evans, E., 2000. Effect of chain length and unsaturation on elasticity of lipid bilayers. Biophys. J. 79 (1), 328–339. https://doi.org/10.1016/S0006-3495(00)76295-3.

Sandre, O., Moreaux, L., Brochard-Wyart, F., 1999. Dynamics of transient pores in stretched vesicles. Proc. Natl. Acad. Sci. U S A 96 (19), 10591–10596. https://doi.org/10.1073/pnas.96.19.10591.

Shigematsu, T., Koshiyama, K., Wada, S., 2014. Molecular dynamics simulations of pore formation in stretched phospholipid/cholesterol bilayers. Chem. Phys. Lipids 183, 43–49. https://doi.org/10.1016/J.CHEMPHYSLIP.2014.05.005.

Shigematsu, T., Koshiyama, K., Wada, S., 2015. Effects of stretching speed on mechanical rupture of phospholipid/cholesterol bilayers: molecular dynamics simulation. Sci. Rep. 5, 15369. https://doi.org/10.1038/srep15369.

Siesjö, B.K., 1993. Basic mechanisms of traumatic brain damage. Ann. Emerg. Med. 22 (6), 959–969. https://doi.org/10.1016/S0196-0644(05)82736-2.

Stoica, Bogdan A, Faden, Alan I, 2010. Cell death mechanisms and modulation in traumatic brain injury. Neurotherapeutics 7 (1), 3–12. https://doi.org/10.1016/j.nurt.2009.10.023.

Terasaki, M., Miyake, K., McNeil, P.L., 1997. Large plasma membrane disruptions are rapidly resealed by Ca^{2+}-dependent vesicle–vesicle fusion events. J. Cell Biol. 139 (1), 63–74. http://www.ncbi.nlm.nih.gov/pmc/articles/PMC2139822/.

Tieleman, D.P., Leontiadou, H., Mark, A.E., Marrink, S.-J.J., 2003. Simulation of pore formation in lipid bilayers by mechanical stress and electric fields. J. Am. Chem. Soc. 125 (21), 6382–6383. https://doi.org/10.1021/ja029504i.

Tieleman, D.P., MacCallum, J.L., Ash, W.L., Kandt, C., Xu, Z., Monticelli, L., 2006. Membrane protein simulations with a united-atom lipid and all-atom protein model: lipid–protein interactions, side chain transfer free energies and model proteins. J. Phys.: Condensed Matter 18 (28), S1221–S1234. https://doi.org/10.1088/0953-8984/18/28/S07.

Tomasini, M.D., Rinaldi, C., Silvina Tomassone, M., 2010. Molecular dynamics simulations of rupture in lipid bilayers. Exp. Biol. Med. 235 (2), 181–188. https://doi.org/10.1258/ebm.2009.009187.

Yuan, J., Lipinski, M., Degterev, A., 2003. Diversity in the mechanisms of neuronal cell death. Neuron 40 (2), 401–413. https://doi.org/10.1016/S0896-6273(03)00601-9.

Zeldovich, J.B., 1943. On the theory of new phase formation; cavitation. Acta Physicochim. URSS 18, 1.

Zhelev, D.V., Needham, D., 1993. Tension-stabilized pores in giant vesicles: determination of pore size and pore line tension. Biochim. Biophys. Acta (BBA) – Biomembr. 1147 (1), 89–104. https://doi.org/10.1016/0005-2736(93)90319-U.

CHAPTER 9

MRE-based modeling of head trauma

Amit Madhukar[a], Martin Ostoja-Starzewski[a,b]

[a]*Department of Mechanical Science & Engineering, University of Illinois at Urbana-Champaign, Urbana, IL, United States* [b]*Beckman Institute and Institute for Condensed Matter Theory, University of Illinois at Urbana-Champaign, Urbana, IL, United States*

9.1 Introduction

The finite element (FE) method is an invaluable tool to determine the mechanical response of brain tissue under impacts or by impulse. The information provided by numerical models can lead to the development of improved diagnostic tools and protective measures to reduce the prevalence of traumatic brain injury. The accuracy of these FE models is highly dependent on the accuracy of the material model. To date most FE models have utilized mechanical properties averaged over large portions of the brain, as reviewed in Madhukar and Ostoja-Starzewski (2019). While simplifying the formulation, this comes at the cost of ignoring potentially significant effects due to heterogeneities present in brain tissue. In this work, we present a high-resolution magnetic resonance elastography (MRE)-based FE model to account for the local differences in mechanical response between different regions of the brain.

The tissues of the brain are heterogeneous, their constitutive response varying from location to location. This is most noticeable for white matter due to the presence of axons with diverse orientations. Overall, the shear stiffness of white matter is 1.2–2.6 higher than that of gray matter (Chatelin et al., 2010). Locally, white matter tracts with highly oriented fibers such as the corpus callosum and corona radiata have material properties vastly different from other regions. Johnson et al. (2013) determined that global white matter was softer on average than either the corpus callosum or the corona radiata. This can be explained by considering the structure of these regions. The corpus callosum is a tight bundle of highly aligned fibers which is expected to provide more structural rigidity than the superficial white matter. The fan-like structure of the corona radiata provides a similar response, though to a lesser extent since the fibers are not as highly aligned. As such the corpus callosum was found to be approximately 11% stiffer than the corona radiata. The brainstem is another structure with a high level of heterogeneity. Arbogast and Margulies (1998) investigated the prevalence of trauma observed in the brainstem after head injuries. They determined that the brainstem was 80–100% stiffer than the cerebrum and concluded this regional stiffness variation is one

Multiscale Biomechanical Modeling of the Brain.
DOI: https://doi.org/10.1016/B978-0-12-818144-7.00005-0
Copyright © 2022 Elsevier Inc. All rights reserved.

reason for the selective vulnerability of this region to rotational motion. FE models that utilize homogenized white matter material properties have no way to resolve these local features.

Clinical studies have shown that damage to the brain in blunt head injuries is primarily confined to the cerebral white matter, with a propensity for lesions in the brainstem and the corpus callosum, that is, regions with highly organized axon tracts (Arbogast and Margulies, 1998; Gennarelli et al., 1982; Gentry et al., 1988a, 1988b; Ng et al., 1994; Arfanakis et al., 2002). These regions serve as vital connection points to other parts of the brain, meaning damage to them is potentially more dangerous. In this work we introduce a heterogeneous material description of white matter structures to our high-resolution FE model to account for the local differences in mechanical response between different regions of the brain.

MRE (Muthupillai et al., 1995) can measure heterogeneity of brain tissue in vivo where the head is excited with shear waves to quantitatively assess the local mechanical properties of brain tissue. MRE is applied as a three-step process beginning by first inducing shear waves in tissue with frequencies ranging from 50 to 500 Hz using an external driver. Second, the waves are imaged using a phase-contrast magnetic resonance imaging (MRI) pulse sequence synchronized with the frequency of the applied vibration. Finally, the mechanical properties of the tissue are estimated by inverting the observed displacements using a viscoelastic material model; such as that presented by Van Houten et al. (2001).

In our recent work (Madhukar and Ostoja-Starzewski, 2020), we have utilized our homogeneous FE model—presented and validated in our previous works (Chen and Ostoja-Starzewski, 2010; Chen et al., 2012; Madhukar et al., 2017)—and introduced a voxel-based heterogeneous material model using results from Johnson et al. (2013). The high-resolution model utilizes tissue properties reconstructed at the same spatial resolution as the MRE acquisition process. The resulting FE mesh has more than sufficient resolution to accurately capture the dynamic shear wave propagation during impacts. To the best of our knowledge, this is the first attempt to include the heterogeneity of brain tissue in a high-resolution FE model.

9.2 Model formulation

9.2.1 MRE acquisition and inversion

To date, many studies have utilized mechanical properties averaged over tissue classes, thus missing potentially significant localized effects. We introduce material heterogeneities to white matter structures in our model using the nonlinear inversion (NLI) technique developed by Johnson et al. (2013). In the study, shear displacements at a frequency of 50 Hz are generated within the subject's head using a remote electromagnetic shaker connected to a custom head cradle. A multishot MRE sequence utilizing spiral readout gradients (Glover, 1999) with periods matching that of the applied shear wave was developed to reduce errors during the inversion step. In total, the imaging volume comprised 20 axial slices of 2 mm thickness covering the ventricles, corpus callosum, and corona radiata resulting in a $2 \times 2 \times 2$ mm^3 isotropic spatial resolution for the reconstructed mechanical properties.

The NLI algorithm (Van Houten et al., 2001) was applied to estimate the material properties of tissue from the measured displacement data using a viscoelastic material model. The inversion is performed by iteratively optimizing the material properties at each voxel location and a Rayleigh damping model is used to represent the material response of brain tissue under time–harmonic conditions (McGarry and Houten, 2008). The Rayleigh damping model introduces the complex-valued shear modulus and density to account for attenuation related to both elastic and inertial forces—where the imaginary shear modulus includes damping effects due to inertial forces, while the imaginary part of the density includes damping related to inertial forces. Inertial damping effects are included here to allow for better characterization of material response when performing the inversion. The notation G' and G'' is used to denote the storage and loss shear moduli, respectively, which are the real and imaginary part of the complex valued shear modulus

$$G = G' + iG'' \tag{9.1}$$

where i is the imaginary unit. Due to the nearly incompressible nature of brain tissue, we assume that λ is very large compared to the shear modulus. The damping ratio

$$\xi := G''/2G' \tag{9.2}$$

can be determined as well which physically describes the level of motion attenuation in the tissue.

Fig. 9.1 presents the distribution of storage and loss moduli along the coronal, sagittal and horizontal planes. The relative stiffness of the corpus callosum and corona radiata can be clearly observed.

9.2.2 Finite element mesh generation

We generate our FE mesh from high-resolution T1-weighted MRI data sets of a single volunteer. Segmentation was performed as outlined in Johnson et al. (2013) into skull, scalp, cerebrospinal fluid, gray matter, and white matter. Features that are below the imaging resolution such as membranes, blood vessels, bridging veins, and draining sinuses are excluded from the present model. Our FE model consists of approximately million elements, each assigned to a different tissue class based on the results of the segmentation. The resulting FE mesh is presented in Fig. 9.2. Due to the nature of the image segmentation, interfaces between tissue types can contain inconsistencies and nonsmooth features such as jagged edges which would cause numerical artifacts. We therefore apply a volume-preserving Laplacian smoothing algorithm as a postprocessing step to improve mesh quality. We ensure traction and displacement continuity at all material interfaces. Our MRI voxel-based approach produces meshes which

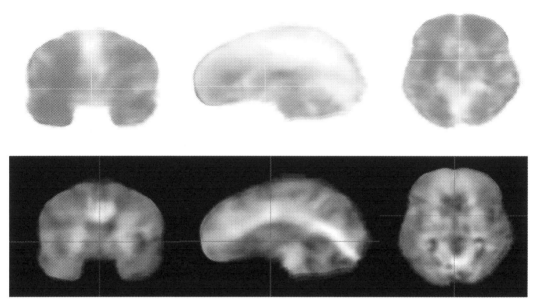

Fig. 9.1: Distribution of loss (top) and storage (bottom) modulus in the FE model. Darker regions indicate higher magnitude of shear modulus.

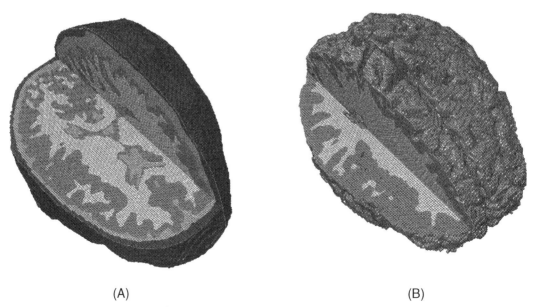

(A) (B)

Fig. 9.2: (A) Voxel-based finite element mesh segmented into skull (red), CSF (yellow), gray matter (gray) and white matter (white). (B) Surface of cerebral cortex with sulci and gyri clearly resolved.

realistically model the complicated folding structure of the cerebral cortex as well as the differentiation of gray and white matter of brain tissues.

9.2.3 Material properties

To keep the MRE image acquisition to a manageable region, we apply heterogeneity to only the white matter regions while other tissues are maintained to be homogeneous. Since the white matter is composed of highly oriented tracts of myelin-sheathed axonal fibers, while gray matter is made up of cell bodies and supporting vascular networks that can be assumed to be isotropic (Johnson et al., 2013; Prange and Margulies, 2002), this assumption is reasonable for an initial implementation.

The material properties utilized in our model are based on data available in the literature presented in Table 9.1. A second-order Ogden model is chosen to model the large deformation of white and gray matter tissues with model parameters given in Table 9.2. Due to limitations in the MRE inversion, we are limited to a locally linear viscoelastic model for the heterogeneous

Table 9.1: Material properties of different tissues used in the FE model.

Tissue	Mass density (kg/m^3)	Bulk modulus K (Pa)	Shear modulus G (Pa)
Skull (Khalil and Hubbard, 1977)	2070	3.61E+9	2.7E+9
Gray matter	1040	Hyperviscoelastic, see Table 9.2	
White matter	1040	Hyperviscoelastic, see Table 9.2	
	Mass density (kg/m^3)	Young's modulus E (Pa)	Poisson ratio
CSF (Wang et al., 2018)	1000	160	0.49

Table 9.2: Hyperviscoelastic material properties as used in Kleiven (2007).

	"Compliant"	"Average"	"Stiff"
μ_1 (Pa)	26.9	53.8	107.6
μ_2 (Pa)	−60.2	−120.4	−240.8
α_1	10.1	10.1	10.1
α_2	−12.9	−12.9	−12.9
G_1 (kPa)	160	320	640
G_2 (kPa)	39	78	156
G_3 (kPa)	3.1	6.2	12.4
G_4 (kPa)	4.0	8.0	16.0
G_5 (kPa)	0.05	0.10	0.20
G_6 (kPa)	1.5	3.0	6.0
β_1 (1/s)	10^6	10^6	10^6
β_2 (1/s)	10^5	10^5	10^5
$\vdots$	$\vdots$	$\vdots$	$\vdots$
β_6 (1/s)	10^1	10^1	10^1

input data. We therefore incorporate the relative stiffness from MRE and scale the nonlinear material parameters between the three material models presented in Table 9.2. More details of the material model are presented in Madhukar and Ostoja-Starzewski (2020).

9.2.4 Experimental verification

Experimental verification is performed using two well-known experimental results: Nahum et al.'s liner impact experiments (Nahum et al., 1977) and Hardy et al.'s brain transnational motion experiments (Hardy et al., 2001, 2007). In Nahum's experiments, intercranial pressure history is measured after an impact on the mid-sagittal plane of the specimen's skull resulting depicted in Fig. 9.3A and the resultant input force time history is given in Fig. 9.3B. We directly compare displacement data during impact utilizing experimental data from Hardy et al.'s experiment collected using neutral density targets (NDT) placed within the specimen's head under linear accelerations.

9.3 Results and discussion

Here we briefly present the main validation results for the new MRE-based model. More information is presented in Madhukar and Ostoja-Starzewski (2020).

A distributed impact load is applied to mid-frontal area of the model in the anterior–posterior direction simulating Nahum's experiments as described in the previous section. The base of the skull is not constrained to allow for rotation under impact. The pressure–time history for the loading is presented in Fig. 9.4 at the coup impact site. We plot the results for both the homogeneous and heterogeneous model. We see that the heterogeneous model predicts

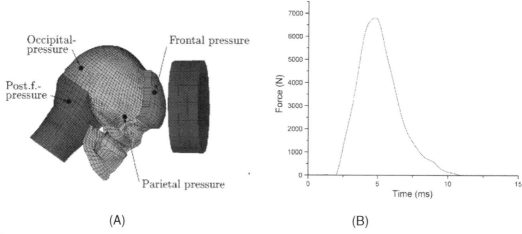

Fig. 9.3: (A) Impact location in Nahum et al.'s experiments (adapted from Kleiven and von Hans, 2002). (B) Input force time history (adapted from Chen and Ostoja-Starzewski, 2010).

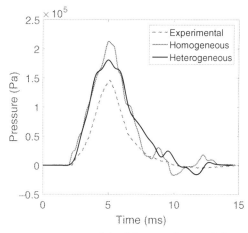

Fig. 9.4: Comparison of heterogeneous model with experimental data from Nahum et al. (1977). We find that the heterogeneous model exhibits response closer to the Nahum data.

responses more closely matching the experimental results, indicating that the inclusion of heterogeneity leads to a more accurate model.

We plot the contours of the von Mises stress distribution on the sagittal plane due to the frontal impact in Fig. 9.5. We see the wave travels inward toward the center of the brain, it attenuates and eventually dissipates before traveling a large distance toward the center of the brain. The attenuation of the stress within the cerebrum is considered by plotting the pressure along the sagittal plane at two time values, as shown in Fig. 9.6. The attenuation is greater in the heterogeneous model as predicted in Madhukar and Ostoja-Starzewski (2020) and is consistent with studies of transient wave propagation in elastodynamics of random media (Nishawala et al., 2016; Zhang and Ostoja-Starzewski, 2020). Pressure history is presented for three distinct points along the sagittal plane for three points within regions of strong heterogeneities, as given in Fig. 9.7. The differences in mechanical properties of these regions are given in Table 9.3. We find that the difference in peak pressure response is proportional to the difference in shear stiffness between the homogeneous and heterogeneous models. However, the time at which these events occur is not significantly affected. This indicates that the pressures in regions of high stiffness within the brain are overestimated in the homogeneous models. In summary, relative to the MRI-based model, the new MRE-based heterogeneous model more accurately predicts the local response within the white matter by taking into account the differences in tissue stiffness of local white matter structures.

The comparison between Hardy's displacement data is presented in Figs. 9.8 and 9.9 for displacements in the x and z directions, respectively, at two NDT locations in the anterior (A) and posterior (P) positions. Quantitative comparisons of the two models are performed by computing the displacement magnitude (excursion) at the NDTs, presented in Table 9.4.

146 Chapter 9

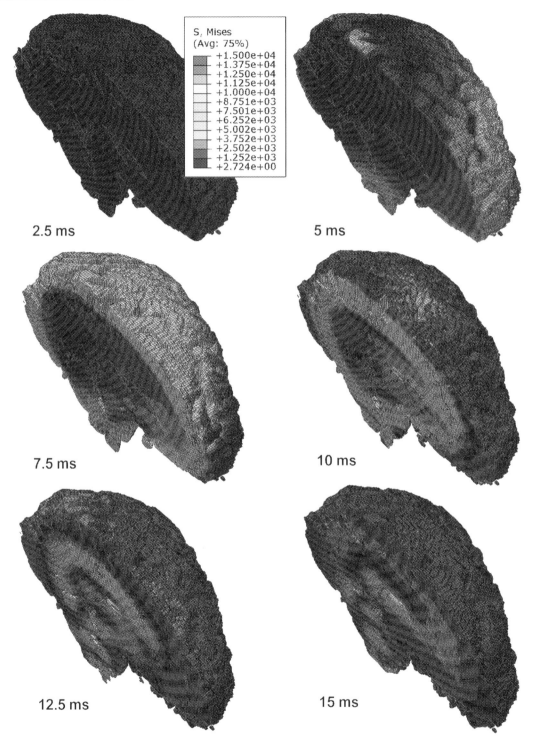

Fig. 9.5: Shear wave propagation due to frontal impact. Notice the attenuation of the wave front as time progresses.

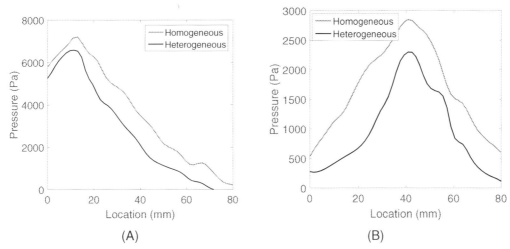

Fig. 9.6: Comparison of attenuation of pressure wave along the sagittal plane between homogeneous and heterogeneous model. (A) At 3 ms and (B) at 7 ms.

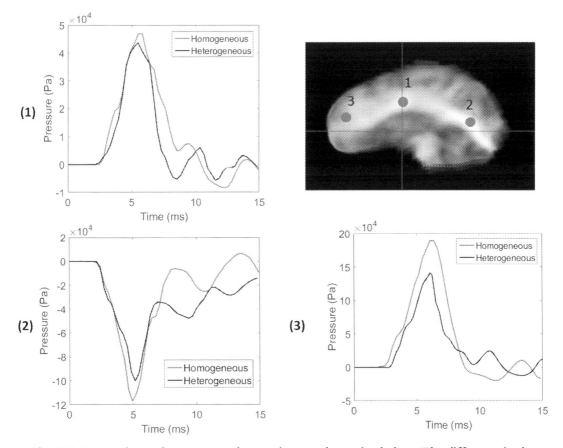

Fig. 9.7: Comparison of pressure at three points on the sagittal plane. The difference in the heterogeneous and homogeneous models is most evident in regions of high relative stiffness.

Table 9.3: Difference of material properties and peak pressure, displacement response for three distinct points (indicated in Fig. 9.7) along the sagittal plane within the white matter.

Location	% Difference in shear modulus (G_∞)	% Difference in peak pressure	% Difference in peak displacement
1	12.11	−9.12	−3.16
2	24.75	−13.91	−6.62
3	18.67	−29.05	−12.96

We find that the inclusion of heterogeneities more consistently and closely predicts the excursion determined experimentally.

Many works in the literature utilize diffusivity measures from diffuse tensor imaging which measures the distribution of axon tracts within white matter to construct anisotropic fiber-based models, for instance in Ji et al. (2015), Giordano and Kleiven (2014), Giordano et al. (2014), and Garimella and Kraft (2017). Our MRE-based heterogeneous

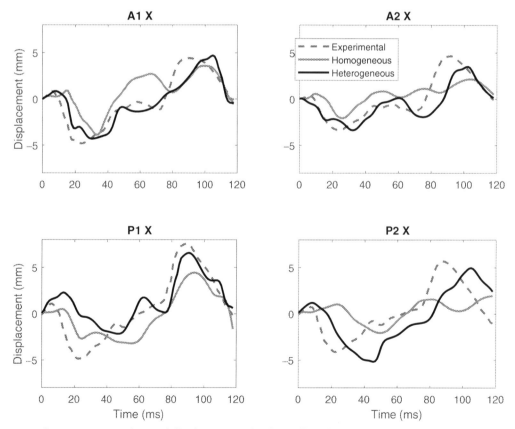

Fig. 9.8: Comparison of displacements in the x direction for two positions in the anterior (A) and posterior (P) to Hardy C383-T1 experiment.

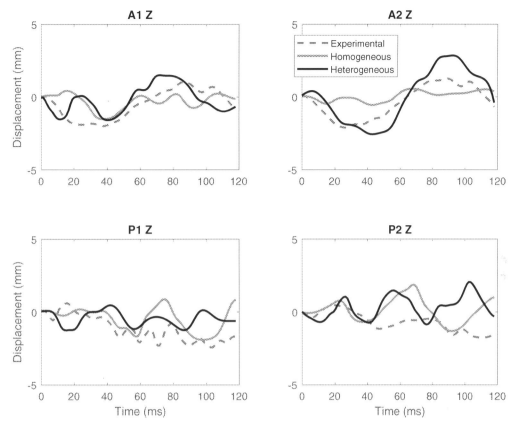

Fig. 9.9: Comparison of displacements in the z direction for two positions in the anterior (A) and posterior (P) to Hardy C383-T1 experiment.

model, while isotropic, is able to incorporate the relative differences between white matter structures. It is insightful to consider the differences between these two approaches. Results by Johnson et al. (2013) show that MRE and diffuse tensor imaging measures correlate well with each other within the corpus callosum since they are both sensitive to the underlying tissue microstructure or, more specifically, axon diameter. More details are presented in Madhukar and Ostoja-Starzewski (2020).

Table 9.4: Total excursions (in mm) for Hardy's C383-T1 experiment compared to the predicted values for two models.

Location	Experiment	Homogeneous	Heterogeneous
A1	9.24	7.47	9.02
A2	8.04	4.22	6.88
P1	12.42	7.66	8.76
P2	9.80	4.01	10.14

9.4 Conclusion

We present a high-resolution, voxel-based heterogeneous FE head model to study transient wave dynamics during traumatic brain injury. We utilize heterogeneous shear modulus determined using a NLI technique from MRE experiments performed by Johnson et al. (2013). While many FE models employ homogenized or averaged, mechanical properties to approximate constitutive relations of brain tissues, our approach allows us to investigate the response due to local structures within the white matter. We note that both the corpus callosum and corona radiata are significantly stiffer than overall white matter, with the corpus callosum exhibiting greater stiffness and lower viscous damping than the corona radiata. These differences are explained by examining the organizational and compositional characteristics of each structure. Incorporating this heterogeneity in our model affects wave propagation within the cerebrum and yields results that more closely match experimental results. We find that local variations in stiffness affect the local mechanical response. For instance, intracranial pressure magnitude following an impact is lower in regions of high local stiffness.

The unique architecture of the human head, consisting of the hard solid skull, the membranes, cerebrospinal fluid, and the viscoelastic brain core, leads to partial conversion of the pressure impact into a shear wave converging toward head's center. However, the shear wave attenuates as it converges inward due to (1) wave damping due to brain tissue viscoelasticity and (2) the heterogeneous brain structure which introduces extra wave scattering. The magnitude of this attenuation is greater for the heterogeneous material model then the homogeneous one.

References

Arbogast, K.B., Margulies, S.S., 1998. Material characterization of the brainstem from oscillatory shear tests. J. Biomech. 31 (9), 801–807.

Arfanakis, K., Haughton, V.M., Carew, J.D., Rogers, B.P., Dempsey, R.J., Elizabeth Meyerand, M., 2002. Diffusion tensor MR imaging in diffuse axonal injury. Am. J. Neuroradiol. 23 (5), 794–802.

Chatelin, S., Constantinesco, A., Willinger, R., 2010. Fifty years of brain tissue mechanical testing: from in vitro to in vivo investigations. Biorheology 47 (5-6), 255–276.

Chen, Y., Ostoja-Starzewski, M., 2010. MRI-based finite element modeling of head trauma: spherically focusing shear waves. Acta Mech. 213 (1-2), 155–167.

Chen, Y., Sutton, B., Conway, C., Broglio, S.P., Starzewski, M.O., 2012. Brain deformation under mild impact: magnetic resonance imaging-based assessment and finite element study. Int. J. Numer. Anal. Model. Ser. B 3 (1), 20–35.

Garimella, H.T., Kraft, R.H., 2017. Modeling the mechanics of axonal fiber tracts using the embedded finite element method. Int. J. Numer. Methods Biomed. Eng. 33 (5), e2823.

Gennarelli, T.A., Thibault, L.E., Hume Adams, J., Graham, D.I., Thompson, C.J., Marcincin, R.P., 1982. Diffuse axonal injury and traumatic coma in the primate. Ann. Neurol. 12 (6), 564–574.

Gentry, L.R., Godersky, J.C., Thompson, B., 1988a. MR imaging of head trauma: review of the distribution and radiopathologic features of traumatic lesions. Am. J. Roentgenol. 150 (3), 663–672.

Gentry, L.R., Godersky, J.C., Thompson, B., Dunn, V.D., 1988b. Prospective comparative study of intermediate-field MR and CT in the evaluation of closed head trauma. Am. J. Roentgenol. 150 (3), 673–682.

Giordano, C., Cloots, R.J.H., Van Dommelen, J.A.W., Kleiven, S., 2014. The influence of anisotropy on brain injury prediction. J. Biomech. 47 (5), 1052–1059.

Giordano, C., Kleiven, S., 2014. Connecting fractional anisotropy from medical images with mechanical anisotropy of a hyperviscoelastic fibre-reinforced constitutive model for brain tissue. J. R. Soc. Interface 11 (91), 20130914.

Glover, G.H., 1999. Simple analytic spiral k-space algorithm. Magn. Reson. Med. 42 (2), 412–415.

Hardy, W.N., Foster, C.D., Mason, M.J., Yang, K.H., King, A.I., Tashman, S., 2001. Investigation of head injury mechanisms using neutral density technology and high-speed biplanar x-ray. Stapp Car Crash J. 45, 337–368.

Hardy, W.N., Mason, M.J., Foster, C.D., Shah, C.S., Kopacz, J.M., Yang, K.H., King, A.I., Bishop, J., Bey, M., Anderst, W., et al., 2007. A study of the response of the human cadaver head to impact. Stapp Car Crash J. 51, 17.

Ji, S., Zhao, W., Ford, J.C., Beckwith, J.G., der, R.P.B., Greenwald, R.M., Flashman, L.A., Paulsen, K.D., McAllister, T.W., 2015. Group-wise evaluation and comparison of white matter fiber strain and maximum principal strain in sports-related concussion. J. Neurotrauma 32 (7), 441–454.

Johnson, C.L., McGarry, M.D.J., Gharibans, A.A., Weaver, J.B., Paulsen, K.D., Wang, H., Olivero, W.C., Sutton, B.P., Georgiadis, J.G., 2013. Local mechanical properties of white matter structures in the human brain. Neuroimage 79, 145–152.

Khalil, T.B., Hubbard, R.P., 1977. Parametric study of head response by finite element modeling. J. Biomech. 10 (2), 119–132.

Kleiven, S., 2007. Predictors for traumatic brain injuries evaluated through accident reconstructions Technical Report, SAE Technical Paper.

Kleiven, S., von Hans, H., 2002. Consequences of head size following trauma to the human head. J. Biomech. 35 (2), 153–160.

Madhukar, A., Chen, Y., Ostoja-Starzewski, M., 2017. Effect of cerebrospinal fluid modeling on spherically convergent shear waves during blunt head trauma. Int. J. Numer. Methods Biomed. Eng. 33 (12), e2881.

Madhukar, A., Ostoja-Starzewski, M., 2019. Finite element methods in human head impact simulations: a review. Ann. Biomed. Eng. 47, 1832–1854. https://doi.org/10.1007/s10439-019-02205-4.

Madhukar, A., Ostoja-Starzewski, M., 2020. Modeling and simulation of head trauma utilizing white matter properties from magnetic resonance elastography. Modelling 1 (2), 225–241.

McGarry, M.D.J., Houten, E.E.W.V, 2008. Use of a Rayleigh damping model in elastography. Med. Biol. Eng. Comput. 46 (8), 759–766.

Muthupillai, R., Lomas, D.J., Rossman, P.J., Greenleaf, J.F., Manduca, A., Ehman, R.L., 1995. Magnetic resonance elastography by direct visualization of propagating acoustic strain waves. Science 269 (5232), 1854–1857.

Nahum, A.M., Smith, R., Ward, C.C., 1977. Intracranial pressure dynamics during head impact Technical Report, SAE Technical Paper.

Ng, H.K., Mahaliyana, R.D., Poon, W.S., 1994. The pathological spectrum of diffuse axonal injury in blunt head trauma: assessment with axon and myelin stains. Clin. Neurol. Neurosurg. 96 (1), 24–31.

Nishawala, V.V., Ostoja-Starzewski, M., Leamy, M.J., Porcu, E., 2016. Lamb's problem on random mass density fields with fractal and Hurst effects. Proc. R. Soc. A: Math. Phys. Eng. Sci. 472 (2196), 20160638.

Prange, M.T., Margulies, S.S., 2002. Regional, directional, and age-dependent properties of the brain undergoing large deformation. J. Biomech. Eng. 124 (2), 244–252.

Van Houten, E.E.W.V., Miga, M.I., Weaver, J.B., Kennedy, F.E., Paulsen, K.D., 2001. Three-dimensional subzone-based reconstruction algorithm for MR elastography. Magn. Resonan. Med. 45 (5), 827–837.

Wang, F., Han, Y., Wang, B., Peng, Q., Huang, X., Miller, K., Wittek, A., 2018. Prediction of brain deformations and risk of traumatic brain injury due to closed-head impact: quantitative analysis of the effects of boundary conditions and brain tissue constitutive model. Biomech. Model. Mechanobiol. 17 (4), 1165–1185.

Zhang, X., Ostoja-Starzewski, M., 2020. Impact force and moment problems on random mass density fields with fractal and hurst effects. Philos. Trans. R. Soc. A 378 (2172), 20190591.

CHAPTER 10

Robust concept exploration of driver's side vehicular impacts for human-centric crashworthiness

A.B. Nellippallil[a,1], P.R. Berthelson[b], L. Peterson[c], Raj K. Prabhu[d]

[a]*Department of Mechanical and Civil Engineering, Florida Institute of Technology, Melbourne, FL, United States* [b]*Center for Applied Biomechanics, University of Virginia, Charlottesville, VA, United States* [c]*Center for Advanced Vehicular Systems (CAVS), Mississippi State University, Mississippi State, MS, United States* [d]*USRA, NASA HRP CCMP, NASA Glenn Research Center, Cleveland, OH, United States*

10.1 Frame of reference

Motor vehicle accidents are a significant contributor toward the global total number of accident-related injuries; producing a reported 1.35 million deaths and up to 50 million injuries annually (Toroyan and Laych, 2015). Within these events, there is an increased incidence of injury to the central nervous system, comprising of the brain and the spinal cord. As such, it is important to understand the mechanisms behind these phenomena to better develop safety and restraint features for consumer vehicles. The most commonly used metric for head injury assessment, head injury criterion (HIC), is used to quantify head injury risk as a fac tor of prolonged linear acceleration within the center of gravity (CG) of the head (Versace, 1971). Additionally, neck injury is often assessed using neck injury criterion (Nij), a measure of injury risk based on forces and moments taken from the C1 vertebra of the test subject (Kleinberger et al., 1998). However, the use of Nij is limited to cases in which the occupant is impacted from the front. To supplement this limitation, other metrics are proposed by slightly

---

[1] An earlier version of this chapter has been published as part of the proceedings of the 2020 ASME International Design Engineering Technical Conferences & Computers and Information in Engineering Conference (IDETC-CIE) (Nellippallil et al., 2020) Nellippallil, A.B., Berthelson, P.R., Peterson, L., and Prabhu, R., 2020, "Head and neck injury based robust design for vehicular crashworthiness," ASME Design Automation Conference, Paper Number IDETC2020- 22539. An updated and longer version of this chapter is under review in the *ASCE-ASME Journal of Risk and Uncertainty in Engineering Systems* (Nellippallil et al., 2021) Nellippallil, A.B., Berthelson, P.R., Peterson, L., and Prabhu, R.K., 2021, "A computational framework for human-centric vehicular crashworthiness design and decision-making under uncertainty," ASCE-ASME J. Risk Uncertainty Eng. Syst. Part B: Mech. Eng., Under review.

Multiscale Biomechanical Modeling of the Brain.
DOI: https://doi.org/10.1016/B978-0-12-818144-7.00002-5
Copyright © 2022 Elsevier Inc. All rights reserved.

154　Chapter 10

tweaking the calculation of Nij, such as, lateral Nij (Soltis, 2001) and Nkm (Schmitt et al., 2010), for use in the other impact cases.

In addition to the development of injury metrics and criteria, a large selection of computational human body models (HBMs) have been created and validated to replace or supplement traditional physical testing of cadaveric or dummy models. These include models to represent isolated organs/body systems (Fice et al., 2011; Panzer et al., 2011; Panzer and Cronin, 2009; Danelson et al., 2021; Takhounts et al., 2003, 2008), dummy models (Noureddine et al., 2002; Canha et al., 2000), and full body HBMs (Iwamoto et al., 2002, 2021a, 2021b; Gayzik et al., 2021a, 2021b). Surrogate model or metamodel-based design studies that are computationally less expensive than the higher-order finite element (FE) simulations are carried out for vehicular design optimization for crashworthiness (Cadete et al., 2005; Zhu et al., 2012; Zhang et al., 2013; Fang et al., 2005; Hamza and Saitou, 2021; Pan et al., 2013) and injury response analysis for motor-vehicle collision scenarios (Prasanna, 2015; Nie et al., 2013; Tay et al., 2021; Wimmer et al., 2015). The authors focus on injury prediction in pedestrian impact scenarios (Nie et al., 2013; Wimmer et al., 2015), and crashworthiness design in front impact scenarios (Pan et al., 2013; Prasanna, 2015; Tay et al., 2021) in these studies. The effect of impact velocity on head and neck injury was studied by a few, however, limiting the study to just one impact orientation (Prasanna, 2015; Mattos et al., 2015; Deng et al., 2013, 2014; Pelenyte-Vyšniauskiene and Jurkauskas, 2007). Understanding the effects of crash variables, such as impact velocity, impact location, and, angle of impact, as well as the interactions between these variables, on the injury response of the occupant's head and neck across collisions with all sides of the vehicle is important. We note that in past studies all potentially hazardous crash orientations (impact locations and angles of impact) and impact velocities are not considered while making design decisions. We also note the absence of studies that take into consideration both head injury and neck injury simultaneously during the design phase. Based on literature search, we recognize in this chapter the need for a study that take into consideration both head injury and neck injury simultaneously and the effects of crash variables, such as, impact velocity, impact location, and angle of impact on these two types of injury in vehicle impact scenarios for design decision making. We recognize that simulation-based models for injury predictions are typically incomplete, inaccurate, and not of equal fidelity. Therefore, these models have uncertainty inherent in them which needs to be managed during design. Our interest is to manage the uncertainty present by identifying design solutions that are relatively insensitive to the sources of uncertainty, defined as robust design (Allen et al., 2006). The robust design types are classified according to the sources of uncertainty as (Nellippallil et al., 2018a): (i) Type I robust design for managing uncertainty due to noise factors, (ii) Type II robust design for managing uncertainty due to control factors, (iii) Type III robust design for managing uncertainty present in models. An approach that takes into account simultaneously Type I, II, and III robust design is required to truly achieve robust solutions.

The chapter is organized as follows. In Section 10.2, we define the human-centric crashworthiness problem. In Section 10.3, we present the adapted concept exploration framework (CEF) for robust concept exploration. In Section 10.4, we carry out the head and neck injury-based robust concept exploration for driver's side vehicular impacts and demonstrate the utility of the framework presented. In Section 10.5, we discuss our way forward. We end the paper with our closing remarks in Section 10.6.

10.2 Problem definition

A human head response under a car crash impact scenario is critical for predicting and modeling traumatic brain injuries, which is one of the major types of injuries and causes of deaths in car accidents. Our interest in this chapter is to predict the patterns of head and neck injury risk using HIC and lateral Nij, respectively, under a set of impact variables. Once these patterns are established from a design perspective, we will have a better knowledge of the effects of the impact variables, their ranges and their interactions in defining head and neck injury. This will help us further in formulating design studies with a focus on managing the head and neck injury risk by controlling the impact variables. The information from such a design study will be of value for vehicular design, road, and traffic design and management, etc. In Fig. 10.1, we show the vehicle model and a full schematic of the orientation of the impact

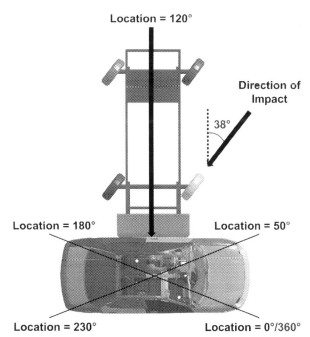

Fig. 10.1: A full schematic of a driver's side car impact with a moving deformable barrier (MDB) at an impact location of 120° and an angle of impact of 38°.

location and angle of impact variable along the driver's side surface as a moving deformable barrier (MDB) impacts the vehicle.

The design parameters considered for the problem defined is explained briefly below.

- *Impact location* (denoted by ω): The impact location spanned from $0°$ to $360°$. This was the location at which a MDB impacts the vehicle model, see Fig. 10.1. An angle from $0°$ to $50°$ corresponded to front location, $50°–180°$ corresponded to the driver's side, $180°–230°$ corresponded to the rear side, and $230°$ to $360°$ corresponded to passenger's side.
- *Angle of impact* (denoted by θ): The angle of impact in this study spanned from $-45°$ to $45°$. This is the angle at which an MDB impacted the vehicle model, see Fig. 10.1. The orientation of these angles will rotate with the MDB as the face of the impacted vehicle changes.
- *Impact velocity* (denoted by v): The impact velocity in this study spans from 10 mph to 45 mph. This is the constant velocity at which the MDB impacts the vehicle model.

Next, we discuss the decision-based robust design framework used to formulate the design problem and carry out robust solution space exploration for the problem defined.

10.3 Adapted CEF for robust concept exploration

The robust design framework presented in this chapter is an adapted from of the CEF presented by Nellippallil and co-authors (Nellippallil et al., 2018a, 2018b). The adapted CEF is used for systematic problem formulation and robust concept exploration to identify *satisficing* robust design solutions. In Fig. 10.2, the computing infrastructure for the adapted CEF is shown.

The modifications include: (i) the shift in focus on developing surrogate models or meta-models (response surface model) using carefully planned design experiments involving the use of costly FE car crash simulations to capture injury risk patterns (predicted responses) and the associated uncertainty involved, (ii) the error margin index (EMI) formulation of Type I, II, and III robust design problem using the response surface mean and prediction interval models developed. The problem formulation and robust concept exploration via the adapted CEF is explained using Steps A to G shown in Fig. 10.2.

Step A: Clarifying the design task (factors, ranges, goals, constraints). In Step A, the design task for the problem is clarified in terms of goals, constraints and problem-specific requirements. The rough design space is generated by identifying the control factors and their ranges.

Step B: Design of experiments. In Step B, design experiments for simulations are planned based on the control and noise factors identified in Step A. The experiments are performed to

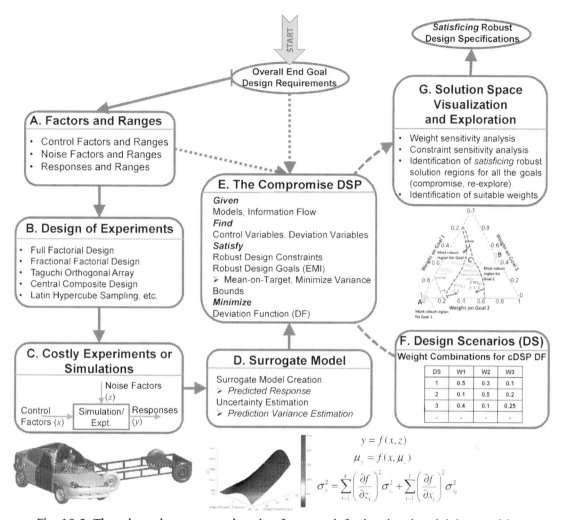

Fig. 10.2: The adapted concept exploration framework for head and neck injury problem formulation and robust concept exploration.

develop surrogate models for the mean response and prediction intervals (based on the variance associated).

Step C: Performing costly experiments/simulations. In Step C, costly experiments or simulations are carried out using a simulation infrastructure based on the design space identified in Step A and design experiments planned in Step B.

Step D: Building surrogate models. In Step D, surrogate models are formulated to map the design space to response performance based on the data generated from the design experiments.

158 Chapter 10

Step E: Formulation of the compromise decision support problem. In Step E, the design problem is formulated using the foundational mathematical construct of the CEF—the compromise decision support problem (cDSP) constructs. The cDSP is a hybrid formulation involving mathematical programming and goal programming. The cDSP and the design constructs involved are anchored in Taguchi's robust design paradigm. Using the cDSP, a designer seeks to explore multiple possible solutions that satisfice the multiple conflicting goals present (see Mistree et al., 1993; Nellippallil et al., 2017, 2019; Fonville et al., 2019).

Step F: Formulating the design scenarios and exercising the cDSP. In Step F, design scenarios are formulated for exercising the cDSP. The cDSP formulated is exercised for each of these design scenarios and results are gathered.

Step G: Solution space visualization and exploration. In Step G, the results obtained after exercising the cDSP for the design scenarios are used to generate the solution space. The solution space generated is further explored and "satisficing robust solutions" are identified.

In Section 10.4, we detail the formulation of the problem using the steps in the adapted CEF presented. Each of the steps is discussed in the context of the head and neck injury problem discussed in this chapter.

10.4 Head and neck injury criteria-based robust design of vehicular impacts

In Sections 10.4.1 to 10.4.5, we discuss in detail the steps involved in formulating the design problem.

10.4.1 Clarification of design task—Step A

In this chapter, we define a boundary and focus on the driver's side impacts. The flexibility of the generic robust design framework presented is such that, an easy formulation of a design problem focusing on the other vehicle impact locations like front, rear, or passenger's side is possible. The driver's side location is defined using *impact location* (ω) as between 50° and 180°. From this range, an ω value of 120° is considered as the location where a direct impact on the driver is possible as shown in Fig. 10.1 resulting in high injury risk. At this location, the impact is directly on the driver's location. We fix the impact location variable at 120° during problem formulation, discussed in Section 10.4. The three control factors identified in this study are impact location between 50° and 180°, the impact angle between −45° and 45°, and impact velocity between 25 mph and 45 mph. Our focus in this problem is on managing model parameter uncertainty present due to the control factors (Type II robust design), and model structure uncertainty present in the predicted responses of HIC and lateral Nij (Type III robust design). For defining the requirements for the design goals, both HIC and lateral Nij responses are studied. Based on a literature search, a value between 800 and 1000 is identified as the injury threshold for the HIC metric (Kleinberger et al., 1998; Henn, 1998). Values greater than the threshold have high chances of head injury risks and are identified as life-threatening by experts (Henn, 1998). Lateral Nij is measured in terms of four metrics, each representing a different loading case within the neck of the occupant: tension-left (Ntl), tension-right (Ntr), compression-left (Ncl), and compression-right (Ncr) (Soltis, 2001). For all these loading cases of lateral Nij, a value of 1.0 is identified as the injury threshold. Values exceeding the injury threshold of 1.0 are deemed to have a significant likelihood of inducing an injury within the cervical spine of the occupant (Soltis, 2001). In the adapted CEF, limits to the requirements/goals are identified. The designer's preference types for the goals are classified as "smaller is better," "nominal is better," or "larger is better." The designer needs to identify an upper requirement limit, both upper and lower requirement limits, and a lower requirement limit, respectively, depending on the designer's preference on the goals, see McDowell et al. (2009) for details. For the head and neck injury goals, the preference type selected is a "smaller is better" type as the requirement is to have as minimum injury risk possible for the different impact conditions. System constraints in terms of lower and upper bounds on the control factors and the head and neck injury functions are defined to bound the design problem.

10.4.2 Design of experiments—Step B

A Latin hypercube sampling procedure is used to create a matrix for the vehicular impact simulations. For carrying out the design of experiments, the three impact variables with the ranges specified, impact location (0°–360°), impact angle (−45° to 45°), and impact velocity (10–50 mph) are selected. The reason for including impact location from 0° to 360° is to obtain the injury responses for all the different car crash scenarios (front, driver's side, rear, and passenger's side). Also, a higher upper range for impact velocity is selected for the experiments. An initial Latin hypercube sampling test matrix of 60 vehicular impact cases spanning the different impact locations is created. An additional 37 cases are added to the initial test matrix to improve the model predictions at critical injury risk regions. The head injury response measured in terms of HIC is recorded for each experiment and is sorted into four groups depending on the location of impact: (1) front impact HIC, (2) driver's side impact HIC, (3) rear impact HIC, and (4) passenger's side impact HIC. The neck injury response measured in terms of lateral Nij is also sorted into groups based on location as carried out for HIC. The groups are further subdivided into five subgroups that include the lateral Nij metric components and the maximum value from all the components as: (1) tension-left (Ntl), (2) tension-right (Ntr), (3) compression-left (Ncl), (4) compression-right (Ncr), and (5) maximum lateral Nij.

10.4.3 Finite element car crash simulations for predicting injury response—Step C

10.4.3.1 Finite element model

The impact scenarios simulated for this study each consisted of three primary FE models. The occupant vehicle model created by the United States National Crash Analysis Center (Zaouk et al., 2000a, 2000b) and modified by researchers at the Center for Advanced Vehicular Systems at Mississippi State University (Horstemeyer et al., 2009; Fang et al., 2005) represented a full-scale 1996 Dodge Neon. Within this vehicle, a fully biofidelic HBM is belted to the driver's seat using a three-point seat belt. The HBM utilized for this study is the Version 4 adult male 50th percentile Total Human Model for Safety (THUMS); created by the Toyota Motor Corporation to represent an adult human male of average height and weight (Iwamoto et al., 2002, 2021a, 2021b). At the start of every crash scenario, the MDB is placed perpendicularly to the impacted surface at the designated impact location. This general FE simulation setup is shown in Fig. 10.3. All car crash cases were simulated using LS-DYNA software (LSTC, Livermore, CA, USA). In-house Python scripts are used to obtain the resultant linear acceleration data from the CG of the head (denoted by the star in Fig. 10.4) of the THUMS model, see Fig. 10.4. Also, various forces and moments from the C1 vertebra of THUMS for each impact scenario are also recorded. The forces and moments recorded include peak positive and negative axial loads (tension and compression), sagittal plane bending moments (flexion and extension), lateral bending moments (left and right), and sagittal plane shear forces (anterior and posterior), depending on the side of vehicle impact, see Fig. 10.4.

10.4.3.2 Head injury metric analysis

The HIC is used in this study to assess the occupant's risk of inducing a head injury under each investigated crash scenario. HIC measures this injury risk in terms of prolonged linear acceleration (Henn, 1998) for durations between 15 and 36 milliseconds (Versace, 1971, Kleinberger et al., 1998). For this study, the 15 milliseconds version of HIC, known as HIC_{15},

Fig. 10.3: Finite element car crash simulation using LS-Dyna (impact variable values: impact velocity = 49.484 mph, impact location = 120°, angle of impact = 38.223°).

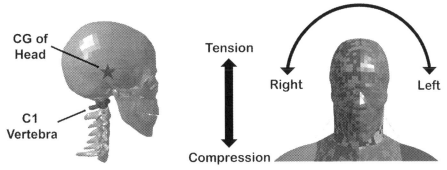

Fig. 10.4: Schematic showing the approach of obtaining the output data from the finite element (FE) model. Accelerations are obtained from the center of gravity (CG) of the head of the THUMS; denoted by the star icon. Additionally, peak positive and negative axial loading forces (tension/compression) and lateral plane bending moments (left/right) are obtained from the C1 vertebra of the THUMS. The assumed orientations of the forces and moments are also shown.

is used for all simulations, as this time duration is the NHTSA standard for 50th percentile test subjects (Kleinberger et al., 1998; Eppinger et al., 2000). The following equation is used for the calculation of HIC_{15} (Versace, 1971),

$$\text{HIC} = \max_{(t_1, t_2)} \left\{ (t_2 - t_1) \left[\frac{1}{t_2 - t_1} \int_{t_1}^{t_2} a(t) \, dt \right]^{2.5} \right\} \qquad (10.1)$$

where a is the resultant linear acceleration (g) measured from the CG of the head of the THUMS model and $(t_2 - t_1)$ is time duration (ms).

10.4.3.3 Neck injury metric analysis

The lateral Nij is used in this study to quantify the injury risk within the neck of the occupant (Kleinberger et al., 1998). A slight modification on Nij, the NHTSA standard neck injury metric for frontal impact cases, lateral Nij is used to measure the injury risk as a factor of axial loads and lateral-plane bending moments near the top of the cervical spine of the occupant. The following equation is used to calculate lateral Nij in each side impact case (Soltis, 2001),

$$\text{Lateral Nij} = \frac{F_Z}{F_{\text{int}}} + \frac{M_X}{M_{\text{int}}} \qquad (10.2)$$

where, F_Z is the axial load (N) and M_X is the lateral-plane bending moment (Nm) (Soltis, 2001); both obtained from the C1 vertebra of the THUMS model. F_{int} and M_{int} are critical intercept values used to normalize the axial load and lateral-plane bending moment, respectively. These values are given in Table 10.1. Four injury metric components are derived from the lateral Nij, each representing a different loading case within the neck of the occupant: tension-left (Ntl), tension-right (Ntr), compression-left (Ncl), and compression-right (Ncr).

162   Chapter 10

Table 10.1: Critical intercept values used to normalize the axial forces and lateral-plane bending moments for the calculation of lateral Nij.

Variable	Intercept value
F_{int} (tension)	6810 N
F_{int} (compression)	6160 N
M_{int} (left)	60 Nm
M_{int} (right)	60 Nm

The orientation assignments for the axial load (tension and compression) values and the lateral-plane bending moment (left and right) values are shown in Fig. 10.4. The greatest of the four components for each impact case is assigned to a separate data set called "maximum lateral Nij." For all cases, a lateral Nij value exceeding our injury threshold of 1 is deemed to have a significant likelihood of inducing an injury within the cervical spine of the occupant (Soltis, 2001).

10.4.4 Building surrogate models—Step D

For building the surrogate models, the SURROGATES Toolbox by Viana (FAC, 2011) is used. The surrogate model functions from SURROGATES Toolbox is used in the MATLAB code to fit a polynomial response surface model for the injury risk metrics as a function of the three variables, *impact velocity ϑ, impact angle θ,* and *impact location ω.* The predicted response surface model is the mean response model and is a function of the most probable responses at new observations. To demonstrate the design approach, in this design study, we fix the *impact location ω* value as 120°, the scenario where the impact is directed at the driver's side. For all the design points, the prediction variance is estimated. Prediction intervals consisting of upper and lower limits within which a new observation may be located with some confidence level is developed. An upper and lower prediction interval limits for the response mean model is estimated using the predicted variance with a $100(1 - \alpha)\%$ confidence level as,

$$\hat{\mu}_{y|x_0} - t_{\frac{\alpha}{2}, n-p} \sqrt{V\left(\hat{\mu}_{y|x_0}\right)} \leq \mu_{y|x_0} \leq \hat{\mu}_{y|x_0} + t_{\frac{\alpha}{2}, n-p} \sqrt{V\left(\hat{\mu}_{y|x_0}\right)} \tag{10.3}$$

where $\hat{\mu}_{y|x_0}$ is the mean response at a point x_0, $V\left(\hat{\mu}_{y|x_0}\right)$ is the variance of $\hat{\mu}_{y|x_0}$. The statistics follow a t distribution with $n - p$ degrees of freedom. In this work, we fit separate response surface models to the upper $(f_1(x_0))$, and lower $(f_2(x_0))$ level prediction intervals identified for the mean model $(f_0(x_0))$. These models are developed as response functions of the impact variables, *impact velocity* and *impact angle.* The mean response and upper prediction interval models for the HIC_{15} response are shown in Fig. 10.5. In Figs. 10.6 and 10.7, the mean response model and upper and lower limits of the prediction interval for the estimation of Ntl and Ncr are shown, respectively. In this chapter, along with HIC_{15}, we pick two lateral

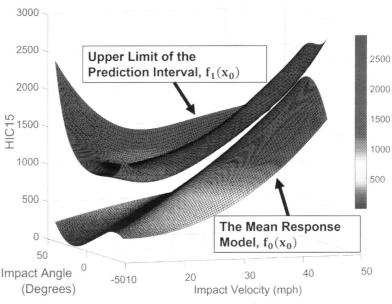

Fig. 10.5: The mean response model and upper limit of the prediction interval for the estimation of HIC_{15} for a given impact location of 120° (lower limit not shown).

Nij components as our goals—Ntl, and Ncr. The reason for picking the components are because of the conflicting nature of the two responses while varying the impact variables, see Figs. 10.6 and 10.7. The red contours in Figs. 10.5–10.7 denote high values of HIC_{15} and lateral Nij components that result in life-threatening situations during vehicle crash scenarios.

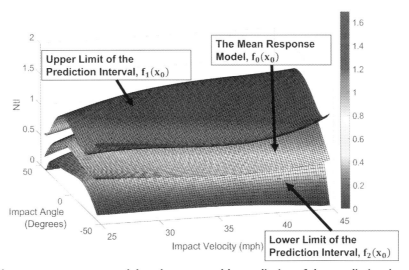

Fig. 10.6: The mean response model and upper and lower limits of the prediction interval for the estimation of Ntl for a given impact location of 120°.

164   Chapter 10

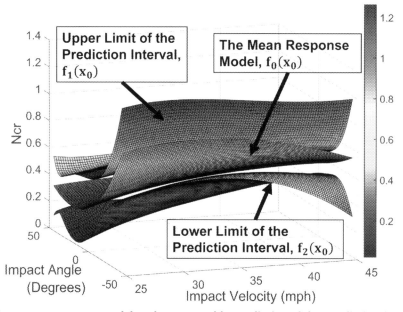

Fig. 10.7: The mean response model and upper and lower limits of the prediction interval for the estimation of Ncr for a given impact location of 120°.

10.4.5 Formulation of robust design cDSP—Step E

To formulate the robust design problem and quantify the safety margin against system failure due to model structure and model parameter uncertainty, the compromise Decision Support Problem construct with the EMI metric is used. The EMI is used for achieving Type I, II, and III robust design in one single formulation. In Fig. 10.8, the formulation of EMI and achieving Type I, II, and III robust design by maximizing EMI for "smaller is better" case is shown.

The designer's interest here is to maximize the EMI as much possible for the multiple goals present. Maximizing EMI involves moving the response mean away from the upper requirement limit (for smaller is the better case) and reducing the response deviation/variation, see Fig. 10.8. The steps involved in formulating the EMI are discussed in detail in Nellippallil et al. (2018a). The steps are briefly presented in this section in Steps A to D.

Step A: Calculation of response variations for all the model levels for variations in design variables.

Let f represents the response functions and x_i the design variables present. Let us assume that there are k uncertainty bounds. The response variations (ΔY_j) for each level is calculated as,

$$\Delta Y_j = \sum_{i=1}^{n} \left| \frac{\partial f_j}{\partial x_i} \right| \Delta x_i \quad (10.4)$$

Robust concept exploration of driver's side vehicular impacts for human-centric   165

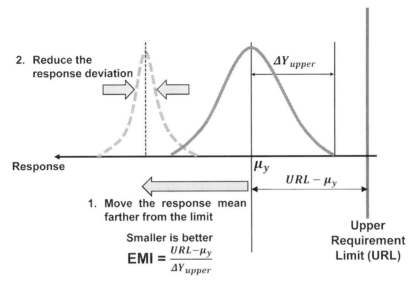

Fig. 10.8: Formulation of EMI and schematic representation of maximization of EMI by moving the mean and reducing the response deviation to achieve Type I, II, and III robust design.

where $j = 0, 1, 2, \ldots, k$ (number of uncertainty bounds). In Fig. 10.9, the response variations for the upper bound, mean, and lower bound response functions are denoted as ΔY_1, ΔY_0, and ΔY_2, respectively.

Step B: Calculation of minimum and maximum responses for variability in design variables and variability around mean response (uncertainty bounds).

Let Y_{max} and Y_{min} be the maximum and minimum responses by considering the variability in design variables and uncertainty bounds around the mean response.

$$Y_{max} = \text{Max}\left[f_j(x) + \Delta Y_j \right] \text{ and} \tag{10.5}$$

$$Y_{min} = \text{Min}\left[f_j(x) - \Delta Y_j \right] \tag{10.6}$$

where $j = 0, 1, 2, \ldots, k$ (number of uncertainty bounds), $f_0(x)$ is the mean response function, and $f_1(x)\ldots f_k(x)$ are the uncertainty bound functions.

Step C: Calculation of upper and lower deviation of response at x.

The upper and lower deviations of responses at x are calculated as,

$$\Delta Y_{upper} = Y_{max} - f_o(x) \text{ and} \tag{10.7}$$

$$\Delta Y_{lower} = f_o(x) - Y_{min} \tag{10.8}$$

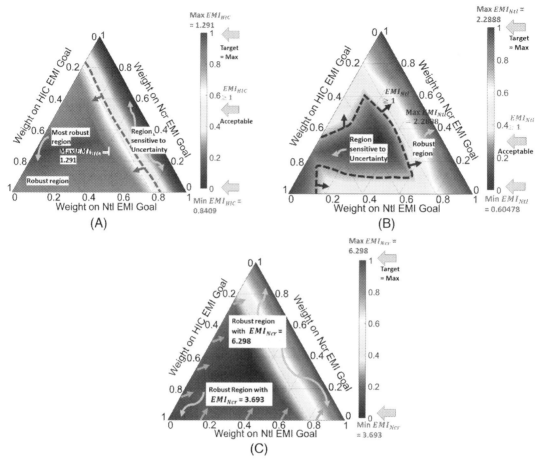

Fig. 10.9: Robust solution space for HIC (A), Ntl (B), and Ncr (C).

Step D: Calculation of EMI.

The EMI is calculated using the mean response function, $f_o(x)$ and the upper and lower deviations, ΔY_{upper}, and ΔY_{lower}. For a "smaller is better" case, the EMI is calculated as,

$$\text{EMI} = \frac{URL - \mu_y}{\Delta Y_{upper}} \quad (10.9)$$

Maximizing the EMI such that EMI ≥ 1 is achieved means that: (1) the response mean is away from the upper/lower requirement limits, and (2) the response deviation due to design variables and models are minimized. An EMI ≤ 1 implies that the upper/lower requirement limits are violated due to uncertainty present in design variables and/or models. Our interest, therefore, using the robust design cDSP formulation is to maximize the individual EMIs of the multiple goals present in the problem. The robust design cDSP for this problem reads as follows:

---

cDSP for head and neck injury based robust design of driver's side impacts

Given
Requirements on the cDSP goals
Maximize EMI for HIC_{15} goal
Maximize EMI for Ntl goal
Maximize EMI for Ncr goal
URL for HIC_{15} goal = 1000
URL for Ntl goal = 1
URL for Ncr goal = 1
System variables, their ranges, and variability (**Table 10.2**)
   Fixed parameters (**Table 10.3**)

Table 10.2: System variables, ranges, and variability.

Sr. no	System variables (X)	Ranges	Variability (Δx)
1	X_1, Impact velocity (ϑ)	25–45 mph	$[\pm 2]$
2	X_2, Impact angle (θ)	$-45°$ to $45°$	$[\pm 1]$

Find
μ_x (mean location of system variables)

Table 10.3: Parameters and values.

Sr. no	Parameter	Value
1	Impact location (ω)	120° (Driver's side)

Deviation Variables
d_i^-, d_i^+ i = 1, 2, 3

Satisfy
System constraints
Constraint for injury risk, Ntr
Ntr (x) ≤ 0.8
Constraint for injury risk, Ncl
Ncl (x) ≤ 0.8
System Goals

Goal 1:
Maximize EMI for yield strength

$$\frac{\mathrm{EMI}_{\mathrm{HIC}}(x)}{\mathrm{EMI}_{\mathrm{Target. HIC}}} + d_1^- - d_1^+ = 1$$

where EMI(x) = $\{f_0(x) - \mathrm{LRL}\}/\{Y_{\min} - f_0(x)\}$

---

168 Chapter 10

where

$$Y_{\min} = \min\left\{\left(f_j(x) - \sum_{i=1}^{n}\left|\frac{\partial f_j}{\partial x_i}\right| \cdot \Delta x_i\right)\right\}$$

Goal 2:
Maximize DCI for tensile strength

$$\frac{\text{EMI}_{\text{Ntl}}(x)}{\text{EMI}_{\text{Target, Ntl}}} + d_2^- - d_2^+ = 1$$

Goal 3:
Maximize DCI for hardness

$$\frac{\text{DCI}_{\text{Ncr}}(x)}{\text{DCI}_{\text{Target, Ncr}}} + d_3^- - d_3^+ = 1$$

where DCI(x) = {$f_0(x) - LRL$}/ΔY

where

$$\Delta Y = \sum_{i=1}^{n}\left|\frac{\partial f_0}{\partial x_i}\right|\Delta x_i$$

Variable bounds
Defined in Table 10.2
Bounds on deviation variables

$$d_i^-, d_i^+ \geq 0 \text{ and } d_i^- * d_i^+ = 0, i = 1, 2, 3$$

Minimize
We minimize the deviation function

$$Z = \sum_{i=1}^{3} W_i\left(d_i^- + d_i^+\right); \sum_{i=1}^{3} W_i = 1$$

We exercise the robust design cDSP for several design scenarios and further explore the solution space to identify robust solutions that satisfice the requirements in the problem. We discuss this in the next section.

10.4.6 Formulating the design scenarios, exercising the cDSP and exploration of solution space—Step E

In order to exercise the cDSP, we select thirteen different design scenarios based on judgment to effectively represent the solution space for visualization and exploration. In each of these scenarios, we assign different weight preferences to each goal in the cDSP, see Table 10.4 for weight preferences assigned on goals for each scenario.

Table 10.4: Design scenarios (DS) and corresponding weights assigned on goals.

DS	W_1	W_2	W_3
1	1	0	0
2	0	1	0
3	0	0	1
4	0.5	0.5	0
5	0.5	0	0.5
6	0	0.5	0.5
7	0.25	0.75	0
8	0.25	0	0.75
9	0.75	0	0.25
10	0.75	0.25	0
11	0	0.75	0.25
12	0	0.25	0.75
13	0.33	0.34	0.33

We execute the cDSP formulated for each of these scenarios and the results in terms of the achieved goal values are documented. Using this information, we create ternary plots to visualize and explore the solution regions of interest, see Fig. 10.9.

In Figs. 10.9, we show via the ternary plots the achieved values of HIC_{15} (Goal 1), Ntl (Goal 2), and Ncr (Goal 3), respectively for all the thirteen scenarios. For Goals 1, 2, and 3, we are interested in achieving high values of EMI_{HIC}, EMI_{Ntl}, and EMI_{Ncr}, respectively. The regions with $EMI_{HIC} \geq 1$, $EMI_{Ntl} \geq 1$, and $EMI_{Ncr} \geq 1$ are identified by the dashed line in ternary space of Fig. 10.9A–C, respectively. Solutions inside these regions are robust under both model structure and model parameter uncertainty. The maximum EMI_{YS}, EMI_{Ntl}, and EMI_{Ncr} are achieved in the red regions and these regions are therefore the most robust regions. The dark blue regions are the relatively least robust in the solution spaces generated. In Fig. 10.9A and B, we see that the blue regions are regions that are sensitive to uncertainty as the EMI values are less than 1.

Since our interest is in identifying *satisficing robust solution* regions for all the multiple conflicting goals, we plot the superposed plot with all the robust solution spaces of interest as shown in Fig. 10.10. The light-yellow region identified in Fig. 10.10 satisfices the robust design requirements identified for all the goals. In Fig. 10.10, we highlight eight points A, B, C, D, E, F, G, and H in the ternary space. Points A, B, C, D, and E lie within the identified superposed robust region. Points F, G and H lie outside this region. On comparing all the points in Table 10.5, we see that Point F is the most robust solution satisfying the HIC_{15} goal, but at the same time the most uncertain solution point for Ntl goal. Similarly, points G and H are the most robust solutions satisfying both Ntl and Ncr goals but the worst for HIC_{15} goal. Since, the goals are conflicting, there is a need to explore the compromised solutions from the superposed region and pick the ones that are robust and best satisfied for all three goals. From Table 10.5, we see that the points A, B, C, D, and E have the same solutions and are

170   Chapter 10

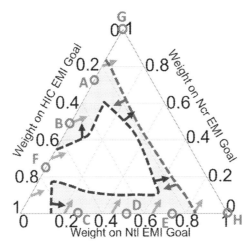

Fig. 10.10: Superposed robust solution spaces.

Table 10.5: cDSP results for points selected.

	Impact variables		EMI values for injury goals		
Sol. pt	X_1, Impact velocity (ϑ)	X_2, Impact angle (θ)	HIC	Ntl	Ncr
A	25	38.04	1.2	1.39	4.23
B	25	38.04	1.2	1.39	4.23
C	25	38.04	1.2	1.39	4.23
D	25	38.04	1.2	1.39	4.23
E	25	38.04	1.2	1.39	4.23
F	30.95	40.34	1.29	0.95	4.09
G	25	45	0.84	2.28	6.29
H	25	45	0.84	2.28	6.29

therefore relatively insensitive to uncertainty. The corresponding EMI values are greater than 1. Thus, we pick these as the best satisficing robust solution meeting all the three goals.

As we can see, the solution points with the lowest impact velocity are the most robust. This occurs because the injury risks associated with low velocity regions are less, thereby, maximizing the numerator of the EMI formulation as the respective mean injury values are farther from the upper requirement limits defined. From the ternary plots, we can also explore regions that have increased chances of causing head and neck injuries.

The utility of the framework presented is that we are able to formulate conflicting multiobjective design problems in a systematic manner and explore satisficing robust solutions that are relatively insensitive to the uncertainty present. The study presented is useful especially for the design of roads and traffic systems taking into consideration the effects of vehicular impact variables on human head and neck injuries. The work will be further extended to include the damage on the vehicle as a response/goal and study the correlations of vehicular damage to the human head and neck injury for variations in impact variables. We discuss this briefly next.

10.5 Future: correlate human brain injury to vehicular damage

Future studies would benefit from a correlation of magnitudes and locations of damage in load bearing structures to human body (our focus is the brain herein) damage during vehicular impacts. Such studies may elucidate structures whose susceptibility to damage renders them unable to optimally absorb impact energy and reduce energy transferred to human occupants in crash events. Damage in ductile metals (automotive structural materials) often manifests as the nucleation, growth, and coalescence of microscopic voids (Puttick, 1959; Rabotnov, 2021). Significant void evolution in these materials can contribute to strain localization, macroscopic crack evolution, and fracture. Voids commonly nucleate at interfacial regions such as grain, subgrain, and inclusion-matrix material boundaries (Puttick, 1959; Barnby, 1967; Gangulee and Gurland, 1967). Void nucleation occurs when sufficient dislocation densities (plasticity) accumulate at boundaries such that localized interfacial decohesion or inclusion fracture mechanisms become energetically favorable to further dislocation motion (Barnby, 1967; Yeh and Liu, 1996; Mishnaevsky Jr et al., 1999; Rincon et al., 2009). Nucleated voids continue to grow in regions of geometrically imposed stress concentrations as plastic deformation continues and these growth rates are strongly influenced by stress triaxiality (McClintock, 1968; Rice and Tracey, 1969; Gurson, 1977). The stress fields around voids can interact depending upon void size, shape, spacing, and orientation resulting in void coalescence by sheeting (fracture of adjoining material ligament between voids) and impingement (diffusion-like growth of voids into one another) mechanisms (Cottrell, 1959; Brown and Embury, 1973; Bourcier and Koss, 1979). Void coalescence leads to transgranular fracture mechanisms ultimately contributing to unstable deformation and fracture of the aggregate material (Puttick, 1959). Void nucleation, growth, and coalescence are strongly correlated to plastic deformation. Other factors strongly influence void evolution rates including stress state triaxiality (McClintock, 1968; Rice and Tracey, 1969; Gurson, 1977), strain rate (Raj and Ashby, 1975), and geometric characteristics and material properties of microstructural constituents (Gangulee and Gurland, 1967).

Plastic equivalent strain could be a suitable first approximation of damage in vehicle structure to human correlation studies because of void evolution mechanisms' dependence upon plastic strain. A more robust approach would consider the coupled effects of plastic strain and stress triaxiality in accordance with traditional void-based damage models (McClintock, 1968; Rice and Tracey, 1969; Gurson, 1977). Contemporary damage models (Tvergaard and Needleman, 1984; Horstemeyer and Gokhale, 1999) that consider microstructure–property effects could be used to calculate void-based damage in vehicular components with even greater accuracy.

172   Chapter 10

However, these physically motivated models require extensive experimental investigations to quantify microstructure property characteristics for calibration purposes.

In our future study planned, we propose to quantify plasticity and metal damage using surrogate models as a function of the impact variables. The plastic strain surrogate model thus developed will be used with the head and neck models presented in this chapter for a multiobjective robust design study. Using this study, we plan to explore correlations between vehicular damage and human injury in terms of the different vehicular crash impact variables. For the human body, we also hope to use the mechano-porosity internal state variable (MPISV) model explained in Chapter 8 of this book.

10.6 Summary

In this chapter, we present a decision-based robust design framework to quantify and manage the impact-based injury risks on occupants for different model-based car crash scenarios. We illustrate the utility of the framework for a human-centric crashworthiness design problem focused on minimizing human head and neck injury by managing the impact variables. New contributions in this chapter from the problem standpoint include the designer's ability to explore satisficing robust solution regions for impact variables when multiple conflicting injury goals and multiple sources of uncertainty are present. Using the framework, a designer is able to formulate conflicting multiobjective design problems in a systematic manner and explore satisficing robust solutions that are relatively insensitive to the uncertainty present. The study presented using the human-centric crashworthiness design is useful especially for the design of roads and traffic systems taking into consideration the effects of vehicular impact variables on human head and neck injuries.

In the future, one can employ the MPISV model described in Chapter 8 of this book to provide a more realistic damage metric for the brain instead of the HIC metric used herein. The robust design methodology has not been employed with the MPISV to date related to brain impacts, so this work would be a contribution to the community.

References

Allen, J.K., Seepersad, C., Choi, H., Mistree, F., 2006. Robust design for multiscale and multidisciplinary applications. J. Mech. Des. 128 (4), 832–843.

Barnby, J., 1967. The initiation of ductile failure by fractured carbides in an austenitic stainless steel. Acta Metall. 15 (5), 903–909.

Bourcier, R., Koss, D., 1979. Ductile Fracture under Multiaxial Stress States between Pairs of Holes. Michigan Technological Univ Houghton Dept of Metallurgical Engineering, Houghton.

Brown, L., Embury, J., 1973. Initiation and growth of voids at second-phase particles, Proceedings of the Conference on Microstructure and Design of Alloys, 1. Institute of Metals and Iron and Steel Insitute, London, pp. 164–169.

Cadete, R.N., Dias, J.P., Pereira, M.S., 2005. Optimization in vehicle crashworthiness design using surrogate models, Sixth World Congresses of Structural and Multidisciplinary Optimization. Rio de Janeiro, Brazil.

Canha, J., DiMasi, F., Tang, Y., Haffner, M., and Shams, T., 2000, "Development of a finite element model of the thor crash test dummy," No. 0148-7191, SAE Technical Paper.

Cottrell, I., 1959, "AH (1959) Theoretical aspects of fracture," B.L. Averbach, O.K. Felbeck, G.T. Hahn, and O.A. Thomas, Fracture. MIT Press,Cambridge, Massachusetts, USA, pp. 20-45.

Danelson, K.A., Gayzik, F.S., Mao, M.Y., Martin, R.S., Duma, S.M., Stitzel, J.D., 2021. Bilateral carotid artery injury response in side impact using a vessel model integrated with a human body model, Proceedings of the Annals of Advances in Automotive Medicine/Annual Scientific Conference. Association for the Advancement of Automotive Medicine, p. 271.

Deng, X., Chen, S.A., Prabhu, R., Jiang, Y., Mao, Y., Horstemeyer, M., 2014. Finite element analysis of the human head under side car crash impacts at different speeds. J. Mech. Med. Biol. 14 (06), 1440002.

Deng, X., Potula, S., Grewal, H., Solanki, K., Tschopp, M., Horstemeyer, M., 2013. Finite element analysis of occupant head injuries: parametric effects of the side curtain airbag deployment interaction with a dummy head in a side impact crash. Accid. Anal. Prevent. 55, 232–241.

Eppinger, R., Kuppa, S., Saul, R., Sun, E., 2000. Supplement: Development of Improved Injury Criteria for the Assessment of Advanced Automotive Restraint Systems: II. *NHTSA*, Washington.

FAC, V., 2011. SURROGATES Toolbox User's Guide, Gainesville, FL, USA, version 3.0 ed., Available at: https://sites.google.com/site/srgtstoolbox/.

Fang, H., Rais-Rohani, M., Liu, Z., Horstemeyer, M., 2005. A comparative study of metamodeling methods for multiobjective crashworthiness optimization. Comput. Struct. 83 (25-26), 2121–2136.

Fang, H., Solanki, K., Horstemeyer, M., 2005. Numerical simulations of multiple vehicle crashes and multidisciplinary crashworthiness optimization. Int. J. Crashworthiness 10 (2), 161–172.

Fice, J.B., Cronin, D.S., Panzer, M.B., 2011. Cervical spine model to predict capsular ligament response in rear impact. Ann. Biomed. Eng. 39 (8), 2152–2162.

Fonville, T.F., Nellippallil, A.B., Horstemeyer, M.F., Allen, J.K., Mistree, F., 2019. A goal-oriented, inverse decision-based design method for multi-component product design, ASME Design Automation Conference.

Gangulee, A., Gurland, J., 1967. On the fracture of silicon particles in aluminum- silicon alloys. AIME Met. Soc. Trans. 239 (2), 269–272.

Gayzik, F.S., Moreno, D.P., Vavalle, N.A., Rhyne, A.C., Stitzel, J.D., 2021. Development of the global human body models consortium mid-sized male full body model, Proceedings of the International Workshop on Human Subjects for Biomechanical Research. National Highway Traffic Safety Administration.

Gayzik, F.S., Moreno, D.P., Vavalle, N.A., Rhyne, A.C., Stitzel, J.D., 2021. Development of a full human body finite element model for blunt injury prediction utilizing a multi-modality medical imaging protocol, Proceedings of the 12th International LS-DYNA User Conference. Dearborn, MI.

Gurson, A.L., 1977. Continuum Theory of Ductile Rupture by Void Nucleation and Growth: Part I—Yield Criteria and Flow Rules for Porous Ductile Media. ASME. J. Eng. Mater. Technol. 99 (1), 2–15.

Hamza, K., Saitou, K., 2021. Vehicle crashworthiness design via a surrogate model ensemble and a co-evolutionary genetic algorithm, Proceedings of the ASME 2005 International Design Engineering Technical Conferences and Computers and Information in Engineering Conference. American Society of Mechanical Engineers, pp. 899–907.

Henn, H.-W., 1998. Crash tests and the head injury criterion. Teach. Math. Its Applic. 17 (4), 162–170.

Horstemeyer, M.F., Gokhale, A.M., 1999. A void–crack nucleation model for ductile metals. Int. J. Solids Struct. 36 (33), 5029–5055.

Horstemeyer, M., Ren, X., Fang, H., Acar, E., Wang, P., 2009. A comparative study of design optimisation methodologies for side-impact crashworthiness, using injury-based versus energy-based criterion. Int. J. Crashworthiness 14 (2), 125–138.

Iwamoto, M., Kisanuki, Y., Watanabe, I., Furusu, K., Miki, K., Hasegawa, J., 2002. Development of a finite element model of the total human model for safety (THUMS) and application to injury reconstruction, Proceedings of the 2002 International Research Council on Biomechanics of Injury. Munich, Germany, 31–42.

Iwamoto, M., Nakahira, Y., Tamura, A., Kimpara, H., Watanabe, I., Miki, K., 2021. Development of advanced human models in THUMS, Proceedings of the Sixth European LS-DYNA Users' Conference, 47–56.

174 Chapter 10

Iwamoto, M., Omori, K., Kimpara, H., Nakahira, Y., Tamura, A., Watanabe, I., Miki, K., Hasegawa, J., Oshita, F., Nagakute, A., 2021. Recent advances in THUMS: development of individual internal organs, brain, small female and pedestrian model, Proceedings of Fourth European LS Dyna Users Conference, 1–10.

Kleinberger, M., Sun, E., Eppinger, R., Kuppa, S., Saul, R., 1998. Development of improved injury criteria for the assessment of advanced automotive restraint systems. NHTSA Docket 4405 (9), 12–17.

Mattos, G.A., Mcintosh, A.S., Grzebieta, R., Yoganandan, N., Pintar, F.A., 2015. Sensitivity of head and cervical spine injury measures to impact factors relevant to rollover crashes. Traffic Inj. Prev. 16 (Suppl. 1), S140–S147.

McClintock, F.A., 1968. A Criterion for Ductile Fracture by the Growth of Holes. ASME. J. Appl. Mech. 35 (2), 363–371.

McDowell, D.L., Panchal, J., Choi, H.-J., Seepersad, C., Allen, J., Mistree, F., 2009. Integrated Design of Multiscale, Multifunctional Materials and Products. Butterworth-Heinemann, New York.

Mishnaevsky Jr, L., Lippmann, N., Schmauder, S., Gumbsch, P., 1999. In-situ observation of damage evolution and fracture in $AlSi_7Mg_{0.3}$ cast alloys. Eng. Fract. Mech. 63 (4), 395–411.

Mistree, F., Hughes, O.F., Bras, B., 1993. Compromise decision support problem and the adaptive linear programming algorithm. Prog. Astronaut. Aeronaut. 150 251-251.

Nellippallil, A.B., Berthelson, P.R., Peterson, L., Prabhu, R., 2020. Head and neck injury based robust design for vehicular crashworthiness, ASME Design Automation Conference Paper Number IDETC2020-22539.

Nellippallil, A.B., Berthelson, P.R., Peterson, L., Prabhu, R.K., 2021. A computational framework for human-centric vehicular crashworthiness design and decision-making under uncertainty. ASCE-ASME J. Risk Uncertainty Eng. Syst. Part B: Mech. Eng. Under review.

Nellippallil, A.B., Mohan, P., Allen, J.K., and Mistree, F., 2018a, "Robust concept exploration of materials, products and associated manufacturing processes." ASME IDETC, Paper Number: DETC2018-85913.

Nellippallil, A.B., Mohan, P., Allen, J.K., Mistree, F., 2019. Inverse thermo-mechanical processing (ITMP) design of a steel rod during hot rolling process, ASME Design Automation Conference.

Nellippallil, A.B., Rangaraj, V., Gautham, B., Singh, A.K., Allen, J.K., Mistree, F., 2018b. An inverse, decision-based design method for integrated design exploration of materials, products, and manufacturing processes. J. Mech. Des. 140 (11), 111403.

Nellippallil, A.B., Rangaraj, V., Gautham, B.P., Singh, A.K., Allen, J.K., Mistree, F., 2017. A goal-oriented, inverse decision-based design method to achieve the vertical and horizontal integration of models in a hot-rod rolling process chain, ASME Design Automation Conference.

Nie, B., Xia, Y., Zhou, Q., Huang, J., Deng, B., Neal, M., 2013. Response surface generation for kinematics and injury prediction in pedestrian impact simulations. SAE Int. J. Transport. Saf. 1 (2), 286–296.

Noureddine, A., Eskandarian, A., Digges, K., 2002. Computer modeling and validation of a hybrid III dummy for crashworthiness simulation. Math. Comput. Modell. 35 (7-8), 885–893.

Pan, F., Zhu, P., Chen, W., Li, C.-z., 2013. Application of conservative surrogate to reliability based vehicle design for crashworthiness. J. Shanghai Jiaotong Univ. (Sci.) 18 (2), 159–165.

Panzer, M.B., Cronin, D.S., 2009. C4–C5 segment finite element model development, validation, and load-sharing investigation. J. Biomech. 42 (4), 480–490.

Panzer, M.B., Fice, J.B., Cronin, D.S., 2011. Cervical spine response in frontal crash. Med. Eng. Phys. 33 (9), 1147–1159.

Pelenyte-Vyšniauskiene, L., Jurkauskas, A., 2007. The research into head injury criteria dependence on car speed. Transport 22 (4), 269–274.

Prasanna, V., 2015. Development of Response Surface Data on the Head Injury Criteria Associated with Various Aircraft and Automotive Head Impact Scenarios MS Thesis. Wichita State University.

Puttick, K., 1959. Ductile fracture in metals. Philos. Mag. 4 (44), 964–969.

Rabotnov, Y.N., 2021. Paper 68: on the equation of state of creep, Proceedings of the Institution of Mechanical Engineers, Conference Proceedings. SAGE Publications Sage UK, London, England 2-117-112-122.

Raj, R., Ashby, M., 1975. Intergranular fracture at elevated temperature. Acta Metall. 23 (6), 653–666.

Rice, J.R., Tracey, D.M., 1969. On the ductile enlargement of voids in triaxial stress fields. J. Mech. Phys. Solids 17 (3), 201–217.

Rincon, E., Lopez, H., Cisneros, M., Mancha, H., 2009. Temperature effects on the tensile properties of cast and heat treated aluminum alloy A319. Mater. Sci. Eng.: A 519 (1-2), 128–140.

Schmitt, K.-U., Niederer, P.F., Muser, M.H., Walz, F., 2010. Trauma Biomechanics-Accidental Injury in Traffic and Sport. Springer Science & Business Media, Springer-Verlag Berlin Heidelberg.

Soltis, S., 2001. An overview of existing and needed neck impact injury criteria for sideward facing aircraft seats, Proceedings of the Third Triennial International Aircraft Fire and Cabin Safety Research Conference. Atlantic City, NJ, 22–25.

Takhounts, E.G., Eppinger, R.H., Campbell, J.Q., Tannous, R.E., Power, E.D., and Shook, L.S., 2003, "On the development of the SIMon finite element head model," SAE Technical Paper.

Takhounts, E.G., Ridella, S.A., Hasija, V., Tannous, R.E., Campbell, J.Q., Malone, D., Danelson, K., Stitzel, J., Rowson, S., and Duma, S., 2008, "Investigation of traumatic brain injuries using the next generation of simulated injury monitor (SIMon) finite element head model," SAE Technical Paper.

Tay, Y.Y., Moradi, R., Lankarani, H.M., 2021. A response surface methodology in predicting injuries to out-of-position occupants from frontal airbags, Proceedings of the ASME 2014 International Mechanical Engineering Congress and Exposition. American Society of Mechanical Engineers V012T015A027-V012T015A027.

Toroyan, T., Laych, K., 2015. Global Status Report on Road Safety 2015. World Health Organization, Geneva.

Tvergaard, V., Needleman, A., 1984. Analysis of the cup-cone fracture in a round tensile bar. Acta Metall. 32 (1), 157–169.

Versace, J., 1971, "A review of the severity index," No. 0148-7191, SAE Technical Paper.

Wimmer, P., Benedikt, M., Huber, P., Ferenczi, I., 2015. Fast calculating surrogate models for leg and head impact in vehicle–pedestrian collision simulations. Traffic Inj. Prev. 16 (Suppl. 1), S84–S90.

Yeh, J.-W., Liu, W.-P., 1996. The cracking mechanism of silicon particles in an A357 aluminum alloy. Metall. Mater. Trans. A 27 (11), 3558–3568.

Zaouk, A., Marzougui, D., Bedewi, N., 2000a. Development of a detailed vehicle finite element model part I: methodology. Int. J. Crashworthiness 5 (1), 25–36.

Zaouk, A., Marzougui, D., Kan, C.-D., 2000b. Development of a detailed vehicle finite element model Part II: material characterization and component testing. Int. J. Crashworthiness 5 (1), 37–50.

Zhang, S., Zhu, P., Chen, W., 2013. Crashworthiness-based lightweight design problem via new robust design method considering two sources of uncertainties. Proc. Inst. Mech. Eng. Part C: J. Mech. Eng. Sci. 227 (7), 1381–1391.

Zhu, P., Pan, F., Chen, W., Zhang, S., 2012. Use of support vector regression in structural optimization: application to vehicle crashworthiness design. Math. Comput. Simul 86, 21–31.

CHAPTER 11

Development of a coupled physical–computational methodology for the investigation of infant head injury

M.D. Jones[a], G.A. Khalid[b], Raj K. Prabhu[c]

[a]Cardiff School of Engineering, Cardiff University, Cardiff, United Kingdom [b]Electrical Engineering Technical College, Middle Technical University, Baghdad, Iraq [c]USRA, NASA HRP CCMP, NASA Glenn Research Center, Cleveland, OH, United States

11.1 Introduction

Up to this point in the book, we have focused on multiscale modeling of the brain from an adult perspective. This chapter focuses the effort on infants. As traumatic brain injuries are a major problem for adults, traumatic brain injury is also a major cause of death or permanent invalidity among the very young. Infancy represents the period of greatest brain fragility and vulnerability; it is also when the brain undergoes its most rapid development. In clinical practice, head-injured infants and young children are difficult to assess since clinicians are too often provided with the only primary signs and symptoms of an unwitnessed incident or a brief third-party description. It is, therefore, difficult to establish a sufficiently detailed understanding of the "cause-and-effect relationship to predict and mitigate potential pathophysiological consequences. Current medical training of the causes and effects of infant head injuries rely on evidence drawn primarily from clinical experience, radiology, postmortem, and population-based epidemiology to inform the diagnostic and management process. Current practice, however, has not and will not provide the necessary breakthrough since there are many age-dependent and biomechanical variables that are poorly understood at the bedside and require further consideration, including impact location geometry, tissue response, mass, velocity, angle, location of head impact and surface response characteristics.

Experimentation on living infants is inconceivable, and postmortem human surrogate (PMHS) experimentation, a logical alternative, is extremely rare, due largely to limited access as a consequence of moral and ethical considerations. A few infant cadavers have been quantitatively tested during short-fall impacts and provide essential data with which subsequent analyses can be validated (Prange et al., 2004; Loyd, 2011). While the PMHS studies provide advancement in infant's head impact understanding, they have many shortcomings: being only four in number

Multiscale Biomechanical Modeling of the Brain.
DOI: https://doi.org/10.1016/B978-0-12-818144-7.00011-6
Copyright © 2022 Elsevier Inc. All rights reserved.

(1, 3, and 11 days and 9 months old), impacted multiple times, from only two fall heights (0.15 m and 0.30 m), at five (only "generally described") locations, at a single angle 90° to a single surface (steel force plate). Furthermore, out of experimental necessity, acceleration, the current key correlated parameter for a head injury, was derived, by calculation by dividing the force, measured at the impact surface by the head mass. This limited global, "rigid body" approximation provides only a single generalized head impact response value, rather than specific regional or localized acceleration responses. Thus, any injury prediction strategies are currently incapable of representing the significant complexities associated with infant head impacts. An overriding limitation is the technical complexity of measuring localized head responses.

Modeling appears a logical way to advance understanding of traumatic head loading, since it offers significant potential to perform simulations and parametric and multivariate analyses. With respect to adult head modeling, significant advances are being made in brain multi-physics. The general properties of the skull (neuro cranium) have been characterized and are becoming better understood. With respect to infants and young children, the skull is not currently well understood, and since the infant skull is so pliable, a model is an essential prerequisite to subsequent brain modeling.

Physical and computational models have been developed to combine known anthropometry and material properties; however, with respect to physical modeling, materials, and manufacturing technologies have proved limiting. Anthropomorphic test devices have been used to act as surrogates for PMHS testing; however, infant head anthropomorphic test devices are based on scaled animal and adult response data and since the pediatric head response is poorly characterized, their specific validity is largely unknown. Research-based physical models each have significant limitations, either failing to adequately represent the anatomical complexity or measuring the head as if it were a "rigid body," the impact being measured at one point, the impact surface. This approach is incapable of providing localized area-specific details with respect to head response.

A paucity of dynamic property, physiological response, and injury threshold data pose a significant limitation. Thus, any impact response of either a physical or computational head model will be dependent on the material properties used to develop them. Consequently, for physical or computational models to work as biofidelic surrogates, their material response properties have to mimic those of a human infant.

In addition to an absence of whole head response data, there are very few material response studies relating specifically to infants. Pediatric cranial bone is, in actuality, a very thin, heterogeneous, and highly curved material with a distinctly orientated fiber pattern. At birth, the cranial bones have a visible fiber orientation, due to the bone trabeculae radiating from a center of ossification, producing dissimilarity in the elastic modulus between tangential and radial fiber orientation. Young's modulus is known to increase with age and the parietal bone, across all ages, is nearly three times stiffer when grain fibers are oriented parallel to the long

Development of a coupled physical–computational methodology for the investigation 179

axis of the specimen (McPherson and Kriewall, 1980a). This variance in the elastic modulus between the parallel and perpendicular specimens confirms significant anisotropic material proprieties (orthotropic appearance) in the immature cranial bone. Margulies and Thibault (2000) performed three-point bending tests on parietal bone specimens from four human infant cadavers with fibers running perpendicular to the long axis of the specimen, at two different strain rates, 2.54 mm/min and 2540 mm/min. The data is not, however, transferable to the larger strain rates in excess of 2.54 m/s. Thus, to determine injury risk and develop effective injury interventions at higher rates, it is crucial that relevant rates be investigated.

To obtain material properties for infant cranial bone, at strain rates more appropriate to domestic scenarios, Coats and Margulies (2006a) assessed the effect of rate dependency of infant cranial bone and sutures at strain rates, 1.58 and 2.82 m/s, respectively. This study was fundamental, since it provided age-specific material properties for infants, which was used in this present study for the development and justification of both physical and finite element (FE) infant head models. With respect to the material properties of pediatric cranial sutures, no significant effect of age or strain rate on the modulus of elasticity, the ultimate stress, and ultimate strain was reported at the coronal suture.

Chatelin et al. (2012) provided the first comparison of human brain tissue at different ages (from 2 months to 50 years) and brain regions. While the tests were performed at nondestructive strain rates, thus limiting the direct relevance of the data to pediatric brain injury, the study confirms that brain tissue response is age-dependent as the adult brain samples were 3–4 times stiffer than the infant and young child at low strain rates.

Computational FE models were developed to investigate skull deformation and how it relates to skull fracture or brain injury in the pediatric population. Thibault and Margulies (1998) conducted an investigation into the effects of the infant cranium and sutures in protecting the brain during impact head injury. Their FE simulations incorporated the tissue response data from their experimental tests, which were performed on samples of human infant cranial bone and porcine bone suture. The authors assumed that the suture properties of infant porcine were the same as a human infant; however, Coats and Margulies (2006) later showed it to be between 22 and 80 times stiffer. This study represented an initial step in understanding the mechanical response of cranial bones subjected to traumatic loading and head injury; hence, defining child head injury thresholds.

Roth et al. (2007) subsequently published the study of an FE-model of a 6-month head constructed from CT images to assess the relative dynamic response of the head during impact and shaking. Furthermore, using the 6-month FE-model as a reference, Roth et al. (2008) reported the effect of scaling by comparing the original model with another 6-month model developed from scaling down an adult FE-model. The research highlights that children's heads should not be considered as scaled adult heads, where the output stress has different values and locations. Like Roth et al. (2008) this model was limited by

180 Chapter 11

skull material properties, which were later reported to be too stiff, and the fact this Fe-model was not validated against PMHS response data. Later, Roth et al. (2009) reported a similar geometric analysis of a 3-year-old child, scaled FE-head model and compared its impact response to the same-aged model, accurately constructed from CT images. Similar results were reported to Roth et al. (2008), highlighting a need for geometrically accurate Fe-models. Roth et al. (2009) further modeled a different FE-head model of a 3-year-old child for reconstructing fall-related accidental scenarios to investigate neurological injury. Twenty-five cases were investigated, composed of two groups: those with no neurological lesions and those with severe lesions. Although, the model's von Mises stresses were a good predictor for neurological lesions, still the mechanical properties of the brain at this age and the biofidelity of the model were unknown. Subsequently, while these pediatric FE-head models provide some indication of head injury mechanics, they are limited, since their validity has not been assessed against experimental pediatric PMHS data. Roth et al. (2010) and Li et al. (2013) validated their models against PMHS data from a similar age-matched group and provided a valuable insight into the mechanics of child head injuries; however, a study limitation is that the material properties of the infant cranial bone were modeled as homogeneous and isotropic, using "mean material properties" from the literature (Margulies and Coats, 2006) rather than the heterogeneous, anisotropic material proprieties reported by McPherson and Kriewall (1980b). Since the cranial bones provide a significant contribution to overall head impact response, to provide the greatest degree of biofidelity, it is deemed imperative that the material properties be accurately represented. Recently, Li et al. (2017) developed three pediatric FE-head models representing a newborn, a 5-month, and a 9-month old child. The cranial bones of these models incorporated the orthotropic appearance of skull bones and a nonlinear constitutive law for soft tissues (scalp, sutures, and dura mater) for infant head modeling. However, while the newborn Fe-model was globally validated against Prange et al. (2004) PMHS response data, no localized bone or suture response values were reported.

In an attempt to address the aforementioned experimental shortcomings, Jones et al. (2017) sought to produce a significant advance in infant head injury investigation by employing a new approach, a "coupled-methodology" combining a physical infant head model, producing "real-world" global, regional, and localized impact response data and the subsequent development and validation of a computer FE head model (Khalid et al., 2019).

Jones et al. (2017) exploited recent advances in 3D object printing and developed and validated a physical head model with a high degree of skull anatomical precision and tissue matched response characteristics. The high speed digital image correlation (HS-DIC) of the physical model for the first time provided regional and localized data, which, in combination with the "global response" value provided an opportunity to develop a geometrically accurate computational FE-head model, utilizing the scan data used during the development of the physical model (McPherson and Kriewall, 1980b; Kriewall, 1982) and the material structures reported in Margulies and Thibault (2000) whose responses were matched to published infant human tissue data (McPherson and Kriewall, 1980b; Kriewall, 1982; Margulies and Thibault, 2000). The infant FE-head computationally replicated the fiber orientation of immature cranial bones and therefore, provided an orthotropic response for each bone. To ensure that the FE-head biofidelity corresponded "globally" with that of Prange et al. (2004), it was validated against the response of the real-world PMHS impact tests. Furthermore, to ensure that this FE-head biofidelity corresponded both "globally" and "locally" with that of Jones et al. (2017) study, the FE-head was subjected to a series of global and local simulations against the response of the Jones's physical model.

The novelty of the approach is that it couples real-world physical modeling with computational models. Providing an opportunity to create previously unachievable levels of "micro" physical modeling precision of not only "boney" structures but also membranous, gel, and vascular structures. The model provides a platform for the subsequent development of advanced pediatric models.

11.2 Methods

11.2.1 Pediatric head development

To create high-resolution 3D computer-aided design (CAD) models of the infant head geometries and achieve the highest level of replication, high-resolution 0.35 mm thickness skull and brain postmortem images were obtained from a 10-day old infant. Mimics Innovation Suite (Materialise, Belgium) was applied to the segmentation of clinical data, from Dicom images, to create the 3D CAD head model. Subsequently, the 3D CAD pediatric head (cranial bones, brain, sutures, and fontanelles) was imported from Mimics to 3-matic, where CAD operations were performed and a successful FE analysis mesh developed and applied to the skull's natural but complex structure and the brain. The final segmented 3D representation of the human infant head was meshed with a second-order tetrahedral mesh, available in Mimics Remesh (3-matic v.10). The nodes of each anatomical structure were tied to those of neighboring structures. Because of complex geometries, this was best achieved by meshing all structures together and sharing surface nodes between adjacent structures. The final segmented 3D representation of the human infant head was meshed with one meshing algorithm available in Mimics, which is a tetrahedral mesh. The tetrahedral mesh algorithm creates a mesh comprised entirely of 10-node tetrahedral focused at anatomical surfaces and interfaces to impart a more realistic shape and contour of the human infant skull and brain. Abaqus Explicit was chosen, since it is both suitable for high impact deformation and it can numerically support the creation of radially anisotropic properties, as shown in Fig. 11.1. To further replicate the physical model (Jones et al., 2017) an oval-shaped pad to represent the physical scalp surrogate was meshed and assembled.

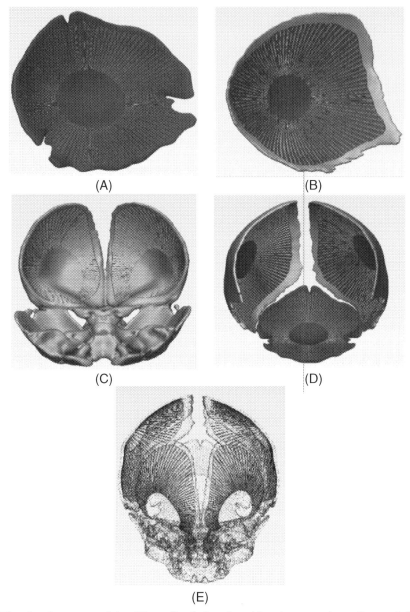

Fig. 11.1: The development of the 3D pediatric head and implementation of material properties by fibrous grooves at the cranial bones model: (A) occipital bone, (B) right parietal bone, (C) left parietal bone, (D) frontal bones, and (E) an assembly of the cranial bones.

11.2.2 Material properties

The infant physical model consists of cranial bones, fontanelles, and sutures (Jones et al., 2017). The cranial cavity was printed using material response matched polypropylene polymers for the frontal, parietal and occipital bones, sutures, and fontanelles. The moduli of

Development of a coupled physical–computational methodology for the investigation **183**

elasticity of the polypropylene polymers were 2300 MPa for the frontal bones, 1400 MPa for the parietal and occipital bones, and 7.2 MPa for the sutures and fontanelles (Yoganandan et al., 2004). A comparison of these three polymeric materials was made with infant tissue mechanical properties, including the variation for different cranial bones. The materials response values were considered to be appropriate for modeling the infant cranium under the experimental impact test condition. Since the constitutive properties of the physical-model cranial bones, brain, sutures, fontanelles and scalp all influence the total system response to mechanical loading, the FE tissue response properties must correspond with similar areas of physical model.

Correspondingly, the cranial bones of this FE model were modeled to be in-plane, orthotropic, and with different elastic moduli in the parallel and perpendicular directions, relative to the fibers. The material constant, used to define orthotropic material, was calculated using a mathematical equation in Abaqus. The material constants used in the FE-model were the elastic moduli, the shear moduli and Poisson's ratio, with a parallel and perpendicular fiber direction and direction perpendicular to both directions, shown in Table 11.1.

The ratio of parietal bone stiffness, between parallel and perpendicular, was 4.2–1.0 and frontal bone stiffness was 1.8–1.0 (McPherson and Kriewall, 1980b). No test data were available for the occipital bones, thus, a ratio of 4.2–1.0 was assumed, equivalent to the parietal bones. The cranial bone material orientation in the FE-head, compared to reality and further replicating the fibrous appearance of the cranial bones in the physical model. To model the brain, a surrogate brain material, gelatin was used in the 3D printed head in accordance with a previous study by Cheng et al. (2010), which conducted a dynamic tensile test and modeled the gelatin as a brain surrogate material. A latex rubber and polyamide micro fleece scalp,

Table 11.1: Of the material contact properties used to define the pediatric infant head FE model (Khalid, 2018).

FE head model structure	Material properties
Cranial bones	$p = 1896 \ kg/m^3$
Occipital bone	$E_1 = 1412 \ MPa, E_2 = 336 \ MPa, E_3 = 336 \ MPa$ $v_{12} = 0.19, v_{13} = 0.045, v_{23} = 0.22, v_{21} = 0.045$ $G_{12} = 281 \ MPa, G_{13} = 312 \ MPa, G_{23} = 137 \ MPa$
Parietal bone	$E_1 = 2201 \ MPa, E_2 = 524 \ MPa, E_3 = 524 \ MPa$ $v_{12} = 0.19, v_{13} = 0.045, v_{23} = 0.22, v_{21} = 0.045$ $G_{12} = 439 \ MPa, G_{13} = 513.3 \ MPa, G_{23} = 214 \ MPa$
Frontal bone	$E_1 = 2201 \ MPa, E_2 = 1222 \ MPa, E_3 = 1222 \ MPa$ $v_{12} = 0.19, v_{13} = 0.11, v_{23} = 0.22, v_{21} = 0.11$ $G_{12} = 627 \ MPa, G_{13} = 750.6 \ MPa, G_{23} = 500 \ MPa$
Sutures and fontanelle	$p = 1025 \ kg/m^3, E = 8.1 \ MPa, v = 0.49$
Brain	$p = 997 \ kg/m^3, E = 0.0272 \ MPa, v = 0.499$
Scalp	$p = 399 \ kg/m^3, E = 0.42 \ MPa, v = 0.42$

representing the elastic and frictional properties (Jones et al., 2016) of the scalp, was modeled and applied to the outer contact surface of the skull, while leaving the skull surface uncovered to permit the application of random speckle pattern for HS-DIC analysis of the physical model for subsequent derivation of local/regional response data (Jones et al., 2017).

11.2.3 Mesh convergence

Meshing the high-resolution CT images produced a large number of elements and required long processing times. Thus, a convergence study was performed to determine the coarsest mesh required to produce accurate results, as shown in Fig. 11.2. A threshold mesh resolution of 20×10^5 was applied, producing a minimal residual error and significantly shorter computational time. The final mesh consisted of 2,626,855 second-order modified tetrahedral elements.

11.2.4 Boundary and loading conditions

Two boundary conditions were considered essential to the biofidelic response of the FE-model, the cerebrospinal fluid (CSF) and the foramen magnum. The CSF was represented by a contact algorithm approach (sliding interface) at the skull-brain interface. The foramen magnum of the FE-head model was developed as a "force free" opening (Coats et al., 2007), thus having no restraints.

The pediatric FE-head skull thickness was assumed to be constant, in accordance with the newborn FE-head reported by Coats et al. (2006) and Roth et al. (2007), reported thicknesses values of 1.2 and 1.4 mm, respectively. The values applied to the present study are in accordance with the 2 mm reported by Li et al. (2013).

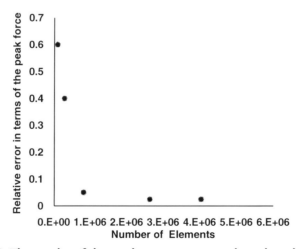

Fig. 11.2: The results of the mesh convergence study and application of boundary and loading conditions (Khalid, 2018).

11.2.5 Global validation of the FE-head against PMHS

The PMHS study of Prange et al. (2004) provided sufficient data to characterize the "global" impact behavior of the pediatric head, which was applied to validating the response of the present infant FE-head by simulation. Eight different impact scenarios were simulated and compared against the kinematic response data from Prange et al. (2004). Two predefined impact velocities were applied (1.72 m/s and 2.43 m/s), corresponding to the drop height of 0.15 m or 0.30 m, respectively. The impact locations in FE simulations were, anterior–posterior, forehead impact, left–right, parietal impact, posterior–anterior, occipital impact, and superior–inferior vertex impacts.

11.2.6 Global, regional, and local validation of the FE-head against the physical model

The physical model (Jones et al., 2017) was subjected to "global" validation against the PMHS data (Prange et al., 2004). Similarly, the computational FE-head was subjected to the same "global" validation. During validation, the physical model was subjected to measurement by HS-DIC, which generated regional and local measurement data. Thus, the FE-head could be subjected to additional validation against the physical-model regional and local data. Provision of regional and local validation of the FE-head, for the first time, provided the opportunity to, in combination with the "global response," significantly improve the understanding of skull impact mechanics. Applying the data from Jones et al. (2017) and measuring the acceleration and deformation at different regions of the head, comparisons were made with the response of the physical model. "Global" validation of the FE-head was achieved. Regional and local validation was possible from the physical-model data, since the HS-DIC system provided the translational impact response, including the distribution of the strain throughout the different structures of the head (Jones et al., 2017), as shown in Fig. 11.3. From the translational acceleration response of the physical model, a series of computational FE simulations were conducted, replicating the different orientations of the head. A correlation was established between the regional and local impact response, to obtain the important features of the physical-model impact tests and the FE analyses.

11.2.7 Statistical analysis

A statistical analysis was performed to assess the agreement between the FE-model prediction and the measured acceleration–time impact curves, in terms of peak values (the differences between the peak values of the physical and FE-model experiments) and the correlation score (CS; difference between the physical and FE-model experiments using the CS measured at 1.72 m/s and 2.43 m/s). For the purpose of global validation, two indicators were used to compare the head response of the FE-head, with both PMHS and physical-model impact response: (a) peak values and (b) CS (Li et al., 2017; Kimpara et al., 2006; De Lange et al., 2005). The first measure used the peak values at discrete spatial points, whereas the second

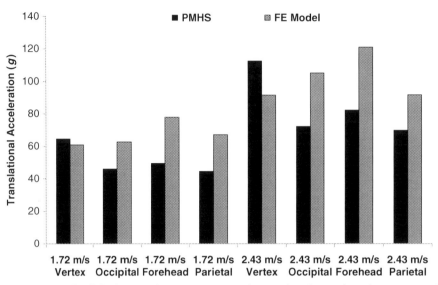

Fig. 11.3: FE model validation against PMHS experimental cadaver data (Prange et al., 2004).

measures used the "full-field" spatial information, signified by the two statistical measures RMS error and CS, between the model and the experiment for model evaluation. The function of the models was evaluated using CS, to assess the agreement between the model prediction and the measured acceleration–time impact curves, in terms of phase, amplitude, and shape (Li et al., 2017; Kimpara et al., 2006; De Lange et al., 2005). The CS values ranged from 0 to 100, which are classified according to a biofidelity rating (Miller et al., 1998). The qualitative index was categorized into five classifications: excellent: 86–100, good: 65–86, fair: 44–65, marginal: 26–44, and unacceptable: 0–26. The error measures were used to calculate a CS value, two discrete-time histories, and four correlation functions were applied. In a similar manner to the physical-model HS-DIC approach (Jones et al., 2017), the local kinematic variables (translational acceleration, duration of impact), head deformation (strain), and stress distribution were measured from discrete points on the skull. The output variables of the head were measured in four directions, anterior–posterior, forehead impact, posterior–anterior, occipital impact, left–right, parietal impact, and superior–inferior vertex impact. All the analyses performed in this study assumed that a 15% difference in a variable's response was a threshold trigger for consideration (Coats et al., 2007; Roth et al., 2007, 2010).

11.3 Results and discussion

11.3.1 Global validation of the FE-head versus the postmortem human surrogate

To ensure that the FE-head biofidelity corresponded with that of PMHS impact experiments (Prange et al., 2004), it was subjected to a series of validation simulations, in terms of the acceleration–time response under impact loading conditions. The impact–acceleration

Development of a coupled physical–computational methodology for the investigation   187

response was obtained by dividing the impact force by the head mass to produce impact translational accelerations and durations. Four different impact locations and the two impact velocities produced a CS equal to, or greater than, 86, according to the biofidelity rating (De Lange et al., 2005).

11.3.2 Global validation of the FE-head versus the physical model

Global validation of the FE-head was achieved by comparing its simulated impact response against the physical model, considered to provide a "real-world" reference, due to its geometrical and material response similarities. The acceleration–time curves generally correlated well, there were no significant differences between acceleration response at different impact locations or velocities, suggesting mechanical similarity. The CSs were all above the threshold level of 86. Differences, expressed as a percentage of peak response, were 2% at the vertex, 27% at the occipital bones, 29% at the frontal bones, and 3% at the parietal bones at 1.27 m/s; compared to the variation in acceleration response at 2.43 m/s, represented by 6% (vertex), 11% (occipital), 3% (frontal) and 9% (parietal). While the variation at 1.27 m/s at both the vertex and the occipital bone was greater than the 15% threshold, difficulties reproducing the exact impact location between the head and the impact surface in the simulation and the physical test provided a potential source of difference and justification for the computational FE approach, since impact positions can be more precisely regulated.

A further reason for differences may in part be due to the fact that the brain and skull are a continuum. Thus, it could be that higher accelerations, away from the point of impact could be a result of the mass of the brain acting directly in compression on bones close to the impact point; such that the "effective mass" of that area of the skull is greater and, therefore, accelerating less, for the same force, than an area where the brain is being moved away from the bone during the impact.

Another difference is that physiologically, CSF facilitates a very low friction coupling between the brain and the skull. The "brain" in the physical model consisted of gel contained in a rubber balloon (Cheng et al., 2010). Though relatively unconstrained, a gel boundary layer would still exist at the inner surface of the balloon; how accurately this would represent the CSF is, at present, unknown. The FE-head may demonstrate a similar limitation, albeit for a different reason, due to the use of tie constraints anchoring the surface of the brain and skull, potentially producing a stiffer response. The total impact durations of the FE-head were 5 milliseconds at 1.72 m/s and 6 milliseconds at 2.43 m/s and are very close to the physical model, which was 6 milliseconds and 7 milliseconds, respectively; a difference of less than 15%. There is no significant difference in impact duration of greater than 15% between the FE-head simulations and the physical-model tests at both impact velocities.

188   Chapter 11

11.3.3 FE-head regional and local validation versus the physical model

Acceleration is considered the most practically achievable measure for establishing a correlation between head impact loading and injury. Measurement can be achieved either by impacting the head against a force plate or instrumenting the head with transducers. To practically consider that an acceleration–time response from an impact force plate represents a head impact response, a "solid body assumption" is essential. However, this assumption can only be considered to hold true while the skull and brain can be considered to behave as a single structure. With respect to the infant head, the loosely associated skull bones, sutures, and fontanelles, are of relatively low mass, compared to the massive brain (Burdi et al., 1969). Thus, the reliance on a single head acceleration value cannot provide a sufficient level of response understanding when considering the movement of the pediatric skull relative to the brain. Given the flexible nature of the pediatric head, it must be an assumption that different areas of the head will accelerate at relatively different rates. The rigid body assumption represents an acceleration "summation" at the point of impact; it is therefore, perhaps more likely to disproportionately represent the acceleration of the most massive part of the structure, that is, the brain. The use of HS-DIC was applied to the physical model for measuring displacements at the skull surface and henceforth, velocity and acceleration directly from different parts of the head, without having to apply the global approximation approach. Subsequent to global validation of the FE-head, with the physical-model impact response data, FE simulation was further applied to investigating the observations of the physical model and supplement the "global approximation" approach. Unique to this present study, the FE-head was applied to providing specific regional and localized response data at different locations on the surface of the skull. The FE-head was subjected to translational head impact, from which local acceleration responses, from different regions of the head, were compared with first the global acceleration response (force/mass) from the FE-head, obtained using the "global approximation" approach; second, the FE-head local impact response was compared and validated against the local response from the physical-model response data from the HS-DIC system (Jones et al., 2017), as shown in Fig. 11.4.

From the comparison of the global and local acceleration response of the FE-head, the relative local translational accelerations, show a variation greater than 15%, and greater than those calculated using the "global approximation" approach. During perpendicular impacts, regional and local accelerations in the skull plates, sutures, and fontanelles were significantly greater than the calculated peak global acceleration response of the FE-head applying the rigid body approach, at the impact velocity of 1.72 m/s by 18% at the vertex, 59% at the occipital bone, 55% at the frontal bone, and 53% at the parietal regions and at the impact velocity 2.43 m/s, 7% (vertex), 28% (occipital), 45% (frontal), and 22% (parietal). There is, therefore, a major observable difference between the FE-head response (skull structures) and the "global" acceleration. The peak acceleration in the FE frontal, occipital,

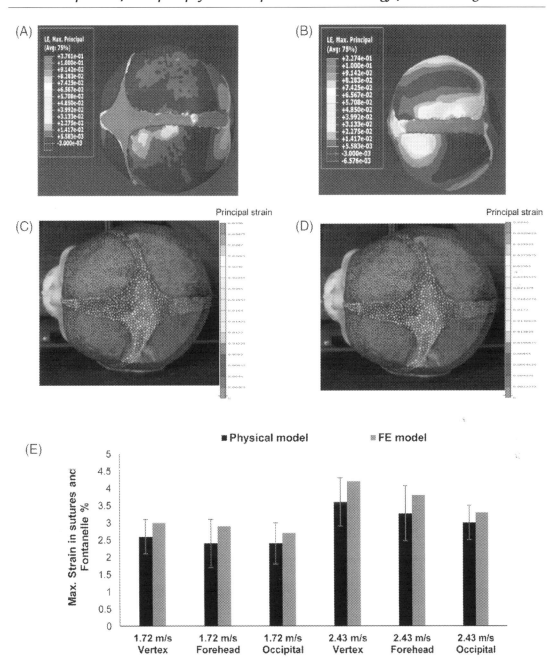

Fig. 11.4: Coronal view of the FE-head at (A) the vertex and (B) the occipital bone impact and the comparative coronal view of physical head deformation for a 2.43 m/s impact onto a metal plate using HS-DIC, (C) 3 milliseconds postimpact and 0.032 mm deformation, (D) 5 milliseconds postimpact and 0.034 mm deformation, (E) maximum-strain of the physical and FE-heads at different impact locations and two impact velocities.

190 Chapter 11

and parietal bones, at 1.27 m/s is nearly three times greater than the "global" response of frontal, occipital, and parietal bones reported by the FE-head. Interestingly, the greatest percentage differences were during occipital impacts, where at 1.72 m/s the difference in the recorded bone acceleration was 59% greater than the "global" value. The smallest increase was observed at 2.43 m/s during the impact onto the vertex, the difference, 7%, was probably a result of the acceleration occurring over a significantly shorter time, which produces a greater peak acceleration than reported by the FE-head using "global approximation" approach.

In contrast, the FE-head calculated accelerations across the entire skull surface; further confirming that the "global approximation" approach (Prange et al., 2004) can be supplemented with regional\local response information, since the cranial bones, sutures and fontanelles move relative to one another. Thus, the FE-head shows agreement with the physical-model study, which reported that acceleration patterns are very sensitive to impact location. It is evident that the local acceleration values from different regions of the FE-head show a good correlation to those of the physical-model–HS-DIC study (Jones et al., 2017). There was no variation in local impact acceleration response greater than 15%, that is, at the impact velocity of 1.72 m/s, 11% at the vertex, 4% at the occiput, 6% at the frontal, and 13% at the parietal regions and at the impact velocity of 2.43 m/s by 12% (vertex), 3% (occiput), 5% frontal, and 11% at the parietal region. As discussed above, this is unsurprising, since the mass of the infant skull bones is relatively small compared with the overall mass of the head (Burdi et al., 1969). The bones are, therefore, likely to come to rest more quickly than the more massive brain. From the comparison of the impact duration between the physical model and present FE-head, a little variation exists between them, when provided with the same input parameters, represented at 1.72 m/s by 3% at the vertex, 11% at the occiput, 13% at the frontal bones, 10% at the parietal regions and at 2.43 m/s by 4% (vertex), 11% (occiput), 10% (frontal), and 8% (parietal); since both have the same material properties and exhibit the same stiffness.

11.3.4 Head deformation

During impact, the majority of the strain was produced within the suture and fontanelle areas of the FE-head of the order of 0.3% and 3%, at 1.72 m/s and 2.43 m/s, respectively. The maximum-strain percentage of the FE-head from the vertex, frontal, and occipital impacts was compared to the physical model (Jones et al., 2017) from the same regions, showing no significant difference between the FE and physical models at both impact velocities. The maximum percentage difference was 4% at 1.72 m/s, and the minimum was 2% at 2.43 m/s. Since both models possess the same mass and the same applied material properties, they demonstrate the same material stiffness during impact.

11.4 Summary

The aim of this study was to develop a coupled physical–computational modeling framework for investigating and improving the biomechanical understanding of pediatric head impact injury mechanics. The study presents the development and validation of a computational FE-head model, adapting and including the shape, structure, and material properties of a physical head model. The FE-head model has been demonstrated to simulate relevant head loading parameters, which are in good agreement with the physical model's global and local acceleration and strain data within the sutures and fontanelles.

References

Burdi, A., Huelke, H., Snyder, R., Lowrey, G., 1969. Infants and children in the adult world of automobile safety design: pediatric and anatomical considerations for design of child restraints. J. Biomech. 2, 267–280.

Chatelin, S., Vappou, J., Roth, S., Raul, J., Willinger, R., 2012. Towards child versus adult brain mechanical properties. J. Mech. Behav. Biomed. Mater. 6, 166–173.

Cheng, J., Howard, I., Rennison, M., 2010. Study of an infant brain subjected to periodic motion via a custom experimental apparatus design and finite element modelling. J. Biomech. 43 (15), 2887–2896.

Coats, B., Margulies, S., 2006. Material properties of human infant skull and suture at high rates. J. Neurotrauma 23 (8), 1222–1232.

Coats, B., Ji, S., Margulies, S., 2007. Parametric study of head impact in the infant. Stapp. Car Crash J. 51, 1–15.

De Lange, R., Vna Rooij, L., Mooi, H., Wismans, J., 2005. Objective biofidelity rating of a numerical human occupant model in frontal to lateral impact. Stapp. Car Crash J. 49, 457–479.

Jones, M., Darwall, D., Khalid, G., Prabhu, R., Kemp, A., Arthurs, O., Theobald, P., 2017. Development and validation of a physical model to investigate the biomechanics of infant head impact. Forensic Sci. Int. 276, 111–119.

Jones, M., Oates, B., Theobald, P., 2016. Quantifying the biotribological properties of forehead skin to enhance head impact simulations. Biosurf. Biotribol. 2 (2), 75–80.

Khalid, G. 2018. A coupled physical-computational methodology for the investigation of short fall related infant head impacts. PhD Thesis, Cardiff University.

Khalid, G., Prabhu, R., Arthurs, O., Jones, MD., 2019. A coupled physical-computational methodology for the investigation of short fall related infant head impact injury. Forensic Sci. Int. 300, 170–186.

Kimpara, H., Nakahira, Y., Iwamoto, M., Miki, K., 2006. Investigation of anteroposterior head–neck responses during severe frontal impacts using a brain–spinal cord complex FE model. Stapp. Car Crash J. 5, 509–544.

Kriewall, T., 1982. Structural, mechanical, and material properties of fetal cranial bone. Am. J. Obstet. Gynecol. 143, 707–771.

Li, X., Sandler, H., Kleiven, S., 2017. The importance of nonlinear tissue modelling in finite element simulations of infant head impacts. Biomech. Model. Mechanobiol. 16 (3), 823–840.

Li, Z., Luo, X., Zhang, J., 2013. Development/global validation of a 6-month-old pediatric head finite element model and application in investigation of drop-induced infant head injury. Comput. Methods Programs Biomed. 112 (3), 309–319.

Loyd, A., 2011. Studies of the Human Head from Neonate to Adult: An Inertial, Geometrical and Structural Analysis with Comparisons to the ATD Head. PhD Thesis, Dissertation, Duke University.

Margulies, S., Coats, B., 2006. Material properties of porcine parietal cortex. J. Biomech. 39, 2521–2525.

Margulies, S., Thibault, K., 2000. Infant skull and suture properties: measurements and implications for mechanisms of pediatric brain injury. J. Biomech. Eng. 122, 364.

McPherson, G., Kriewall, T., 1980a. Fetal head molding: an investigation utilizing a finite element model of the fetal parietal bone. J. Biomech. 13 (1), 9–16.

192   Chapter 11

McPherson, G., Kriewall, T., 1980b. The elastic modulus of fetal cranial bone: a first step towards an understanding of the biomechanics of fetal head molding. J. Biomech. 13 (1), 9–16.

Miller, R., Margulies, S., Leoni, M., Nonaka, M., Chen, X., Smith, D., Meaney, D. 1998, Finite Element Modeling Approaches of Predicting Injury in an Experimental Model of Severe Diffuse Axonal Injury, Report No.: 0768002931.

Prange, M., Luck, J., Dibb, A., Van Ee, C., Nightingale, R., Myers, B., 2004. Mechanical properties and anthropometry of the human infant head. Stapp. Car Crash J. 48, 279.

Roth, S., Raul, J.-.S., Ludes, B., Willinger, R., 2007. Finite element analysis of impact and shaking inflicted to a child. Int. J. Legal Med. 121 (3), 223–228.

Roth, S., Raul, J., Willinger, R., 2008. Biofidelic child head FE model to simulate real world trauma. Comput. Methods Programs Biomed. 90 (3), 262–274.

Roth, S., Raul, J.S., Willinger, R., 2010. Finite element modelling of pediatric head impact: global validation against experimental data. Comput. Methods Programs Biomed. 99 (1), 25–33.

Roth, S., Vappou, J., Raul, J., Willinger, R., 2009. Child head injury criteria investigation through numerical simulation of real world trauma. Comput. Methods Programs Biomed. 93 (1), 32–45.

Thibault, K., Margulies, S., 1998. Age-dependent material properties of the porcine cerebrum: effect on pediatric inertial head injury criteria. J. Biomech. 31 (12), 1119–1126.

Yoganandan, N., Derosia, J., Humm, J., 2004. An improved method to calculate paediatric skull fracture threshold, *National Highways Traffic Safety Administration*, 23rd International Technical Conference on the Enhanced Safety of Vehicles (ESV), 1–8.

CHAPTER 12

Experimental data for validating the structural response of computational brain models

A. Alshareef, J.S. Giudice, D. Shedd, K. Reynier, T. Wu, M.B. Panzer

School of Engineering, University of Virginia, Charlottesville, VA, United States

12.1 Introduction

Traumatic brain injuries (TBIs) are one of the most common yet least understood injuries to the body. According to the World Health Organization, there has been an increase in global TBI incidences, and it is expected to become the third leading cause of death by 2020 (Meaney et al., 2014). In the United States, an estimated 1.7 million TBIs occur annually, and TBI is a contributing factor in one-third of all injury-related deaths (Taylor et al., 2017).

The study of brain biomechanics and the associated models to predict injury is important for understanding and mitigating brain injuries. In the past two decades, various computational models of the brain have been developed to assess brain injury risk and to evaluate safety gear. As discussed in previous chapters, finite element (FE) brain models allow for investigation into brain mechanics that is not possible with dummies or physical models. They have been vital to reconstructing real-world impacts, assessing injury risk and safety gear, and developing brain injury criteria based on head impact kinematics (Gabler et al., 2016, 2017; Sanchez et al., 2017; Takhounts et al., 2013). There is, however, a need for the improvement of the biofidelity of these models under impact conditions relevant to diffuse TBI, such as concussion or diffuse axonal injury.

Improvements in computational capabilities and the generation of datasets of human brain deformation have allowed for the development of numerous FE brain models. Since 2001, there have been at least 16 models developed (Giudice et al., 2019), 12 within the past 5 years (Fig. 12.1). Given the importance of these models in influencing standards of safety and product development across multiple industries, it is essential to validate the brain deformation predicted by these models using human brain motion under repeatable loading conditions that are causative of injury. However, the FE models in the literature are evaluated against a small set of available datasets of brain deformation, which are limited in their ability to fully validate model results.

Multiscale Biomechanical Modeling of the Brain.
DOI: https://doi.org/10.1016/B978-0-12-818144-7.00009-8
Copyright © 2022 Elsevier Inc. All rights reserved.

193

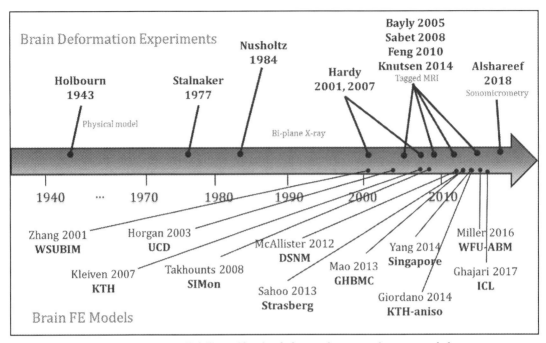

Fig. 12.1: The availability of brain deformation experiments and the development of FE brain models (Alshareef, 2019).

Various techniques have been utilized to study in vivo and in situ human brain motion in response to motion of the head. Most of the FE models in the literature are validated based on two datasets available for brain models: the Hardy brain motion datasets (Hardy et al., 2001, 2007) and the preceding brain pressure datasets by Nahum et al. (1977) and Trosseille et al. (1992). While the pressure datasets provide a reasonable metric to verify the correct implementation of the brain materials, they are not useful as a validation dataset for brain deformation nor are they useful from an injury standpoint. The Hardy dataset provides validation data for human brain deformation under impact conditions, but there are many limitations associated with the study.

There have been recent developments in the acquisition of brain deformation and brain mechanics using novel imaging techniques to aid in the structural validation of computational brain models. A recent study by the University of Virginia quantified in situ brain deformation using sonomicrometry in cadaveric specimens subjected to rotational loading (Alshareef et al., 2018). The technique provided TBI-relevant and repeatable loading conditions for the creation of a dataset specifically for the purpose of brain model validation. Additionally, studies by Washington University and others have used tagged magnetic resonance imaging (MRI) and magnetic resonance elastography (MRE) (see Chapter 9 of this book) to measure in vivo human brain deformation and material properties at low accelerations (Bayly et al., 2005, 2021; Gomez et al., 2018; Knutsen et al., 2020; Okamoto et al., 2019). These studies demonstrate that the measurement of brain deformation is an emerging science, with challenges associated with the experimental procedures used. This chapter will focus on the importance of validating computational models with all the datasets currently available by outlining the techniques and their advantages and limitations.

12.2 Methods

12.2.1 Experimental brain pressure measurements

In efforts to elucidate the relationship between brain injury and intracranial dynamics, early experiments focused on measuring intracranial pressure in the human cranium. The prominent datasets used to evaluate the pressure response of computational brain models during head impact are experiments by Nahum (1977) and Trosseille (1992). Nahum and Smith (1977) performed a series of sagittal head impacts on seated cadaveric specimens with the objective of correlating brain pressures to head kinematic injury criteria, including the head injury criterion (Versace, 1971). The reported coup intracranial pressures ranged from 23 kPa to 505 kPa for head linear acceleration of 155–397 g. Trosseille et al. (1992) conducted sagittal impact tests on seated cadavers with repressurized vasculature. The reported coup intracranial pressures ranged from 8 kPa to 88 kPa for head linear acceleration of 12–102 g.

The brain pressure experiments found a correlation between head linear acceleration and intracranial pressure. This result is expected due to the high bulk modulus and incompressible nature of the brain (Meaney et al., 2014; Shuck and Advani, 1972), with peak pressures (200–500 kPa) causing very small dilational strains (1e−4) (Bradshaw and Morfey, 2001). In the closed environment of the skull, the brain experiences a pressure gradient aligned with the direction of the acceleration field, which is consistent with hydrostatic pressure theory. When this acceleration field is produced by a direct impact to the head, there is a positive pressure in the brain at the site of impact (sometimes referred to as the *coup pressure*), a negative pressures on the opposite side of the brain (sometimes referred to as the *contracoup pressure*, and a linear gradient of pressure in between (Ommaya et al., 1971). At the time of the experiments by Nahum et al. and Trosseille et al., the impact biomechanics community was focused on assessing brain injury using linear head kinematic injury metrics. In particular, the head injury criterion and the Gadd severity index, which both predict injury based on linear acceleration, were used in assessments of automobiles and sports equipment (Hutchinson et al., 1998; Versace, 1971). The association of brain pressure in response to linear acceleration during impacts to the head supported the use of intracranial pressure as an injury metric and for FE model validation. However, brain pressure and linear acceleration were never correlated to diffuse brain injury or its associated mechanisms (Ommaya et al., 1971; Gennarelli et al., 1972; Ommaya, 1984; Gennarelli, 1993). Subsequently, experimental and computational studies have supported a shift away from brain pressure and linear acceleration to a focus on rotational head kinematics that cause shearing of the brain (Gabler et al., 2016; Gennarelli et al., 1972; Takhounts et al., 2013).

196 Chapter 12

12.2.2 Experimental brain deformation measurements

Various techniques have been utilized to study in situ and in vivo human brain deformation in response to motion of the head, including X-ray, sonomicrometry, and MRI. This section will outline five techniques: high-speed X-ray imaging, dynamic ultrasound, sonomicrometry, tagged MRI (tMRI), and MRE. A discussion will highlight their advantages, limitations, and applicability to structural validation of computational brain models.

12.2.2.1 High-speed X-ray

One approach for investigating in situ brain motion has been high-speed X-ray imaging of radio-opaque markers implanted in the brains of postmortem human surrogates (PMHS). Stalnaker et al. (1977) used lead markers to quantify brain motion during the repressurization of the vasculature and ventricles and showed that the coupling between the brain and skull increased with the increased pressure. Nusholtz et al. (1984) injected a neutral-density radio-opaque gel into the brain to measure brain motion in PMHS using high-speed cineradiography. Frontal impacts using a padded linear impactor on the specimen resulted in head linear accelerations ranging from 25 to 450 g. Minimal brain distortion was observed during these tests, except in a specimen that also sustained a skull fracture.

After these early tests using X-ray, the focus of the brain deformation field shifted toward using kinematic sensors to measure brain motion. Trosseille et al. (1992) used accelerometers implanted in the brains of cadavers to conduct validation tests for an FE model. A subsequent study by Hardy et al. (1997) used similar triaxial neutral density accelerometers designed to move with the brain tissue and measure tissue motion. The measured acceleration was then compared to skull-mounted accelerometers to compare relative motion, with 3–5 mm of peak displacement observed in the brain. Although the method was validated against X-ray measurements of brain motion, the accelerometers did not directly measure brain displacement, and integration errors were introduced in the displacement calculations.

In the early 2000s, the accelerometer method was abandoned with improvements in digital X-ray imaging technology. A large dataset of in situ human brain deformation was generated by Hardy et al. (2001, 2007). The studies used high-speed biplanar X-ray to track the three-dimensional (3D) motion of neutrally-dense targets made from tin granules embedded in polystyrene tubing and were implanted in columnar and cluster arrays in the brains of PMHS head–neck specimens. A depiction of the experiments is shown in Fig. 12.2. The head–neck specimens were inverted and subjected to frontal, occipital, and coronal impacts. The impacts were imparted on the head using a padded linear impactor which caused resultant head linear accelerations ranging from 38 to 291 g, rotational velocities ranging from 4 to 30 rad/s, and head rotational accelerations ranging from 2370 to 24,206 rad/s^2 (Hardy et al., 2007). The markers in the brain were observed to follow figure-eight patterns with peak-to-peak excursions as high as 13.4 mm. The observed deformation was largest in the inferior regions of the brain.

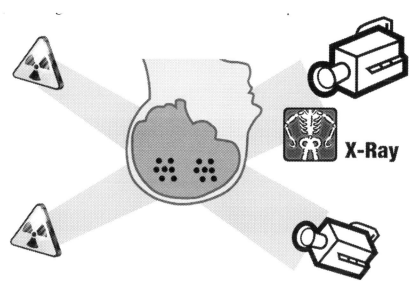

Fig. 12.2: The biplane X-ray methodology has been the primary technique to collect brain deformation for brain FE models. Small NDTs are implanted in the brain (black circles) and tracked using X-ray.

The biplane X-ray method and Hardy et al. dataset provided important experimental targets for the validation of brain FE models, but the experiments contained certain limitations that restrict their applicability for model assessment (Alshareef et al., 2018). The X-ray method requires line-of-sight between the emitter, specimen, and cameras, which limits the mounting hardware, the number and location of embedded targets, and the direction of loading. There are also limitations associated with camera resolution and image distortion. The impact loading used in the Hardy et al. (2001, 2007) dataset resulted primarily in linear acceleration, but FE models used to predict diffuse injuries to the brain require validation for rotational loading. Lastly, the dataset included approximately 62 tests conducted using varied tests conditions and impact speeds. While nearly every human FE brain model in the literature uses the Hardy dataset as the primary validation target, only 3–5 tests from the full dataset are utilized due to missing kinematic data, inability to track targets, or experimental variation and errors (Zhou et al., 2018). A digital archive of the data is not available, and the data used for FE validation is typically digitized from publications, possibly introducing more errors in the analysis. Since the goal of these experiments was not to explicitly collect data for FE validation, there have been inconsistencies in accounting for marker initial position, geometry differences between the model and specimens, and exact boundary conditions of the impacts.

12.2.2.2 Dynamic ultrasound

Dynamic ultrasound imaging has been used with limited success to study surface motion with high sampling rates. Mallory et al. (2015) conducted low-severity sagittal head rotations (2 rad/s, 120–140 rad/s^2) on repressurized cadavers and measured brain deformation

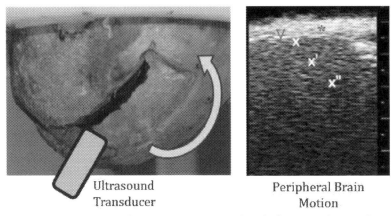

Ultrasound Transducer   Peripheral Brain Motion

Fig. 12.3: The dynamic ultrasound tests use a conventional ultrasound transducer to capture peripheral and meningeal brain motion (figures adapted from Mallory et al. 2015).

using B-mode ultrasound at the surface of the dura following a craniectomy (Fig. 12.3). The study was useful in identifying relative localized motion between the brain and surrounding meningeal membranes but was limited by the penetration depth of ultrasound waves, two-dimensional (2D) tracking of motion, and the substantial disruption of the brain–skull boundary condition. Additionally, the tests were conducted at low rates of head motion because of spatiotemporal errors from the fast sampling rate and out-of-plane motion. Dynamic ultrasound has not been used to evaluate FE model response, but the technique is being refined (Mallory et al., 2018) and has the potential to provide validation data for meningeal and peripheral brain motion.

12.2.2.3 Sonomicrometry

A novel methodology using sonomicrometry was recently developed by Alshareef et al. (2018) as an alternative to high-speed radiography. Sonomicrometry uses ultrasound wave time-of-flight to dynamically measure distances between pairs of small piezoelectric crystals implanted within a medium. Sonomicrometry does not have line-of-sight limitations, which allows for tracking of a large number of crystals in the brain, and it allows for testing under multiple directions and loading conditions for each specimen. An overview of the technique and experiments is shown in Fig. 12.4.

This methodology was first demonstrated on a human cadaveric head–neck specimen, unembalmed and never frozen, that was instrumented with an array of 32 sonomicrometry crystals embedded in the head. The specimen was subjected to dynamic rotation tests that were applied about the three principal directions of rotation (sagittal, coronal, axial), with angular velocity pulses ranging from 20 to 40 rad/s, with durations of 30–60 milliseconds and angular accelerations from 600 to 5500 rad/s^2. The rotation severities were chosen to represent mild to moderate risk of concussion, and were applied in a manner easily translatable to computational models. The sonomicrometry and experimental techniques were able to reliably and

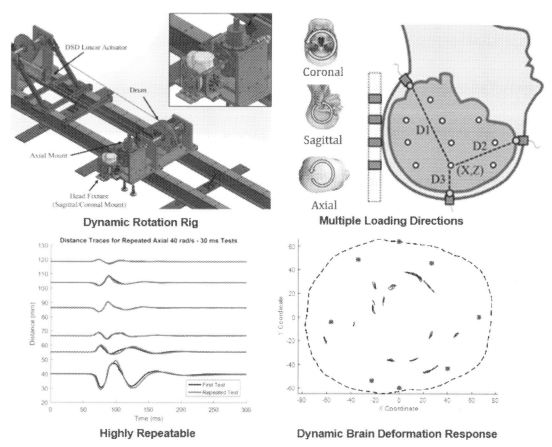

Fig. 12.4: Depiction of the sonomicrometry brain deformation dataset. The dynamic rotation rig (top, left) is used to induce pure head rotation for multiple loading severities and directions. The inserted crystals (top, right) are used to track dynamic distances (Alshareef et al., 2020a), which are found to be highly repeatable (bottom, left). The results reliably capture brain deformation at up to 24 points inside the brain (bottom, right).

repeatedly capture 3D dynamic in situ whole brain deformation during the dynamic head rotation tests. The experiments captured the 3D motion of the brain in a highly repeatable manner, with spatial and temporal precision sufficient to fully capture both the short-term and long-term transient mechanical response of the brain in these tests. The peak-to-peak deformation of the brain ranged from 2 to 23 mm, with axial rotations inducing the largest deformation. The peak-to-peak deformation of the brain increased with increasing angular velocity and decreasing pulse duration (Alshareef et al., 2018).

With the objective of generating controlled and repeatable experimental measurements of brain deformation for FE model validation, the technique was expanded to a total of six specimens. The study generated a dataset of dynamic brain deformation collected under various kinematic conditions to investigate the relationship between brain deformation magnitude and the characteristics of the applied head loading. The final dataset includes six specimens, each tested under 12 kinematic loading conditions, for approximately 5000 dynamic brain motion curves that can be used to evaluate the deformation response of FE models. The dataset also includes 3D computed tomography images and anatomical MRI images of the specimens, as well as a digital archive of the six degree-of-freedom head kinematics, crystal initial position, and 3D displacement of every crystal for every test (Alshareef et al., 2020b).

While previous human brain deformation experiments conducted a breadth of tests and severities with different specimens (Hardy et al., 2007), the sonomicrometry dataset is the first to produce data for multiple loading directions and severities for the same set of specimens. The novelty of this dataset allows for comparisons for the same point in the brain across different angular velocity, loading duration, and rotation direction. It also allows FE model developers to compare the model response to a representative range of specimens under various conditions, minimizing ambiguity associated with population variance and the directional dependence of brain deformation. The availability of this digital sonomicrometry dataset of human brain deformation allows for thorough and rigorous validation of FE models. The plethora of data allows developers to use a subset of the data to calibrate model or material property parameters, while reserving a set of data to validate the calibrated model response.

The sonomicrometry dataset is limited in certain experimental and hardware aspects. While high rate rotational loading provides relevant data to validate models, the technique is conducted in head–neck cadaveric specimens (tested within 72 hours postmortem), and the effects relative to an in vivo test are unknown. The sonomicrometry acquisition was also limited to 24 crystals embedded in the brain tissue due to a limitation in the number of instrumented channels.

12.2.2.4 Tagged magnetic resonance imaging

MRI has also been used as an imaging modality to measure in vivo human brain deformation during noninjurious impact conditions. The method uses synthetic "tag lines" in MRI images, generated through sinusoidal image sequences, to track deformation (Bayly et al., 2005). The tag lines appear as solid black lines on the MRI image and move with the tissue as it deforms. Since tag lines have a short lifetime (approximately 1 second), multiple images are required to generate a dynamic deformation map with adequate temporal resolution. Additionally, multiple images are required to generate tag lines in orthogonal directions to be able to acquire 2D and 3D motion (Fig. 12.5).

In the first reported used of tMRI for the brain, Bayly et al. (2005) collected two 2D brain deformation in volunteers subjected to a sagittal head motion (2–3 g). Sabet et al. (2008) and Feng et al. (2010) expanded on this work with a rotational loading test and analysis of brain motion relative to the skull. Knutsen et al. (2014), Chan et al. (2018), and Gomez et al. (2018) have further refined this technique with optimized 3D tagging sequences, improved analysis of

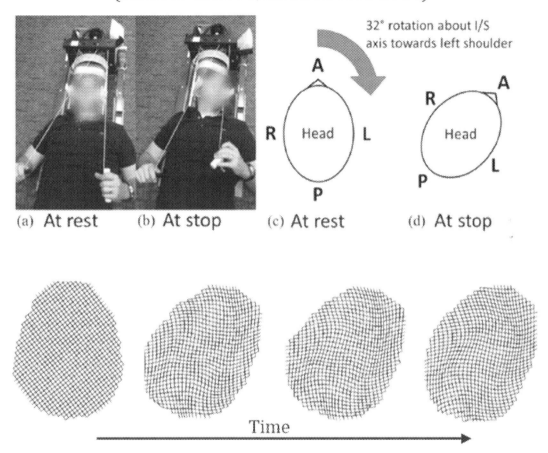

Fig. 12.5: The tMRI technique has been used to collect in vivo brain deformation for volunteers subjected to "head drop" and "mild rotation" conditions (left). The technique requires multiple repetitions to collect tag lines in three orthogonal directions to measure 3D brain deformation (right). Images are adapted from Sabet et al. (2008) and Knutsen et al. (2014).

deformation and strain, and an analysis of a dataset of 34 volunteers under mild head rotation (300 rad/s^2, 40 milliseconds) (Chan et al., 2018; Gomez et al., 2018; Knutsen et al., 2014).

The tMRI method has been effective in measuring dynamic deformation of the human brain in living volunteers. Additionally, the technique has an advantage in its spatial resolution, with tag lines approximately every 8 mm. This allows for strain calculations that are not possible with sonomicrometry and biplane X-ray, where the spatial resolution of the discrete markers is larger than what is needed for accurate strain calculation. However, there are limitations to tMRI that inhibit its use for study of high-rate, injurious loading conditions. One, the volunteer subjects must be tested multiple times in the exact same loading condition to capture each "frame" of the motion using tMRI for a single tag direction. Thus, the induced head motion must be very repeatable, and small variations can alter the results. Second, the imaging sampling rate is approximately 15–50 Hz, which is not high enough to capture the large dynamic deformation of the brain relevant to brain injury (300–1000 Hz required). The sampling rate could be increased to approximately 150 Hz, but the tests would require many more repetitions. Third, the tests are conducted on human volunteers, and the loading was not large enough to cause deformations that can be used to validate the strain response of an injurious loading scenario using an FE model. Extrapolation of tMRI results into the injurious loading regime is not credible given the nonlinear and viscoelastic nature of brain tissue. However, extension of the tMRI technique to animal or cadaveric human models can potentially capture brain deformation at injurious levels with enough kinematic repetitions.

12.2.2.5 Magnetic resonance elastography

As discussed in Chapter 9, MRE is used to measure deformation of the brain on the micron level to measure the stiffness of the tissue. The technique works by generating shear waves in the brain using an induced harmonic motion, depicted in Fig. 12.6, and tracking the propagation of the waves using a conventional MRI scanner with modified sequences that include a "motion-encoding gradient" (Hiscox et al., 2016). The head motion is usually induced by an external driver that can generate harmonic vibrations (5–50 μm, 10–100 Hz). The magnitude and speed of the shear waves along with the anatomical image can then be used to generate a stiffness map, or elastogram, of the brain (Hiscox et al., 2018). Several models exist to transform the shear wave information to elastic or viscoelastic material properties, each with associated assumptions and limitations.

MRE has recently been used to measure brain motion in human volunteers (Badachhape et al., 2018; Johnson et al., 2013; Kurt et al., 2019; Okamoto et al., 2019). The technique is

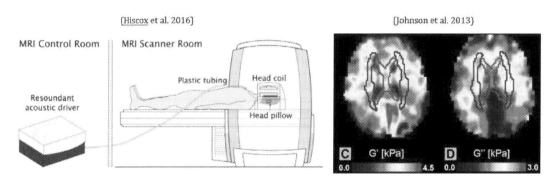

Fig. 12.6: MRE uses external harmonic motion (left) to measure the speed and magnitude of shear waves in the brain to determine the stiffness of the tissue (images adapted from Hiscox et al., 2018 and Johnson et al., 2013).

relatively new in the field, with recent refinements in animal models and human experiments. There are ongoing efforts to derive mechanical properties and motion estimations in a cohort of human volunteers, with varying ages and sexes, that can be used to estimate the material properties of FE models (Bayly et al., 2021). A benefit of MRE also lies in the harmonic motion used to generate the shear waves, which can be used to measure the modal response of the human brain. Modal response or resonance frequency of the brain has never been used for validation of FE brain models and should be investigated in validating models in conjunction with low and high rate loading.

Like tMRI, MRE provides a useful tool for studying in vivo brain motion but is limited to low-rate loading. The mechanics of brain, like other soft tissues, changes at various strain thresholds. At the loading rate and magnitude of deformation of MRE, the brain is likely behaving elastically. While the method provides important information about relative stiffness among brain regions, future studies are needed to elucidate the relationship between MRE results and high-rate loading.

12.3 Challenges and limitations

There are many challenges associated with the collection of experimental brain biomechanics data. Historically, there has been a dearth of available data for brain FE model developers, but there has been a recent surge in available data collected using novel imaging methods. As described in the various methods used, the common theme is the difficulty associated with accurately and reliably measuring brain deformation under loading conditions relevant to TBI. There are tradeoffs with each method, as depicted in Fig. 12.7. The

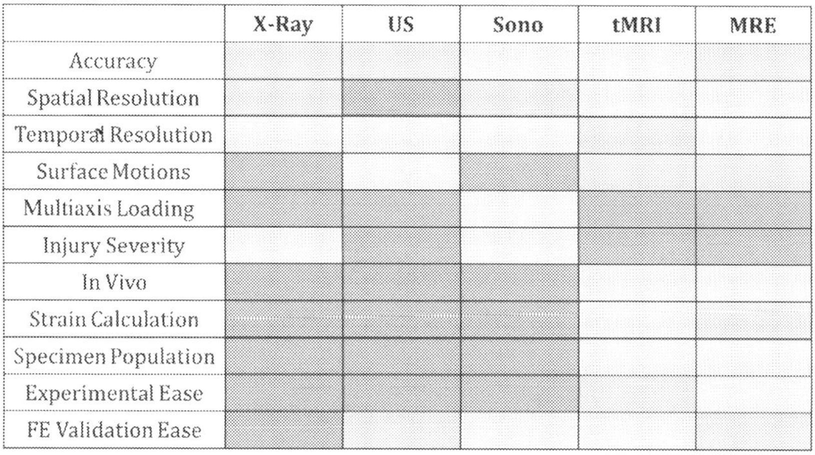

	X-Ray	US	Sono	tMRI	MRE
Accuracy					
Spatial Resolution					
Temporal Resolution					
Surface Motions					
Multiaxis Loading					
Injury Severity					
In Vivo					
Strain Calculation					
Specimen Population					
Experimental Ease					
FE Validation Ease					

Fig. 12.7: Overview of the advantages and disadvantages of each method (or associated) dataset utilized to measure brain deformation. Green indicates a proven result or technique. Orange indicates an acceptable level or unproven aspect. Red indicates a disadvantage of the method.

in vivo studies prevent the use of high-rate loading in human volunteers, while the in situ studies include complex and unique testing environments. The in vivo studies also allow for a wide range of subjects (e.g., age, sex, and anthropometry), while the in situ studies typically include older cadaveric specimens. Animal studies have not been reviewed in this publication, but could be an alternative experimental model for in vivo, high-rate deformation using the sonomicrometry and MRI methods. This model, however, presents different issues with kinematic scaling and the translation of an injury in an animal to that of a human (Wu et al., 2020; Wu et al., 2021).

None of the methods can individually complete our understanding of human brain mechanics or provide definitive validation tools for FE models. A combination of the available datasets must be utilized in a standardized, consistent manner in the modeling community. Currently, there is no consensus on the best methods for brain FE model development. Although many of the models are validated using the same single dataset of human brain deformation, the strain results can vary significantly among models (Giudice et al., 2019; Miller et al., 2017). A recent analytical review of FE brain models suggests that the factors affecting the output of these models include material properties, geometry differences, the size, type, and quality of the mesh, and FE parameters such as hourglass control (Giudice et al., 2019). These differences, in addition to different choices of material properties and what anatomical structures are included, make it difficult to compare strain results across models or to physical experiments (Giudice et al., 2019). Additionally, the models are often simulated under conditions that vary from the head kinematic impacts they were validated against (direct impact versus rotational loading), raising questions about the applicability of the models and what conclusions can be drawn from the results.

To be able to make significant conclusions about tissue-level deformation, the biomechanical response of the models must match as closely as possible to real-world human brain deformation. While most of the brain FE models in the literature use a small subset of the Hardy dataset for validation, there is a lack of consistency in how the models are evaluated. Miller et al. (2017) performed evaluations of several FE brain models in comparison to a subset of the Hardy dataset. The Miller study found that the models yielded a cross-correlation score (Gehre et al., 2009) between 0.26 and 0.41. The study did not account for differences in specimen geometry and picked nodal points that were closest to the absolute marker location. Displacement was only compared in the plane of testing (x–z or y–z) with each test in the Hardy dataset conducted on a different specimen, preventing the validation of the models across different impact directions using the same specimen or dataset. There is also inconsistency on the best metric to evaluate the model results in comparison to experimental brain deformation. The data is 4D (x, y, z motion versus time), and some metrics like root mean square error, cross-correlation score, and correlation score (Ganpule et al., 2017; Giordano and Kleiven, 2016; Kimpara et al.,

2011) are used to evaluate the model objectively. The parameters used in these metrics, as well as the choice of metric, have not been evaluated.

Model validation has been conducted in the literature using brain motion relative to the skull. An underlying assumption exists that matching the brain motion (relative to the skull) will lead to match in tissue strain. While certain techniques like sonomicrometry are unable to yield strain, tMRI allows for the calculation of strain after spatial interpolation. Strain can be a useful validation tool, but often includes more error in the added calculations than brain motion (relative to physical models) (Gomez et al., 2018). A recent study suggested that a validation of brain deformation does not necessitate a match of the strain values (Zhou et al., 2018). Computational models that are validated for the same dataset of brain motion (by comparing nodal displacement relative to skull motion) can yield varying strain results. While these results suggest that a combination of factors, including mesh size and model boundary conditions, can lead to this disparity, it is important to validate models to experimental strain results. A combination of validation using high-rate brain motion results, as well as strain results from in vivo tests should utilized in FE model validation.

12.4 Summary and future perspectives

Several experimental methods have been used to investigate the mechanics of the brain to better understand brain injury mechanisms and to provide data to validate brain computational models. While early datasets of intracranial pressure were used to evaluate model response, they are not correlated to diffuse strain injury mechanisms. A brain deformation dataset by Hardy et al. (2001, 2007) has historically been used to validate brain FE models under head impact, but the dataset includes limitations associated with loading conditions and X-ray image. Recent improvements in biomechanical imaging techniques, including sonomicrometry, tagged MRI, and MRE have allowed a new set of experiments aimed at generating data for FE model validation. There are advantages and disadvantages to each method and dataset, and a combination of the available is needed to validate models at TBI-relevant loading conditions for a range of severities, target populations, and strain outputs. These techniques are still being refined, but there has been a recent surge in data available to brain FE model developers. The data will allow the creation of biofidelic models, aiding in the creation of safety gear, the investigation of injury mechanisms, and analysis using subject-specific models.

The calibration and development of brain FE models requires a standardized approach for evaluation and validation. There are many parameters and boundary conditions that are chosen for a model, which can affect model responses and any conclusions relevant to injury or safety. A consensus on the best approach for model development, the datasets used for evaluation, and the objective metric to evaluate the deformation results is needed to ensure model biofidelity and accuracy.

206 Chapter 12

References

Alshareef, A., 2019. Deformation of the Human Brain Under Rotational Loading. University of Virginia, Charlottesville, VA. https://doi.org/10.18130/v3-yq95-t439.

Alshareef, A., Giudice, J.S., Forman, J., Salzar, R.S., Panzer, M.B., 2018. A novel method for quantifying human in situ whole brain deformation under rotational loading using sonomicrometry. J. Neurotrauma 35, 780–789.

Alshareef, A., Giudice, J.S., Forman, J., Shedd, D.F., Wu, T., Reynier, K.A., Panzer, M.B., 2020a. Application of trilateration and Kalman filtering algorithms to track dynamic brain deformation using sonomicrometry. Biomed. Signal Process. Control 56, 101691.

Alshareef, A., Giudice, J.S., Forman, J., Shedd, D.F., Reynier, K.A., Wu, T., Sochor, S., Sochor, M.R., Salzar, R.S., Panzer, M.B., 2020b. Biomechanics of the Human Brain During Dynamic Rotation of the Head. J. Neurotrauma 37 (13), 1546–1555.

Badachhape, A.A., Okamoto, R.J., Johnson, C.L., Bayly, P.V., 2018. Relationships between scalp, brain, and skull motion estimated using magnetic resonance elastography. J. Biomech. 73, 40–49.

Bayly, P.V., Cohen, T.S., Leister, E.P., Ajo, D., Leuthardt, E.C., Genin, G.M., 2005. Deformation of the human brain induced by mild acceleration. J. Neurotrauma 22, 845–856.

Bayly, P.V., Alshareef, A., Knutsen, A.K., Upadhyay, K., Okamoto, R.J., Carass, A., Butman, J.A., Pham, D.L., Prince, J.L., Ramesh, K.T., et al., 2021. MR Imaging of Human Brain Mechanics In Vivo: New Measurements to Facilitate the Development of Computational Models of Brain Injury. Ann. Biomed. Eng. 1–16.

Bradshaw, D.R.S., and Morfey, C.L. (2001). Pressure and shear responses in brain injury models, SAE Technical Paper.

Chan, D.D., Knutsen, A.K., Lu, Y.-C., Yang, S.H., Magrath, E., Wang, W.-T., Bayly, P.V., Butman, J.A., Pham, D.L., 2018. Statistical characterization of human brain deformation during mild angular acceleration measured in vivo by tagged magnetic resonance imaging. J. Biomech. Eng. 140, 101005.

Feng, Y., Abney, T.M., Okamoto, R.J., Pless, R.B., Genin, G.M., Bayly, P.V., 2010. Relative brain displacement and deformation during constrained mild frontal head impact. J. R. Soc. Interface 7, 1677–1688.

Gabler, L.F., Crandall, J.R., Panzer, M.B., 2016. Assessment of kinematic brain injury metrics for predicting strain responses in diverse automotive impact conditions. Ann. Biomed. Eng. 44, 3705–3718.

Gabler, L.F., Joodaki, H., Crandall, J.R., Panzer, M.B., 2017. Development of a single-degree-of-freedom mechanical model for predicting strain-based brain injury responses. J. Biomech. Eng. 140 (3), 031002-1-031002-13.

Ganpule, S., Daphalapurkar, N.P., Ramesh, K.T., Knutsen, A.K., Pham, D.L., Bayly, P.V., Prince, J.L., 2017. A three-dimensional computational human head model that captures live human brain dynamics. J. Neurotrauma 34, 2154–2166.

Gehre, C., Gades, H., Wernicke, P., 2009. Objective rating of signals using test and simulation responses, 21st International Technical Conference on the Enhanced Safety of Vehicles. Stuttgart, Germany.

Gennarelli, T.A., 1993. Mechanisms of brain injury. J. Emerg. Med. 11, 5–11.

Gennarelli, T.A., Thibault, L.E., and Ommaya, A.K. (1972). Pathophysiologic responses to rotational and translational accelerations of the head. SAE Technical Paper.

Giordano, C., and Kleiven, S. (2016). Development of an unbiased validation protocol to assess the biofidelity of finite element head models used in prediction of traumatic brain injury. SAE Technical Paper.

Giudice, J.S., Zeng, W., Wu, T., Alshareef, A., Shedd, D.F., Panzer, M.B., 2019. An analytical review of the numerical methods used for finite element modeling of traumatic brain injury. Ann. Biomed. Eng. 47 (9), 1855–1872.

Gomez, A.D., Knutsen, A.K., Xing, F., Lu, Y.-C., Chan, D., Pham, D.L., Bayly, P., Prince, J.L., 2018. 3-D measurements of acceleration-induced brain deformation via harmonic phase analysis and finite-element models. IEEE Trans. Biomed. Eng. 66, 1456–1467.

Hardy, W.N., Foster, C.D., King, A.I., Tashman, S., 1997. Investigation of brain injury kinematics: introduction of a new technique. ASME Appl. Mech. Div. Publ. AMD 225, 241–254.

Hardy, W.N., Foster, C.D., Mason, M.J., Yang, K.H., King, A.I., Tashman, S., 2001. Investigation of head injury mechanisms using neutral density technology and high-speed biplanar X-ray. Stapp. Car Crash J. 45, 337–368.

Hardy, W.N., Mason, M.J., Foster, C.D., Shah, C.S., Kopacz, J.M., Yang, K.H., King, A.I., Bishop, J., Bey, M., Anderst, W., et al., 2007. A study of the response of the human cadaver head to impact. Stapp. Car Crash J. 51, 17.

Hiscox, L.V., Johnson, C.L., Barnhill, E., McGarry, M.D., Huston 3rd, J., Van Beek, E.J., Starr, J.M., Roberts, N., 2016. Magnetic resonance elastography (MRE) of the human brain: technique, findings and clinical applications. Phys. Med. Biol. 61, R401.

Hiscox, L.V., Johnson, C.L., McGarry, M.D., Perrins, M., Littlejohn, A., van Beek, E.J., Roberts, N., Starr, J.M., 2018. High-resolution magnetic resonance elastography reveals differences in subcortical gray matter viscoelasticity between young and healthy older adults. Neurobiol. Aging 65, 158–167.

Hutchinson, J., Kaiser, M.J., Lankarani, H.M., 1998. The head injury criterion (HIC) functional. Appl. Math. Comput. 96, 1–16.

Johnson, C.L., McGarry, M.D., Gharibans, A.A., Weaver, J.B., Paulsen, K.D., Wang, H., Olivero, W.C., Sutton, B.P., Georgiadis, J.G., 2013. Local mechanical properties of white matter structures in the human brain. Neuroimage 79, 145–152.

Kimpara, H., Nakahira, Y., Iwamoto, M., Rowson, S., Duma, S., 2011. Head injury prediction methods based on 6 degree of freedom head acceleration measurements during impact. Int. J. Automot. Eng. 2, 13–19.

Knutsen, A.K., Magrath, E., McEntee, J.E., Xing, F., Prince, J.L., Bayly, P.V., Butman, J.A., Pham, D.L., 2014. Improved measurement of brain deformation during mild head acceleration using a novel tagged MRI sequence. J. Biomech. 47, 3475–3481.

Knutsen, A.K., Gomez, A.D., Gangolli, M., Wang, W.-T., Chan, D., Lu, Y.-C., Christoforou, E., Prince, J.L., Bayly, P.V., Butman, J.A., 2020. In vivo estimates of axonal stretch and 3D brain deformation during mild head impact. Brain Multiphys. 1, 100015.

Kurt, M., Wu, L., Laksari, K., Ozkaya, E., Suar, Z.M., Lv, H., Epperson, K., Epperson, K., Sawyer, A.M., Camarillo, D, 2019. Optimization of a multifrequency magnetic resonance elastography protocol for the human brain. J. Neuroimaging 29 (4), 440–446.

Mallory, A., Kang, Y.-S., Herriott, R., Rhule, H., Donnelly, B., Moorhouse, K., 2015. High-Frequency B-Mode Ultrasound for the Measurement of Intracranial Motion in Head Rotation.

Mallory, A., Donnelly, B., Liu, J., Bahner, D., Moorhouse, K., Dupaix, R., 2018. Addressing spatiotemporal distortion of high-speed tissue motion in B-mode ultrasound. Biomed. Phys. Eng. Express 4, 057003.

Meaney, D.F., Morrison, B., Bass, C.D., 2014. The mechanics of traumatic brain injury: a review of what we know and what we need to know for reducing its societal burden. J. Biomech. Eng. 136, 021008.

Miller, L.E., Urban, J.E., Stitzel, J.D., 2017. Validation performance comparison for finite element models of the human brain. Comput. Methods Biomech. Biomed. Engin. 20, 1273–1288.

Nahum, A.M., Smith, R., and Ward, C.C. (1977). Intracranial pressure dynamics during head impact, SAE Technical Paper.

Nusholtz, G.S., Lux, P., Kaiker, P., and Janicki, M.A. (1984). Head impact response—skull deformation and angular accelerations, SAE Technical Paper.

Okamoto, R.J., Romano, A.J., Johnson, C.L., Bayly, P.V., 2019. Insights into traumatic brain injury from MRI of harmonic brain motion. J. Exp. Neurosci. 13, 1179069519840444.

Ommaya, A.K., 1984. The Head: Kinematics and Brain Injury Mechanisms. Elsevier, Amsterdam.

Ommaya, A.K., Grubb Jr, R.L., Naumann, R.A, 1971. Coup and contre-coup injury: observations on the mechanics of visible brain injuries in the rhesus monkey. J. Neurosurg. 35, 503–516.

Sabet, A.A., Christoforou, E., Zatlin, B., Genin, G.M., Bayly, P.V., 2008. Deformation of the human brain induced by mild angular head acceleration. J. Biomech. 41, 307–315.

Sanchez, E.J., Gabler, L.F., McGhee, J.S., Olszko, A.V., Chancey, V.C., Crandall, J., Panzer, M.B., 2017. Evaluation of head and brain injury risk functions using sub-injurious human volunteer data. J. Neurotrauma 34 (16), 2410–2424.

Shuck, L.Z., Advani, S.H., 1972. Rheological response of human brain tissue in shear. J. Basic Eng. 94, 905–911.

Stalnaker, R.L., Melvin, J.W., Nusholtz, G.S., Alem, N.M., and Benson, J.B. (1977). Head impact response, SAE Technical Paper.

Takhounts, E.G., Craig, M.J., Moorhouse, K., McFadden, J., Hasija, V., 2013. Development of brain injury criteria (BrIC). Stapp. Car Crash J. 57, 243.

Taylor, C.A., Bell, J.M., Breiding, M.J., Xu, L., 2017. Traumatic brain injury–related emergency department visits, hospitalizations, and deaths—United States, 2007 and 2013. MMWR Surveill. Summ. 66, 1–16.

Trosseille, X., Tarriere, C., Lavaste, F., Guillon, F., and Domont, A. (1992). Development of a FEM of the human head according to a specific test protocol, SAE Technical Paper.

Versace, J. (1971). A review of the severity index, SAE Technical Paper.

Wu, T., Antona-Makoshi, J., Alshareef, A., Giudice, J.S., Panzer, M.B., 2020. Investigation of cross-species scaling methods for traumatic brain injury using finite element analysis. J. Neurotrauma. 37 (2), 410–422.

Wu, T., Hajiaghamemar, M., Giudice, J.S., Alshareef, A., Margulies, S.S., Panzer, M.B., 2021. Evaluation of tissue-level brain injury metrics using species-specific simulations. J. Neurotrauma. 38, 1879–1888.

Zhou, Z., Li, X., Kleiven, S., Shah, C.S., Hardy, W.N., 2018. A reanalysis of experimental brain strain data: implication for finite element head model validation. Stapp. Car Crash J. 62, 293–318.

CHAPTER 13

A review of fluid flow in and around the brain, modeling, and abnormalities

R. Prichard, M. Gibson, C. Joseph, W. Strasser

Liberty University, Lynchburg, VA, United States

13.1 Introduction

The previous chapters have focused on the solid mechanics and materials aspect of human brain. In this chapter we focus on the fluid aspects of the brain, in particular the effects of water. The brain contains about 77% water with the rest being lipids (fats), blood vessels/blood, proteins, soluble organic materials, carbohydrates, and inorganic salts. Hence, the largest volume of material type in the brain is water! There are three basic categories of water in the brain: free flowing water, intercellular water, and water in the blood.

Basic brain function is related to water operating within the brain. The space between neurons occupied by glial cells is very narrow, and neuronal excitation requires motion of chemical ion concentration via water. Also, ionic transmembrane diffusion requires water to maintain ion homeostasis during neuronal activity. Third, water transport across plasma membranes in blood vessels is important to brain function. In each of these cases, the varying and complex multiscale aspects of water flow require analysis for us to understand the chemical–mechanical–electrical interactions of brain function.

13.2 Flow anatomy

13.2.1 Ventricular system

The brain's ventricular system was discovered by either Aristotle or Herophilos around 300 BC. However, it was not discovered for another two millennia that the four ventricle chambers contained fluid. This fluid, now known as *cerebrospinal fluid* (CSF), consists of ultrafiltered blood plasma. CSF supplies the brain with oxygen and nutrients and removes waste (Louveau et al., 2015, 2017). The venous and glymphatic subsystems (Brinker et al., 2014), which are parts of the ventricular system, work in concert to supply nutrients and remove waste. The arterial system carries nutrients into the central nervous system through four main vessels (right and left internal carotid arteries anteriorly, and right and left vertebral arteries posteriorly). From there, blood flows through smaller divisions of arteries until they

Multiscale Biomechanical Modeling of the Brain.
DOI: https://doi.org/10.1016/B978-0-12-818144-7.00015-3
Copyright © 2022 Elsevier Inc. All rights reserved.

reach the capillary level. The capillaries facilitate the exchange of oxygen for CO_2 and the delivery of vital nutrients such as glucose, amino acids and lipids.

13.2.2 Ventricles and subarachnoid space

Fig. 13.1 shows the brain's four ventricles: two lateral ventricles, the third ventricle, and the fourth ventricle, which contain about 25 mL of CSF in total (Sakka et al., 2011). The passages connecting the ventricles are termed foramina; one intraventricular foramen (or foramen of Monro) connects each lateral ventricle to the third ventricle, and the *cerebral aqueduct* (or *aqueduct of Sylvius*) connects the third ventricle to the fourth ventricle. As shown in Fig. 13.2, the fourth ventricle also connects to the central canal of the spinal cord, and it supplies CSF to the *subarachnoid space* (SAS).

The SAS contains approximately 150 mL of CSF (Sakka et al., 2011), which helps absorb minor cranial impacts but offers only minimal protection against larger impacts. CSF flows into the SAS from the fourth ventricle via three channels: two *lateral apertures* (or *foramina of Luschka*) and the *median aperture* (or *foramen of Magendie*), which connect with the subarachnoid cisterns surrounding the pons and medulla oblongata. More details on flow pathways are addressed later.

Fluid-filled spaces surrounding arteries, known as *perivascular* or *Virchow Robin spaces*, carry CSF from the SAS into the brain. The glymphatic system consists of these perivascular

Fig. 13.1: Anterior (left) and oblique (right) views of a 3D model of the ventricular system, composed of lateral ventricles (blue), interventricular foramina (cyan), third ventricle (yellow), cerebral aqueduct (red), and fourth ventricle (purple). Green indicates the connection to the spinal canal. "Human Ventricular system colored and animated" from Wikipedia and created by Anatomography (lifesciencedb.jp/bp3d), licensed under CC BY-SA 2.1 JP.

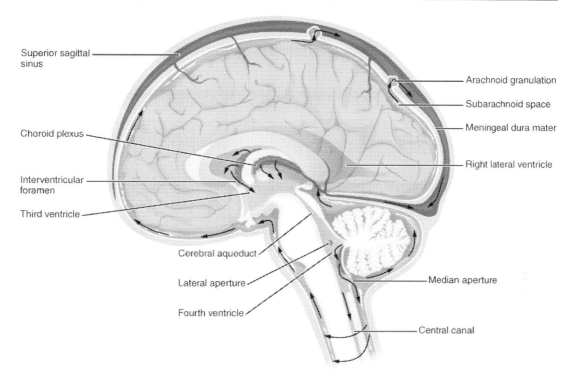

Fig. 13.2: Cross-section of a human brain illustrating the CSF space. *"Cerebrospinal Fluid Circulation"* from OpenStax Anatomy and Physiology, licensed under CC BY 4.0.

spaces and the interstitial spaces within the brain's tissue. The CSF in the perivascular spaces is driven by the advective forces of respiration and the arterial pulse wave (Louveau et al., 2015, 2017; Brinker et al., 2014; Plog and Nedergaard, 2018). In contrast, the interstitial fluid, a local name for CSF within small gaps in the brain's tissue, travels much more slowly by diffusion (Smith et al., 2017; Abbott, 2004; Holter et al., 2017). The interstitial fluid empties into the perivenous space, where it mixes with CSF (Brinker et al., 2014).

13.3 Characteristic numbers

Several dimensionless numbers are important for determining and characterizing the behavior of flow within the CSF space.

13.3.1 Reynolds number

The Reynolds number represents the ratio of inertial to viscous forces within a fluid and is used to indicate the laminar or turbulent nature of a flow. It is defined as the following:

$$\mathrm{Re} = \frac{\rho v L}{\mu} \qquad (13.1)$$

212   Chapter 13

where ρ, v, and μ are, respectively, the fluid's density, velocity, and dynamic viscosity; and L is a characteristic length. In this case, the characteristic length would be the diameter of a CSF passage. Due to the low velocities and small length scales seen in the CSF space, it has been widely reported that flow is laminar. In a study, Gupta *et al.* (Gupta et al., 2010) found the maximum Reynolds number in the cranial SAS to be 386, and Kurtcuoglu similarly observed purely laminar flow in the cerebral aqueduct with Re = 340 (Kurtcuoglu et al., 2007). Depending on geometry and system stability, laminar behavior can extend up to a Reynolds number of 3000, so these are well within the laminar range. Kurtcuoglu further reports (Miller, 2011), "In its entire domain, CSF flow is laminar and does not require any turbulence modeling." However, anatomical abnormalities such as Chiari malformation may result in transitional flows (Støverud et al., 2016; Jain et al., 2017), and some authors suggest that even healthy brains may exhibit transitional flow characteristics (Jain et al., 2017).

13.3.2 Womersley number

Another important parameter to characterize the flow of CSF is the Womersley number. The Womersley number closely relates to the Reynolds number, as it represents the ratio of *transient* inertial forces to viscous forces. It is used to determine whether a pulsatile flow has time to develop a parabolic profile in each pulse, that is, whether viscous effects are significant. The Womersley number, α, is defined as the following:

$$\alpha = L\sqrt{\frac{\omega\rho}{\mu}} \tag{13.2}$$

where ω is the angular frequency of the flow. Typically, $\alpha = 1$ is chosen as the threshold between developed and undeveloped flow. In the CSF space, α is typically significantly greater than one. Womersley numbers found in several studies are listed in Table 13.1.

13.3.3 Péclet number

The Péclet number can be important to characterize the transport of solutes by advective or diffusive means, a topic of great interest due to the debate around the glymphatic hypothesis

Table 13.1: Womersley numbers found in several studies.

Study	α
Martin et al., 2016	4–12
Gupta et al., 2009	15
Gholampour, 2018	3
Sweetman et al., 2011	4.2–8

(discussed later) related to the brain. Accordingly, it is the ratio of the rate of advection to the rate of diffusion, and it is defined as the following,

$$\text{Pe} = \frac{Lv}{D} \quad (13.3)$$

where D is the *mass diffusion coefficient*, an experimentally determined quantity with units of m^2/s. Diffusion is said to dominate if Pe $\ll$ 1 (Holter et al., 2017). The Péclet number ranges drastically throughout the brain depending on the location and the solute in question. Holter et al. calculated Péclet numbers as low as approximately 0.005 for Na$^+$ in the interstitial space (Holter et al., 2017), and Kurtcuoglu et al. calculated Péclet numbers in the third ventricle ranging from 46.6 to 77,600 (Kurtcuoglu et al., 2007).

13.4 Common brain flow abnormalities

Several disorders affect the geometry and therefore the CSF flow dynamics within the central nervous system. The causes of these disorders are not well known, although some can be congenital. Before addressing flow anomalies, a closer look at the small-scale flow is required. As shown in Fig. 13.3, astrocytes are star-shaped cells that occupy a large portion of the space between the neurons and the blood vessels in the brain. They are the support structure for the neurons, providing them with nutrients, proteins, and cholesterol. The astrocytes also connect to the capillaries via their "feet," which contain aquaporin water channels (Abbott, 2004;

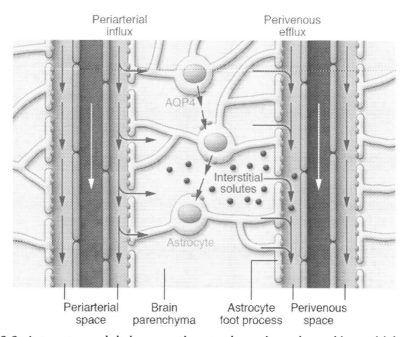

Fig. 13.3: Astrocytes and their connections to the perivascular and interstitial spaces. Reproduced from Louveau et al. (2017), with permission. Copyright American Society for Clinical Investigation, 2017.

Bering, 1952; Igarashi et al., 2014; Papadopoulos and Verkman, 2013). These channels allow them to regulate fluid pressure and transport nutrients and waste products.

13.4.1 Misfolded proteins

In most cases, the aquaporin channels are more than capable of keeping up with the waste products produced by the brain, but this system can be overwhelmed. One of the most common sources of interference is accumulation of misfolded brain proteins, specifically intracellular hyperphosphorylated tau and beta amyloid (Ringstad et al., 2017; Da et al., 2018). Particularly, traumatic brain injuries (TBIs) have resulted in hyperphosphorylated tau protein. Normally, tau protein maintains the proper separation between the fast axonal transport tracks (Fig. 13.4), analogous to railroad ties separating the tracks. These transport tracks move metabolites, proteins, and even mitochondria down the axons and waste products back to the cell body for processing. However, when the tau proteins misfold, transport fails. The axon "dies back," and the proteins begin to pile up within the cell body and precipitate. Both the soluble form and the precipitated form are fatally toxic to the cell. They hijack the synthesis mechanism, producing more defective proteins, and they get transported transsynaptically to previously unaffected cells. Thus, the neurodegenerative process spreads to interconnected regions of the brain. Beta amyloid is another misfolded protein that accumulates in Alzheimer disease. Beta amyloid tends to precipitate in the interstitial spaces, impeding flow needed to clear waste products, but by itself has not been shown to be fatally toxic. As a consequence of

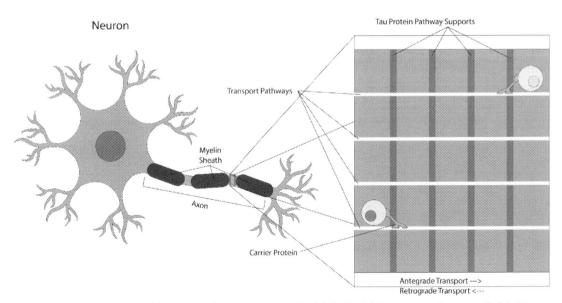

Fig. 13.4: A diagram of fast axonal transport tracks (right) within a neuron's axon (left). Tau proteins (vertical red bars) maintain spacing between transport tracks (horizontal yellow bars).

A review of fluid flow in and around the brain, modeling, and abnormalities **215**

brain trauma—particularly repetitive trauma—significant accumulation of hyperphosphory-lated misfolded tau occurs with similar consequences to neurodegenerative diseases such as Alzheimer's disease (Da et al., 2018).

13.4.2 Injury

Many things can impede natural fluid flow with varying levels of severity. Mild interference in both intracellular and extracellular flow can cause diffuse swelling with disruption in normal physiology as well as a reduction in waste removal. Moderate interference will also cause more severe localized swelling, and it carries the potential for permanent injury due to loss of cells via impact, tract disruption, and ischemia. In addition to the consequences of moderate interference, severe interference will rupture blood vessels, leading to potentially lethal hemorrhage into the brain. Such interference is supposed to be prevented by a properly designed helmet in sports or for police and firefighters. Being able to design such a helmet is but one of the motivations of modeling the fluid flow through the brain. Severe TBI causes acute shearing of white matter tracks. Astrocytes swell and retract the aquaporin channels from the vascular endothelium much like a castle under siege raises its drawbridge. Without these flow channels, the cells lose their ability to regulate the flow of fluid within the interstitial space, leading to further brain swelling (Sullan et al., 2018). The myriad of tiny blood vessels within the brain substance can rupture or leak, leading to internal bleeding in the brain along with even more swelling. Like a stroke, the swelling reduces normal perfusion of blood to the affected areas and thus acutely increases the brain cell loss. Depending on the severity of injury, swelling subsides over days or weeks, and the remaining astrocytes express the aquaporin channels within the capillary end feet. The degree of atrophy or brain shrinkage is a function of the extent of brain cell loss due to apoptosis. The long term effect of brain trauma is formation of misfolded hyperphosphorylated tau. Over ensuing years, this leads to cell death (see Fig. 13.4), and progressive dementia so called chronic traumatic encephalopathy.

13.4.3 Reduced arterial pulsatility

Aging and damage to arterial vessels diminish compliance of the vascular walls. This may be caused by stiffening of the *elastica interna*, the elastic layer of the blood vessel walls. Without proper arterial compliance, the ability of arterial pulsation to move CSF is compromised (Da et al., 2018). Fig. 13.5 illustrates the difference between a healthy vessel and a diseased vessel.

13.4.4 Hydrocephalus

Hydrocephalus is a condition where excess fluid accumulates in the brain under either normal or increased pressure. As fluid pressure in the brain increases, so does the stiffness of the brain's tissue. Similarly, reduced pressure decreases the stiffness. This effect should be considered when modeling the brain. Normal pressure hydrocephalus is related to brain

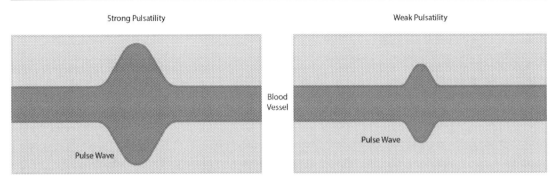

Fig. 13.5: Strong pulsatility in a healthy vessel (left) compared to weak pulsatility in a diseased vessel (right).

atrophy, where fluid simply fills the additional empty space. Further discussion of this topic is beyond the scope of this text. Hydrocephalus associated with high intracranial pressures can be caused by CSF flow obstruction within the ventricular system or by blockage of the normal resorptive channels in the SAS (Ringstad et al., 2017). In the latter circumstance, those channels in the cribriform plate, arachnoid granulations, and the spinal nerve root exit zones become clogged with blood products from prior subarachnoid bleeding or infection. Treatment of the latter may require surgical spinal fluid drainage on a permanent basis. Hydrocephalus is also accompanied with enlargement of the ventricles. While a drainage procedure can remedy the symptoms, the ventricles remain enlarged, which affects CSF flow patterns.

13.4.5 Chiari malformation

The cerebellum normally sits completely within the skull above the foramen magnum. In the Chiari malformation, cerebellar tonsils protrude into the spinal canal. This obstruction reduces subarachnoid CSF flow between the brain and the spine, thus increasing the pressure. Chiari malformation can cause hydrocephalus, headaches, syringomyelia, and syringobulbia.

13.4.6 Syringomyelia and syringobulbia

Syringomyelia is a disorder in which a fluid-filled cyst known as a syrinx forms within the spinal cord. This obstruction can affect CSF flow dynamics (Clarke et al., 2013). Similarly, syringobulbia consists of a cyst extending to the brainstem.

13.5 Boundary conditions for models
13.5.1 General comments

CSF is a clear, Newtonian fluid with minimal proteins and solutes. With properties similar to those of water, it constantly flows through the ventricular system, SAS surrounding the brain, and spinal column. In addition, water flows through the perivascular spaces and diffuses through the interstitial space (Brinker et al., 2014). Fluid flow is primarily driven by advective forces from vascular pulsation, respiratory dynamics that alter venous resistance, and perfusion of blood.

13.5.2 Cardiac flow

Pulsation of vasculature within the cranium displaces CSF into the spinal canal (Miller, 2011), causing an oscillatory flow pattern in sync with the heartbeat. Although there is some disagreement about the exact value, the magnitude of this flow is very low: Czosnyka (Miller, 2011) reports that only about 450 μL of intracranial CSF is displaced in a single cardiac cycle (the *stroke volume*). At the foramen magnum, Yiallourou et al. (2012) reported an average stroke volume of around 760 μL for healthy subjects. This discrepancy could be because of measurement error, differences between subjects, or different measurement locations.

13.5.3 Respiratory flow

In addition to the cardiac pulsation, respiratory motion causes periodic flow (Linninger et al., 2016). While this has been the subject of less investigation than cardiac pulsation, interest in the phenomenon is growing. Vinje et al. (2019) found that respiratory flow was much more significant than cardiac flow, with an average aqueductal stroke volume (*cardiac* stroke volume through the cerebral aqueduct) of 99.3 μL and an average aqueductal respiratory volume (*respiratory* stroke volume through the cerebral aqueduct) of 482 μL. Because the respiratory cycle has a longer period than the cardiac cycle, the difference in peak flow rates was not as dramatic—290 μL/s cardiac versus 320 μL/s respiratory. Respiratory flow is not entirely symmetric: a net upward flow has been observed, and it is balanced by corresponding increased venous drainage out of the head (Dreha-Kulaczewski et al., 2017).

13.5.4 Circulatory flow

Apart from the oscillatory flow from cardiac and respiratory influences, there may be a small circulatory portion of flow caused by secretion and absorption of CSF in different parts of the ventricular system. The classical conceptual model stating the presence of this circulatory flow has long been accepted, but recent research has cast doubt on it. In a comprehensive literature review, Chikly and Quaghebeur (2013) concluded that the classical model is deeply flawed. They state: "From more recent research, there is relatively little convincing, in vivo evidence to support the traditional model of the production, circulation, and reabsorption of CSF. The traditional model is seemingly based on faulty research and misinterpretations of that research, and this hypothesis is now increasingly being challenged."

218 Chapter 13

13.5.4.1 Production—classical model

The classical model holds that production mostly occurs in the choroid plexus at a rate of 350–370 μL/min (Kaczmarek, 1997), and it also occurs diffusely in the brain's parenchyma at an estimated 120 μL/min (Linninger et al., 2009). However, Kurtcuoglu cautions that the methods used to arrive at these numbers have not been validated (Miller, 2011).

13.5.4.2 Absorption—classical model

The traditional viewpoint is that CSF is absorbed in several places—the sagittal sinus (via the arachnoid granulations), lymphatic pathways, and possibly also within the parenchyma (Gupta et al., 2010). Some authors propose the 1/3–1/3–1/3 absorption model: One third via the arachnoid granulations in the midline Parietal dura, one third via the cribriform plate at the floor of the nasal structures, and one third via the spinal nerve roots (Louveau et al., 2015, 2017; Brinker et al., 2014; Bakker et al., 2016). Another recent in vivo study by de Leon et al. also found that CSF is cleared through the cribriform plate to lymphatic vessels contained within the nasal turbinates, which drain into the deep cervical lymph nodes (de Leon et al., 2017). According to Kurtcuoglu, little is certain about the amount of fluid absorbed by each sink. While equilibrium dictates that the absorption rate must equal the production rate, there is no agreement on the absorption rates at each site or even the relative importance of various sites (Miller, 2011).

13.5.4.3 New model

Whereas the classical model states that CSF is produced and absorbed at separate, distinct locations, recent evidence denies the existence of significant circulatory flow. The competing theory that seems to have the most support is that osmotic gradient and active transport of solute causes uniform secretion of CSF from capillaries throughout all the parenchyma. Similarly, it holds that CSF is both absorbed by capillaries in the parenchyma as well as being incorporated in perivascular flow (Chikly and Quaghebeur, 2013; Klarica et al., 2013; Buishas et al., 2014).

13.5.4.4 Diffusive versus advective transport

One source of disagreement is the relative importance of diffusion and advection at transporting nutrients and waste through the CSF. The traditional viewpoint has been that diffusion was the primary mechanism by which solute traveled through the ventricular system (Smith et al., 2017). However, a recent hypothesis termed the *glymphatic hypothesis* ("glymphatic" being a portmanteau of "glial" and "lymphatic") claims that advective flow is the more important factor (Iliff et al., 2012; Jessen et al., 2015). The subject is hotly debated, with compelling evidence on both sides. The viewpoint with most support is somewhere in the middle—both mechanisms are important depending on the region (perivascular compartment versus interstitial space) and molecular size (Vardakis et al., 2017).

Smith et al. (2017) studied mice and found that transport of tracer molecules was independent of molecular size in the paravascular spaces, but it was strongly dependent on molecular size in the brain parenchyma. From these results, they concluded that advection plays a significant role in paravascular spaces but not in the parenchyma. Holter et al. (2017) agree, arguing that advective flow is an important in the paravascular spaces but that diffusion is the driving force in the interstitial space. This is likely because velocities are very low in the interstitial spaces (around 10 nm/s; Holter et al., 2017) but much higher in the paravascular spaces. Kurtcuoglu (Miller, 2011) found that advection is dominant in the third ventricle but that the lower Péclet numbers in the anterior and posterior recess indicate diffusion could be dominant.

13.5.5 Intracranial pressure

Intra Cranial Pressure (ICP) readings are a valuable source of boundary condition data for modeling studies. ICP varies periodically with cardiac and respiratory cycles, and Czosnyka et al. (Miller, 2011) describe additional low-frequency fluctuations known as *B waves*, which have a period ranging from 20 seconds to 2 minutes. The ICP of a healthy brain is typically around 500 Pa, as shown by data from Linninger et al. in Table 13.2. They also found that pressure is nearly six times higher for hydrocephalic brains.

13.6 Brain measurement and imaging

13.6.1 Magnetic resonance imaging

Magnetic resonance imaging (MRI) is used to capture two aspects of the central nervous system: geometry and flow characteristics. Several MRI sequences are generally used to accomplish this: spin echo, field echo, and gradient echo MRI are used to capture the anatomy of the brain, and Phase Contrast MRI (PC MRI) is used to measure fluid velocity and pressure.

13.6.2 Spin/field/gradient echo MRI

MRI techniques are typically able to image the brain's structure with an isotropic resolution in the range of 0.8–1.0 mm (Clarke et al., 2013; Yiallourou et al., 2012; Heidari Pahlavian et al., 2015), although Sigmund et al. achieved an isotropic resolution of as high as 0.18 mm

Table 13.2: ICP data from Linninger et al., 2007.

	Normal			Hydrocephalic		
	Parenchyma	Lateral ventricle	SAS	Parenchyma	Lateral ventricle	SAS
ICP (Pa)	637–664	490–517	489–525	2983–3051	2822–2889	2922–2890

220 Chapter 13

(180 μm) using gradient echo MRI on a powerful (7 T) machine (Sigmund et al., 2012). While this spatial resolution would be enough to capture the arachnoid granulations, whose average diameter is approximately 300 μm (Gupta et al., 2009), it is not enough to image the arachnoid trabeculae. These filaments may substantially affect flow through the SAS where they are found (Gupta et al., 2009, 2010,), but their diameter is only between 5 and 7 μm (Killer, 2003). Furthermore, MRI has a finite temporal resolution. Just as a photograph of a moving subject can blur, motion during an MRI will degrade the quality of the image. Due to arterial pulsation and respiratory motion, the brain is constantly moving (Miller, 2011). While the *spatial* resolution attained by Sigmund et al. (2012) may be sufficient to image the arachnoid granulations, the *temporal* resolution must also be sufficiently high to avoid blurring.

13.6.3 Phase contrast MRI

PC MRI allows visualization of the fluid velocity field within the CSF space. As described by Linninger et al., "PC MRI... uses a velocity encoding (VENC) gradient to generate signal contrast between flowing and stationary hydrogen atoms" (Linninger et al., 2016).

Several variants of PC-MRI exist. In 2D-1dir PC-MRI, normal velocities are measured passing through a plane of interest. This technique is incrementally improved in 2D-3dir PC-MRI, wherein velocities along a plane of interest are measured in all three principal directions. A larger distinction is found in a newer method referred to as 4D PC-MRI. In 4D PC-MRI, velocities in all three principal directions are measured across a 3-dimensional volume (Bollache et al., 2016). This technique is growing in popularity, but it is disadvantaged by a comparatively lower resolution than either 2D-1dir or 2D-3dir. It seems that 2D is preferred for finding accurate flowrates for use as boundary conditions, and 4D is preferred for validating flow patterns found in numerical models.

13.6.4 MRI limitations

All variations of PC-MRI have limitations. For one, the small magnitude and pulsatile nature of CSF flow makes accurate measurement a challenge (Linninger et al., 2016). Furthermore, PC-MRI can only measure a finite range of velocities. When a scan is conducted, a velocity encoding value must be set. Flow velocities greater than this value will produce aliasing artifacts (anomalies in the image), and velocities significantly less can produce a poor signal-to-noise ratio (Linninger et al., 2016). This can prevent PC-MRI from detecting intricate flow structures such as vortices (Heidari Pahlavian et al., 2015). Additionally, in 2D MRI it is necessary to determine the plane perpendicular to the flow direction to convert velocities to volumetric flow rates, but the irregular geometry of the ventricular system makes this difficult (Linninger et al., 2016). This typically restricts flow measurements to the cerebral aqueduct and the cervical spinal canal, which have relatively simple geometry and exhibit mostly axial flow (Miller, 2011).

13.6.5 Pressure monitoring

ICP has traditionally been measured with probes inserted through the skull. Because of the invasiveness of this process, it can be difficult to obtain ICP data for healthy subjects. However, noninvasive ICP measurement techniques are being put into practice. In 2000, Alperin et al. published a paper demonstrating accurate measurement of ICP using PC-MRI (Alperin et al., 2000). While this method calculates ICP from flow rate, thus relying on the accuracy of a known relationship between pressure drop and velocity, the authors demonstrated strong correlation between noninvasively measured ICP and invasively measured ICP with an R-squared of 0.965. Recently, another method was demonstrated which uses a mathematical model to compute ICP from blood pressure and velocity measurements obtained with ultrasonography (Fanelli et al., 2019). This method predicted ICP with a standard deviation of error of 5.1 mm Hg (680 Pa), which is quite large relative to the typical mean value.

13.6.6 MRI segmentation

MRI produces raw imagery which may be in the format of a 3D image or, more commonly, a series of 2D images referred to as *slices*. In either case, the imagery must be subjected to a process known as segmentation, in which each area of the image is classified according to tissue type. These classifications are usually white matter, gray matter, and CSF (Despotović et al., 2015). The results of the segmentation process are shown in Fig. 13.6.

Consequently, each slice must be manually segmented before the boundaries of each slice can be stitched together to form a 3D model (Despotović et al., 2015). Not only is this process

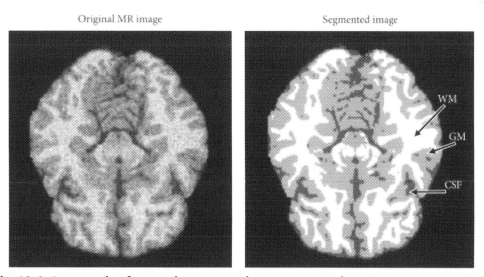

Fig. 13.6: An example of magnetic resonance image segmentation. *GM*, gray matter; *WM*, white matter. Reproduced from Despotović et al. (2015), licensed under CC BY 3.0.

labor-intensive, its repeatability is poor. Martin et al. performed a study wherein four operators each segmented two sets of cervical spine MRI images—one of a healthy patient, and one of a patient with Chiari malformation I. After performing a numerical study on the resulting models, they calculated the peak flow to have a coefficient of variance of up to 17% for the healthy patient and 51% for the Chiari malformation I patient (Martin et al., 2016). Fig. 13.7 shows the inconsistencies seen in this study. Despotović et al. write that "manual segmentation is prone to errors, highly subjective, and difficult to reproduce (even by the same expert)." Ideally, segmentation would be automatically performed on 3D imagery. Automating the segmenting process would avoid operator error, and using 3D imagery would prevent inaccuracies from stitching together 2D slices (Despotović et al., 2015). Nonetheless, most studies have used manual segmentation of 2D slices. While automatic segmentation software exists, Kurtcuoglu wrote in 2011 that its quality when applied to MR images was too low (Miller, 2011), possibly because of the noisiness of MR images (Despotović et al., 2015). While techniques have been proposed specifically for automatically segmenting the CSF

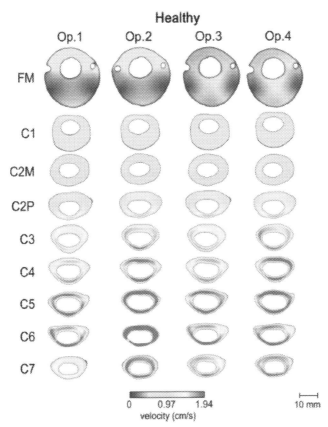

Fig. 13.7: Velocity profiles along the cervical spine illustrating interoperator segmentation inconsistency (Martin et al., 2016). Reproduced with permission, Copyright Biomedical Engineering Society 2015.

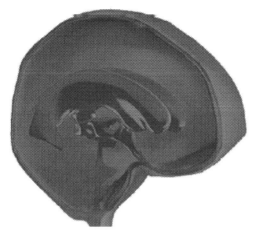

Fig. 13.8: Model of cerebrospinal fluid space used by Gholampour (2018). Licensed under CC BY 4.0.

space (Linninger et al., 2016), they do not appear to be in common use. The geometry of a study that utilized automatic segmentation is shown in Fig. 13.8. Generally, more recent studies tend to use automatic segmentation, so it seems the technique is gaining popularity. A table containing a list of computational studies and their segmentation method can be found in Appendix A.

13.7 Flow modeling

While the ability to study the ventricular system was limited before the advent of modern imaging technologies, Monro (Linninger et al., 2009) created the first mathematical model in 1783. His primitive lumped-parameter (or *compartment*) model had two compartments—brain and blood—and set the direction for modeling attempts over the next two centuries. Over time, lumped-parameter models grew in complexity, adding more compartments to independently model various structures within the brain and even to capture the Fluid–Structure Interaction (FSI) between the brain's fluids and its tissue. A recent model by Linninger et al. (2009) has 19 compartments to account for the ventricular system, the vascular system, and the parenchyma. However, more sophisticated techniques have almost completely replaced lumped-parameter models for understanding flow in the brain.

Computational Fluid Dynamics (CFD) is a computational method that is used to model flow in complex geometries like the brain. CFD breaks a domain into small cells in much the same way as a digital image is broken into pixels. The technique then solves the governing partial differential equations (such as Navier–Stokes) over each individual cell. CFD is an important tool for understanding flow within the brain because it allows calculation of parameters that are difficult or impossible to obtain using other methods. In particular, the limited resolution

224 Chapter 13

Appendix A: Summary of studies of abnormal CSF flow. Adapted from Linninger et al. (2016) with permission.

Reference	Focus of study	Micro-anatomy	Domain	Segmenta-tion	Scope	Experi-mental validation
Global models of the CNS						
Tangen et al., 2015	Studied impact of spinal microanatomy on complex flow profiles and on drug dispersion.	Trabeculae, nerve roots	3D CNS	Mimics	3D CFD (moving walls)	PC MRI
Hsu et al., 2012	Quantified effect of CSF pulsations on drug-distribution rate and spread.	Nerve roots	2D CNS	Mimics; SAS and spinal cord tissue segmented manually	3D CFD (wall condition uncertain)	Cine MRI
Howden et al., 2011	Examined pulsa-tile ventricles and rigid spinal SAS and predicted CSF peak velocity magnitude.	Trabeculae, porous	3D CNS	Mimics	3D CFD (rigid walls)	NA
Sweetman and Linninger, 2011	Predicted CSF flow fields in entire CNS with pulsatile boundaries.	None	3D CNS	Mimics (manual)	3D FSI	PC MRI
Cranial models						
Vinje et al., 2019	Studied compara-tive importance of cardiac and respiratory flow in idealized cylindrical and patient-specific models.	None	Aqueduct	VMTK	3D CFD	PC MRI
Gholampour, 2018	Compared flow in healthy patients to flow in hydrocepha-lus patients over time after shunts were placed.	None	Ventricles, cranial SAS	Mimics	3D FSI	PC MRI
Vardakis et al., 2017	Transport of solute through parenchy-mal tissue was stud-ied with poroelastic model.	None	Paren-chyma	Unknown	3D CFD	Mice study

A review of fluid flow in and around the brain, modeling, and abnormalities 225

Reference	Focus of study	Micro-anatomy	Domain	Segmenta-tion	Scope	Experi-mental validation
Global models of the CNS						
Madhukar et al., 2017	Studied effects of CSF modeling on shear waves in blunt head trauma.	None	Cranium	Unknown	3D FEA	Tagged MRI
Thalakotu-nage and Sethaput, 2016	Computed CSF velocity in cerebral aqueduct for two healthy subjects and one hydrocephalic subject.	None	Aqueduct	3D Slicer	3D FSI	NA
Buishas et al., 2014	Predicted water transport between the compartments of the cranial vault, parenchyma, vascu-lature, and CSF as driven by osmotic pressure and Starling forces.	None	Compart-mental model cra-nial SAS, cerebral vascula-ture, and ECS	NA	Lumped-parameter	None
Sweetman et al., 2011	Compared healthy and hydrocephalic model with FSI to measured data.	None	3D cra-nial SAS	Mimics	3D FSI (ADINA)	PC MRI
Hadzri et al., 2011	Predicted CSF flow and pressure in normal and stenosed aqueduct.	None	Third ventricle, aqueduct	AMIRA	3D CFD (rigid walls)	NA
Gupta et al., 2010	Predicted veloc-ity and pressure distributions in the cranial SAS.	Trabeculae, porous	Cranial SAS	AMIRA (manual)	3D CFD (rigid walls)	PC MRI
Linninger et al., 2009	Compared healthy and hydrocephalic MRI data to simu-lated CSF velocity values; predicted resorption influ-ence on ventricular enlargement.	None	Cranial SAS	Mimics 11.0	3D FSI (ADINA)	PC MRI

(Continued)

226 Chapter 13

Appendix A: (Cont'd)

Reference	Focus of study	Micro-anatomy	Domain	Segmenta-tion	Scope	Experi-mental validation
Global models of the CNS						
Kurtcuoglu et al., 2007	Made subject-specific CSF flow predictions and examined aqueduct flow jet and recirculation.	None	Third ventricle, aqueduct	Manual	3D CFD (rigid walls)	MRI velocity
Kurtcuoglu et al., 2007	Investigated relative importance of advection and diffusion in mass transport in the third ventricle.	None	Third ventricle	Manual	3D CFD (rigid walls)	MRI velocity
Linninger et al., 2007	Predicted CSF flow patterns and pressure gradients in healthy and hydrocephalus patients for CSF spaces and porous brain parenchyma.	None	Cranial SAS, porous parenchyma	Mimics	2D and 3D CFD (rigid walls)	MRI velocity
Kurtcuoglu et al., 2004	Predicted CSF flow in simplified ventricular geometries.	None	Ventricles	NA—idealized model	3D CFD, moving walls	MRI velocity
Linninger et al., 2005	Quantified pulsatile CSF and approximate parenchyma deformation.	None	Ventricles	Unknown	1D FSI, 2D CFD (rigid walls), physical model	Canine ICP and human PC MRI
Haslam and Zamir, 1998	Examined pulsatile flow in elliptical cross sections.	None	Simple ellipse	NA—idealized model	Mathematical model	NA
Jacobson et al., 1996	Computed pressure drop in an hourglass cylinder to predict pressure drop.	None	Simple cylinder	NA—mathematical model	3D CFD (rigid walls)	NA
Spinal models						
Lloyd et al., 2019	Studied effects of cardiac pulsatory waveform on perivascular flow.	None	Axisymmetric annulus	NA	Axisymmetric CFD (rigid walls)	NA
Lloyd et al., 2017	Subarachnoid pressures were compared in Chiari patients and healthy subjects.	None	Cervical	Manual	Axisymmetric CFD (rigid/ moving walls)	PC MRI

A review of fluid flow in and around the brain, modeling, and abnormalities   227

Reference	Focus of study	Micro-anatomy	Domain	Segmentation	Scope	Experimental validation
			Global models of the CNS			
Jain et al., 2017	Examined transitional hydrodynamics in one healthy subject and two Chiari patients.	None	Cervical	VMTK	3D CFD (rigid walls)	NA
Støverud et al., 2016	Compared flow in Chiari patients and healthy subject.	None	Crani-ocervical	VMTK and manual	3D CFD (rigid walls)	2D PC MRI
Martin et al., 2016	Analyzed reliability of CFD results based on variation in MRI segmentation.	None	Cervical	Manual	3D CFD (rigid walls)	4D PC MRI
Pahlavian et al., 2015	Studied impact of tissue motion on flow in a Chiari patient.	None	Cervical	Manual	3D CFD (moving walls and rigid walls)	NA
Cheng et al., 2014	Predicted effect of FSI of spinal cord on fluid flow.	None	Cervical	Manual	3D CFD (rigid walls), 3D FSI	PC MRI
Heidari Pahlavian et al., 2014	Assessed impact of nerve roots and denticulate ligaments on CSF flow patterns.	Nerve roots, denticulate ligaments	Cervical	Manual	3D CFD (rigid walls)	4D PC MRI
Clarke et al., 2013	Compared fluid mechanics in spinal canal of Chiari patients with and without syringomyelia	None	Cervical	Manual (SURFdriver)	3D CFD (rigid walls)	PC MRI
Yiallourou et al., 2012	Compared 4D MRI flow profiles with subject-specific CFD.	None	Cervical	Manual	3D CFD (rigid walls)	4D PC MRI
Rutkowska et al., 2012	Assessed cyclic CSF flow in Chiari patients before and after decompression surgery.	None	Cervical	VMTK; some SAS segmented manually	3D CFD (rigid walls)	PC MRI
Linge et al., 2010	Predicted spatial CSF velocities for the full cardiac cycle.	None	Crani-ocervical	NA—idealized model	3D CFD (rigid walls)	PC MRI
Roldan et al., 2009	Compared flow profiles in Chiari patient and healthy subject.	None	Cervical	Mimics	BEM (rigid walls)	2D PC MRI

(Continued)

228   Chapter 13

Appendix A: (Cont'd)

Reference	Focus of study	Micro-anatomy	Domain	Segmenta-tion	Scope	Experi-mental validation
Global models of the CNS						
Gupta et al., 2009	Predicted pressure and flow profiles for full cardiac cycle and observed no net flow.	Trabeculae, porous	Crani-ocervical	AMIRA (manual)	3D CFD (rigid walls)	MRI velocity
Stockman, 2007	Studied impact of microanatomy on flow patterns with lattice Boltzmann simulations.	Trabeculae, nerve roots	Annular model of spine	NA	Mathematical model	MRI-de-rived flow profile
Loth et al., 2000	Studied flow in short segments of the spine and impact of annular geometric variation.	None	Axial seg-ments	Manual	2D CFD (rigid walls)	MRI peak velocity

Domains are subject specific, unless otherwise noted. Abbreviations: *CFD*, computational fluid dynamics; *CNS*, central nervous system; *CSF*, cerebrospinal fluid; *ECS*, extracellular space; *FEA*, finite-element analysis; *FSI*, fluid–structure interaction; *ICP*, intracranial pressure; *MRI*, magnetic resonance imaging; *NA*, not available; *PC MRI*, phase-contrast magnetic resonance imaging; *SAS*, subarachnoid spaces.

of MRI prevents insight into small-scale effects such as turbulence. Assuming it is properly validated by comparison to experimental results, CFD allows a fine-grained understanding of flow throughout the entire nervous system. An example is provided in Fig. 13.9, which shows calculated velocity magnitude and ICP at different parts of the cardiac cycle.

Intrathecal administration, wherein drugs are injected into the CSF, is a field of growing interest. It is important to study CSF flow patterns to understand how intrathecally administered drugs diffuse. Also, it is believed that abnormal flow patterns may play a role in neurodegenerative diseases like Alzheimer's (Gupta et al., 2010; de Leon et al., 2017), so CFD studies of CSF flow could help combat such diseases. Simulating flow fields within the brain also gives a greater understanding of abnormalities of the brain such as hydrocephalus, syringomyelia, and Chiari malformation (Linninger et al., 2009).

While CFD techniques began to be used in other fields in the mid-20th century, lumped-parameter models remained the dominant method to model the brain until the turn of the century. This is because the largest barrier to simulating the brain is not computational; rather, the greatest difficulty lies in imaging the brain to capture its geometry and boundary conditions. MRI made this possible, and the first CFD study of the ventricular system was published by

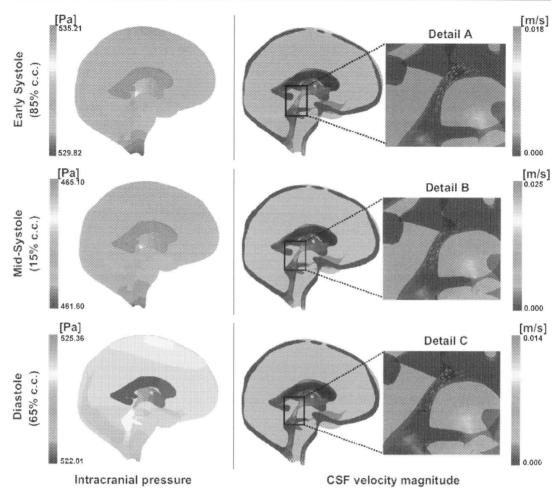

Fig. 13.9: Computational fluid dynamics results from Sweetman et al. (2011) showing cerebrospinal fluid velocity and intercranial pressure at three points in the cardiac cycle. Reproduced with permission, Copyright Elsevier 2011.

Jacobson et al. in 1996. Their study only modeled a tiny portion of the system—the cerebral aqueduct (Jacobson et al., 1996). Since then, CFD has almost completely supplanted lumped-parameter models for studying flow in the ventricular system. While lumped-parameter models can give a general idea of flow in the brain, they are of little practical value (Gupta et al., 2009). They are sufficient to determine simple parameters, such as pressure drop across a structure within the ventricular system, but CFD must be used for specific information (Miller, 2011). Figs. 13.10 and 13.11 show examples of the fine-grained data available from CFD studies. Table 13.3 shows a summary of CFD studies through 2016, while a much more comprehensive table resides in Appendix A.

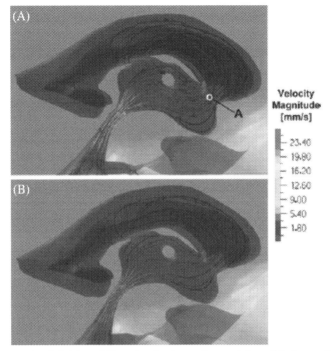

Fig. 13.10: Cerebrospinal fluid pathlines in the lateral ventricles, third ventricle, and cerebral aqueduct (Sweetman and Linninger, 2011). Reproduced with permission, Copyright Biomedical Engineering Society 2010.

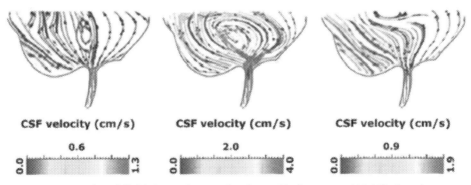

Fig. 13.11: Computational fluid dynamics results from Gholampour (2018) showing a vortex in the inferior section of the third ventricle over a single pulsation. Licensed under CC BY 4.0.

However, CFD methods are far from perfect, and the results are difficult to validate due to the limited scope of MRI data. Furthermore, CFD is extremely computationally expensive—one study of the cerebrospinal junction required nearly one billion computational cells (Jain et al., 2017). If one considers the time requirement for reaching equilibrium in pulsatile flow

Table 13.3: Summary of computational fluid dynamics (CFD) studies, adapted from Linninger et al. (2016), licensed under Creative Commons.

Author	Technique	Geometry	Tissue motion	Arachnoid trabeculae	Nerve roots
Gupta et al., 2009	CFD, anisotropic porous media	3D subject-specific	No	Yes	No
Stockman, 2007	CFD, Lattice Boltzmann	2D idealized	No	Yes	Yes
Roldan et al., 2009	3D rigid wall CFD	3D subject-specific	No	No	No
Linge et al., 2010	3D rigid wall CFD	3D idealized	No	No	No
Loth et al., 2000	3D rigid wall CFD	2D concentric ellipse based on subject	No	No	No
Rutkowska et al., 2012	3D rigid wall CFD	3D patient specific	No	No	No
Bertram, 2009	Numerical model/ wave propagation	2D idealized axisymmetric, tapered tubes	Yes	No	No
Cirovic, 2009	Numerical model/ wave propagation	2D concentric tube with constant diameter	Yes	No	No
Carpenter et al., 2003; Elliott et al., 2009; Cirovic and Kim, 2012	Numerical model/ wave propagation	1D coaxial, fluid-filled, elastic tubes	Yes	No	No
Elliott et al., 2011	Two multiple-compartment hydraulic circuit models	1D coaxial, fluid-filled, permeable tubes	No	No	No
Linninger et al., 2009	Fluid-structure interaction	Multi compartment model of intracranial dynamics	Yes	No	No
Bilston et al., 2006	CFD	2D axisymmetric, cylindrical model	Yes	No	No

in dynamic systems (Strasser and Battaglia, 2017), an unsteady model with one billion cells could take months to complete on a modest supercomputer.

13.7.1 CFD simplifications: rigid walls

Several major simplifications are commonly seen in CFD studies. Perhaps the most significant is the assumption of rigid walls. While the brain's parenchyma and surrounding structures deform from the pressures and flow patterns of the CSF, many studies treat the tissue as rigid and immobile. Some studies have improved accuracy by modeling the walls as moving rigid boundaries. Pahlavian et al. studied flow in the spinal cord of a patient with Chiari malformation, and they found moving boundaries changed peak velocity by 60% compared to a rigid-walled model (Pahlavian et al., 2015). However, the best method of capturing the behavior is FSI modeling, which uses CFD and related techniques to model realistic motion of both CSF and tissue. It was not until recently that FSI simulations began

232 Chapter 13

to grow in popularity, allowing an accurate look at the coupled behavior of the solid tissue and the liquid CSF (Gholampour, 2018; Sweetman and Linninger, 2011). Jacobson's pioneering CFD study was repeated using a FSI model, and pressure drop decreased by 37% (Miller, 2011). However, while FSI is often beneficial, it is not universally necessary. A recent study of the spinal SAS found that predicted pressures were almost unaffected compared to a rigid-walled CFD study, though velocity profiles did show significant differences (Cheng et al., 2014).

13.7.2 CFD simplifications: microstructures

As previously discussed, the brain contains microstructures that can significantly affect flow patterns. Not only are these structures and others like them difficult to image, geometrically modeling micron-scale features is impractical. Consequently, simplifying assumptions are necessary.

13.7.2.1 CFD simplifications: arachnoid granulations

While little is known about the flow rates through the various CSF drainage paths, arachnoid granulations are suspected to be a primary sink. The arachnoid granulations are often simplified as one-way differential pressure valves with a hydraulic conductivity of 92.5 μL/min/mm Hg/cm^2, a value which was measured in an in vitro experiment (Gupta et al., 2010).

13.7.2.2 CFD simplifications: arachnoid trabeculae

Many choose to simply ignore the arachnoid trabeculae (Miller, 2011; Sweetman et al., 2011; Heidari Pahlavian et al., 2015; Bollache et al., 2016), but they have a significant effect on flow through the SAS. Tangen et al. suggest that the trabeculae increase pressure drop across the SAS by two times and that they substantially change flow patterns by creating vorticity (Tangen et al., 2015). This resistance to flow damps brain acceleration (Gupta et al., 2009), which is crucial to model when considering TBI. To take the trabecular network into account, some authors use an anisotropic porous medium model (Gupta et al., 2010; Sweetman and Linninger, 2011). This model, developed by Gupta et al., generalizes the trabeculae as straight circular cylinders connecting two parallel plates (Gupta et al., 2009). Given this simplification, flow results, and a porosity value from literature, they were able to calculate longitudinal and transverse permeability as a function of the separation of the two plates. However, they point out that trabeculae can vary greatly in location and structure, so this model leaves room for improvement.

13.7.2.3 CFD simplifications: other structures

The CSF space houses other microstructures such as pillars and nerve roots, and many of these remaining structures have been completely neglected in CFD studies.

13.8 Literature gap

While substantial literature exists concerning the flow of the CSF, there remain several gaps in knowledge, particularly when considering the different length scales of analysis. Before studies continue under the assumptions of the classical model, the circulation of CSF must be investigated to quantify production and absorption levels for use in CFD boundary conditions. Second, refining and putting into practice automatic (and higher resolution) segmentation tools is important so that geometries will be more consistent and detailed. It has been demonstrated that the rigid wall assumption can greatly affect flow patterns (Cheng et al., 2014), so more attention also needs to be given to FSI modeling of the CSF space. FSI studies have been done using lumped-parameter models and 2D models, but only one comprehensive 3D study exists (Sweetman and Linninger, 2011). Moreover, hardware and software advancements are necessary so that unsteady billion-cell simulations can be produced within a reasonable timeframe. Lastly, one critical application of flow modeling that has remained largely untouched is blunt trauma-induced TBI modeling. While much research exists concerning blast-induced TBI, only one study concerning blunt head trauma could be found (Madhukar et al., 2017). The CDC estimated TBIs contributed to the deaths of 56,800 people in the United States in 2014 (Centers for Disease Control and Prevention, 2014), and modeling the head using FSI techniques would be a valuable tool in preventing injuries.

References

Abbott, N.J., 2004. Evidence for bulk flow of brain interstitial fluid: significance for physiology and pathology. Neurochem. Int. 45 (4), 545–552.

Alperin, N.J., Lee, S.H., Loth, F., Raksin, P.B., Lichtor, T., 2000. MR-intracranial pressure (ICP): a method to measure intracranial elastance and pressure noninvasively by means of MR imaging: baboon and human study. Radiology 217 (3), 877–885.

Bakker, E.N., Bacskai, B.J., Arbel-Ornath, M., Aldea, R., Bedussi, B., Morris, A.W., Weller, R.O., Carare, R.O., 2016. Lymphatic clearance of the brain: perivascular, paravascular and significance for neurodegenerative diseases. Cell. Mol. Neurobiol. 36 (2), 181–194.

Bering, E.A., 1952. Water exchange of central nervous system and cerebrospinal fluid. J. Neurosurg. 9 (3), 275–287.

Bertram, C.D., 2009. A numerical investigation of waves propagating in the spinal cord and subarachnoid space in the presence of a syrinx. J. Fluids Struct. 25 (7), 1189–1205.

Bilston, L.E., Fletcher, D.F., Stoodley, M.A., 2006. Focal spinal arachnoiditis increases subarachnoid space pressure: a computational study. Clin. Biomech. (Bristol Avon) 21 (6), 579–584.

Bollache, E., van Ooij, P., Powell, A., Carr, J., Markl, M., Barker, A.J., 2016. Comparison of 4D flow and 2D velocity-encoded phase contrast MRI sequences for the evaluation of aortic hemodynamics. Int. J. Cardiovasc. Imaging 32 (10), 1529–1541.

Brinker, T., Stopa, E., Morrison, J., Klinge, P., 2014. A new look at cerebrospinal fluid circulation. Fluids Barriers CNS 11 (1), 10.

Buishas, J., Gould, I.G., Linninger, A.A., 2014. A computational model of cerebrospinal fluid production and reabsorption driven by starling forces. Croat. Med. J. 55 (5), 481–497.

Carpenter, P.W., Berkouk, K., Lucey, A.D., 2003. Pressure wave propagation in fluid-filled co-axial elastic tubes part 2: mechanisms for the pathogenesis of syringomyelia. J. Biomech. Eng. 125 (6), 857–863.

234 Chapter 13

Centers for Disease Control and Prevention, 2014, Surveillance Report of Traumatic Brain Injury-Related Emergency Department Visits, Hospitalizations, and Deaths. Centers for Disease Control and Prevention, United States. https://www.cdc.gov/traumaticbraininjury/pdf/TBI-Surveillance-Report-FINAL_508.pdf

Cheng, S., Fletcher, D., Hemley, S., Stoodley, M., Bilston, L., 2014. Effects of fluid structure interaction in a three dimensional model of the spinal subarachnoid space. J. Biomech. 47 (11), 2826–2830.

Chikly, B., Quaghebeur, J., 2013. Reassessing cerebrospinal fluid (CSF) hydrodynamics: a literature review presenting a novel hypothesis for CSF physiology. J. Bodywork Mov. Ther. 17 (3), 344–354.

Cirovic, S., 2009. A coaxial tube model of the cerebrospinal fluid pulse propagation in the spinal column. J. Biomech. Eng. 131 (2), 021008.

Cirovic, S., Kim, M., 2012. A one-dimensional model of the spinal cerebrospinal-fluid compartment. J. Biomech. Eng. 134 (2), 021005. https://pubmed.ncbi.nlm.nih.gov/22482672/.

Clarke, E.C., Fletcher, D.F., Stoodley, M.A., Bilston, L.E., 2013. Computational fluid dynamics modelling of cerebrospinal fluid pressure in Chiari malformation and syringomyelia. J. Biomech. 46 (11), 1801–1809.

Da, S.M., Louveau, A., Vaccari, A., Smirnov, I., Cornelison, R.C., Kingsmore, K.M., Contarino, C., Onengut-Gumuscu, S., Farber, E., Raper, D., 2018. Publisher correction: functional aspects of meningeal lymphatics in ageing and Alzheimer's disease. Nature 564 (7734) E7–E7.

de Leon, M.J., Li, Y., Okamura, N., Tsui, W.H., Saint-Louis, L.A., Glodzik, L., Osorio, R.S., Fortea, J., Butler, T., Pirraglia, E., Fossati, S., Kim, H.-J., Carare, R.O., Nedergaard, M., Benveniste, H., Rusinek, H., 2017. Cerebrospinal fluid clearance in Alzheimer disease measured with dynamic PET. J. Nucl. Med. 58 (9), 1471–1476.

Despotović, I., Goossens, B., Philips, W., 2015. MRI segmentation of the human brain: challenges, methods, and applications. Comput. Math. Methods Med. 2015, 1–23.

Dreha-Kulaczewski, S., Joseph, A.A., Merboldt, K.-D., Ludwig, H.-C., Gärtner, J., Frahm, J., 2017. Identification of the upward movement of human CSF *in vivo* and its relation to the brain venous system. J. Neurosci. 37 (9), 2395–2402.

Elliott, N.S.J., Lockerby, D.A., Brodbelt, A.R., 2009. The pathogenesis of syringomyelia: a re-evaluation of the elastic-jump hypothesis. J. Biomech. Eng. 131 (044503), 874–882. https://pubmed.ncbi.nlm.nih.gov/20833093/.

Elliott, N.S.J., Lockerby, D.A., Brodbelt, A.R., 2011. A lumped-parameter model of the cerebrospinal system for investigating arterial-driven flow in posttraumatic syringomyelia. Med. Eng. Phys. 33 (7), 874–882.

Fanelli, A., Vonberg, F.W., LaRovere, K.L., Walsh, B.K., Smith, E.R., Robinson, S., Tasker, R.C., Heldt, T., 2019. Fully Automated, Real-Time, Calibration-Free, Continuous Noninvasive Estimation of Intracranial Pressure in Children, American Association of Neurological Surgeons, United States, 11. https://thejns.org/pediatrics/view/journals/j-neurosurg-pediatr/24/5/article-p509.xml.

Gholampour, S., 2018. FSI simulation of CSF hydrodynamic changes in a large population of non-communicating hydrocephalus patients during treatment process with regard to their clinical symptoms. PLoS One 13 (4), e0196216.

Gupta, S., Soellinger, M., Boesiger, P., Poulikakos, D., Kurtcuoglu, V., 2009. Three-dimensional computational modeling of subject-specific cerebrospinal fluid flow in the subarachnoid space. J. Biomech. Eng. 131 (2), 021010.

Gupta, S., Soellinger, M., Grzybowski, D.M., Boesiger, P., Biddiscombe, J., Poulikakos, D., Kurtcuoglu, V., 2010. Cerebrospinal fluid dynamics in the human cranial subarachnoid space: an overlooked mediator of cerebral disease. I. Computational model. J. R. Soc. Interface 7 (49), 1195–1204.

Hadzri, E.A., Osman, K., Kadir, A., Mohmmad, R., Abdul Aziz, A., 2011. Computational investigation on CSF flow analysis in the third ventricle and aqueduct of sylvius. IIUMEJ 12 (3). https://journals.iium.edu.my/ejournal/index.php/iiumej/article/view/158.

Haslam, M., Zamir, M., 1998. Pulsatile flow in tubes of elliptic cross sections. Ann. Biomed. Eng. 26 (5), 780–787.

Heidari Pahlavian, S., Bunck, A.C., Loth, F., Shane Tubbs, R., Yiallourou, T., Robert Kroeger, J., Heindel, W., Martin, B.A., 2015. Characterization of the discrepancies between four-dimensional phase-contrast magnetic resonance imaging and in-silico simulations of cerebrospinal fluid dynamics. J. Biomech. Eng. 137 (5), 051002.

Heidari Pahlavian, S., Yiallourou, T., Tubbs, R.S., Bunck, A.C., Loth, F., Goodin, M., Raisee, M., Martin, B.A., 2014. The impact of spinal cord nerve roots and denticulate ligaments on cerebrospinal fluid dynamics in the cervical spine. PLoS One 9 (4), e91888.

Holter, K.E., Kehlet, B., Devor, A., Sejnowski, T.J., Dale, A.M., Omholt, S.W., Ottersen, O.P., Nagelhus, E.A., Mardal, K.-A., Pettersen, K.H., 2017. Interstitial solute transport in 3D reconstructed neuropil occurs by diffusion rather than bulk flow. Proc. Natl. Acad. Sci USA 114 (37), 9894–9899.

Howden, L., Giddings, D., Power, H., Vloeberghs, M., 2011. Three-dimensional cerebrospinal fluid flow within the human central nervous system. DCDS-B 15 (4), 957–969.

Hsu, Y., Hettiarachchi, H.D.M., Zhu, D.C., Linninger, A.A., 2012. The frequency and magnitude of cerebrospinal fluid pulsations influence intrathecal drug distribution: key factors for interpatient variability. Anesth. Analg. 115 (2), 386–394.

Igarashi, H., Tsujita, M., Kwee, I.L., Nakada, T., 2014. Water influx into cerebrospinal fluid is primarily controlled by aquaporin-4, not by aquaporin-1: 17O JJVCPE MRI study in knockout mice. Neuroreport 25 (1), 39.

Iliff, J.J., Wang, M., Liao, Y., Plogg, B.A., Peng, W., Gundersen, G.A., Benveniste, H., Vates, G.E., Deane, R., Goldman, S.A., Nagelhus, E.A., Nedergaard, M., 2012. A paravascular pathway facilitates CSF flow through the brain parenchyma and the clearance of interstitial solutes, including amyloid. Sci. Transl. Med. 4 (147) 147ra111-147ra111.

Jacobson, E.E., Fletcher, D.F., Morgan, M.K., Johnston, I.H., 1996. Fluid dynamics of the cerebral aqueduct. Pediatr. Neurosurg. 24 (5), 229–236.

Jain, K., Ringstad, G., Eide, P.-K., Mardal, K.-A., 2017. Direct numerical simulation of transitional hydrodynamics of the cerebrospinal fluid in Chiari I malformation: the role of cranio-vertebral junction: direct numerical simulation of transitional hydrodynamics of the cerebrospinal fluid in Chiari malformation type I—the role of cranio-vertebral junct. Int. J. Numer. Meth. Biomed. Engng. 33 (9), e02853.

Jessen, N.A., Munk, A.S.F., Lundgaard, I., Nedergaard, M., 2015. The glymphatic system: a beginner's guide. Neurochem. Res. 40 (12), 2583–2599.

Kaczmarek, M., 1997. The Hydromechanics of Hydrocephalus: Steady-State Solutions for Cylindrical Geometry, Springer, New York, 30. https://pubmed.ncbi.nlm.nih.gov/9116602/.

Killer, H.E., 2003. Architecture of arachnoid trabeculae, pillars, and septa in the subarachnoid space of the human optic nerve: anatomy and clinical considerations. Br. J. Ophthalmol. 87 (6), 777–781.

Klarica, M., Miše, B., Vladić, A., Radoš, M., Orešković, D., 2013. 'Compensated hyperosmolarity' of cerebrospinal fluid and the development of hydrocephalus. Neuroscience 248, 278–289.

Kurtcuoglu, V., Poulikakos, D., Ventikos, Y., 2004. Computational modeling of the mechanical behavior of the cerebrospinal fluid system. J. Biomech. Eng. 127 (2), 264–269.

Kurtcuoglu, V., Soellinger, M., Summers, P., Boomsma, K., Poulikakos, D., Boesiger, P., Ventikos, Y., 2007. Computational investigation of subject-specific cerebrospinal fluid flow in the third ventricle and aqueduct of sylvius. J. Biomech. 40 (6), 1235–1245.

Kurtcuoglu, V., Soellinger, M., Summers, P., Poulikakos, D., Boesiger, P., 2007. Mixing and modes of mass transfer in the third cerebral ventricle: a computational analysis. J. Biomech. Eng. 129 (5), 695.

Linge, S.O., Haughton, V., Løvgren, A.E., Mardal, K.A., Langtangen, H.P., 2010. CSF flow dynamics at the craniovertebral junction studied with an idealized model of the subarachnoid space and computational flow analysis. AJNR Am. J. Neuroradiol. 31 (1), 185–192.

Linninger, A.A., Sweetman, B., Penn, R., 2009. Normal and hydrocephalic brain dynamics: the role of reduced cerebrospinal fluid reabsorption in ventricular enlargement. Ann. Biomed. Eng. 37 (7), 1434–1447.

Linninger, A.A., Tangen, K., Hsu, C.-Y., Frim, D., 2016. Cerebrospinal fluid mechanics and its coupling to cerebrovascular dynamics. Annu. Rev. Fluid Mech. 48 (1), 219–257.

Linninger, A.A., Tsakiris, C., Zhu, D.C., Xenos, M., Roycewicz, P., Danziger, Z., Penn, R., 2005. Pulsatile cerebrospinal fluid dynamics in the human brain. IEEE Trans. Biomed. Eng. 52 (4), 557–565.

Linninger, A.A., Xenos, M., Sweetman, B., Ponkshe, S., Guo, X., Penn, R., 2009. A mathematical model of blood, cerebrospinal fluid and brain dynamics. J. Math. Biol. 59 (6), 729–759.

Linninger, A.A., Xenos, M., Zhu, D.C., Somayaji, M.R., Kondapalli, S., Penn, R.D., 2007. Cerebrospinal fluid flow in the normal and hydrocephalic human brain. IEEE Trans. Biomed. Eng. 54 (2), 291–302.

236 Chapter 13

Lloyd, R.A., Fletcher, D.F., Clarke, E.C., Bilston, L.E., 2017. Chiari malformation may increase perivascular cerebrospinal fluid flow into the spinal cord: a subject-specific computational modelling study. J. Biomech. 65, 185–193.

Lloyd, R.A., Stoodley, M.A., Fletcher, D.F., Bilston, L.E., 2019. The effects of variation in the arterial pulse waveform on perivascular flow. J. Biomech. 90, 65–70.

Loth, F., Yardimci, M.A., Alperin, N., 2000. Hydrodynamic modeling of cerebrospinal fluid motion within the spinal cavity. J. Biomech. Eng. 123 (1), 71–79.

Louveau, A., Plog, B.A., Antila, S., Alitalo, K., Nedergaard, M., Kipnis, J., 2017. Understanding the functions and relationships of the glymphatic system and meningeal lymphatics. J. Clin. Invest. 127 (9), 3210–3219.

Louveau, A., Smirnov, I., Keyes, T.J., Eccles, J.D., Rouhani, S.J., Peske, J.D., Derecki, N.C., Castle, D., Mandell, J.W., Lee, K.S., 2015. Structural and functional features of central nervous system lymphatic vessels. Nature 523 (7560), 337.

Madhukar, A., Chen, Y., Ostoja-Starzewski, M., 2017. Effect of cerebrospinal fluid modeling on spherically convergent shear waves during blunt head trauma. Int. J. Numer. Meth. Biomed. Eng. 33 (12), e2881.

Martin, B.A., Yiallourou, T.I., Pahlavian, S.H., Thyagaraj, S., Bunck, A.C., Loth, F., Sheffer, D.B., Kröger, J.R., Stergiopulos, N., 2016. Inter-operator reliability of magnetic resonance image-based computational fluid dynamics prediction of cerebrospinal fluid motion in the cervical spine. Ann. Biomed. Eng. 44 (5), 1524–1537.

Miller, K., 2011. Biomechanics of the Brain. Springer New York, New York, NY.

Pahlavian, S.H., Loth, F., Luciano, M., Oshinski, J., Martin, B.A., 2015. Neural tissue motion impacts cerebrospinal fluid dynamics at the cervical medullary junction: a patient-specific moving-boundary computational model. Ann. Biomed. Eng. 43 (12), 2911–2923.

Papadopoulos, M.C., Verkman, A.S., 2013. Aquaporin water channels in the nervous system. Nat. Rev. Neurosci. 14 (4), 265.

Plog, B.A., Nedergaard, M., 2018. The glymphatic system in central nervous system health and disease: past, present, and future. Annu. Rev. Pathol.: Mech. Dis. 13, 379–394.

Ringstad, G., Vatnehol, S.A.S., Eide, P.K., 2017. Glymphatic MRI in idiopathic normal pressure hydrocephalus. Brain 140 (10), 2691–2705.

Roldan, A., Wieben, O., Haughton, V., Osswald, T., Chesler, N., 2009. Characterization of CSF hydrodynamics in the presence and absence of Tonsillar Ectopia by means of computational flow analysis. AJNR Am. J. Neuroradiol. 30 (5), 941–946.

Rutkowska, G., Haughton, V., Linge, S., Mardal, K.-A., 2012. Patient-specific 3D simulation of cyclic CSF flow at the craniocervical region. AJNR Am. J. Neuroradiol. 33 (9), 1756–1762.

Sakka, L., Coll, G., Chazal, J., 2011. Anatomy and physiology of cerebrospinal fluid. Eur. Ann. Otorhinolaryngol. Head Neck Dis. 128 (6), 309–316.

Sigmund, E.E., Suero, G.A., Hu, C., McGorty, K., Sodickson, D.K., Wiggins, G.C., Helpern, J.A., 2012. High-resolution human cervical spinal cord imaging at 7 T. NMR Biomed. 25 (7), 891–899.

Smith, A.J., Yao, X., Dix, J.A., Jin, B.-J., Verkman, A.S., 2017. Test of the 'glymphatic' hypothesis demonstrates diffusive and aquaporin-4-independent solute transport in rodent brain parenchyma. eLife 6, e27679.

Stockman, H.W., 2007. Effect of anatomical fine structure on the dispersion of solutes in the spinal subarachnoid space. J. Biomech. Eng. 129 (5), 666–675.

Støverud, K.-H., Langtangen, H.P., Ringstad, G.A., Eide, P.K., Mardal, K.-A., 2016. Computational investigation of cerebrospinal fluid dynamics in the posterior cranial fossa and cervical subarachnoid space in patients with Chiari I malformation. PLoS One 11 (10), e0162938.

Strasser, W., Battaglia, F., 2017. The effects of pulsation and retraction on non-Newtonian flows in three-stream injector atomization systems. Chem. Eng. J. 309, 532–544.

Sullan, M.J., Asken, B.M., Jaffee, M.S., DeKosky, S.T., Bauer, R.M., 2018. Glymphatic system disruption as a mediator of brain trauma and chronic traumatic encephalopathy. Neurosci. Biobehav. Rev. 84, 316–324.

Sweetman, B., Linninger, A.A., 2011. Cerebrospinal fluid flow dynamics in the central nervous system. Ann. Biomed. Eng. 39 (1), 484–496.

Sweetman, B., Xenos, M., Zitella, L., Linninger, A.A., 2011. Three-dimensional computational prediction of cerebrospinal fluid flow in the human brain. Comput. Biol. Med. 41 (2), 67–75.

Tangen, K.M., Hsu, Y., Zhu, D.C., Linninger, A.A., 2015. CNS wide simulation of flow resistance and drug transport due to spinal microanatomy. J. Biomech. 48 (10), 2144–2154.

Thalakotunage, H.A., Sethaput, T., 2016. Quantification of CSF velocity through the narrowest point in aqueduct of sylvia for normal and normal pressure hydrocephalus patient by CFD analysis. Int. J. Pharm. Pharm. Sci. 8 (2), 52.

Vardakis, J.C., Guo, L., Chou, D., Tully, B.J., Vindedal, G.F., Jensen, V., Thoren, A.E., Pettersen, K.H., Omholt, S.W., Ottersen, O.P., Nagelhus, E.A., Ventikos, Y., 2017. Poroelastic Modelling of CSF Circulation Via the Incorporation of Experimentally-Derived Microscale Water Transport Properties, International Conference on Computational and Mathematical Biomedical Engineering, United States, 4. https://discovery.ucl.ac.uk/id/eprint/1573538/1/CMBE17_CFD2.pdf.

Vinje, V., Ringstad, G., Lindstrøm, E.K., Valnes, L.M., Rognes, M.E., Eide, P.K., Mardal, K.-A., 2019. Respiratory influence on cerebrospinal fluid flow – a computational study based on long-term intracranial pressure measurements. Sci. Rep. 9 (1), 9732.

Yiallourou, T.I., Kröger, J.R., Stergiopulos, N., Maintz, D., Martin, B.A., Bunck, A.C., 2012. Comparison of 4D phase-contrast MRI flow measurements to computational fluid dynamics simulations of cerebrospinal fluid motion in the cervical spine. PLoS One 7 (12), e52284.

CHAPTER 14

Resonant frequencies of a human brain, skull, and head

T.R. Fonville[a], S.J. Scarola[a], Y. Hammi[b], Raj K. Prabhu[c], Mark F. Horstemeyer[a]

[a]*Liberty University, Lynchburg, VA, United States* [b]*Mississippi State University, Mississippi State, MS, United States* [c]*USRA, NASA HRP CCMP, NASA Glenn Research Center, Cleveland, OH, United States*

14.1 Introduction

When considering the mechanical behavior of a structure, one must consider both the strength and stiffness related to frequency. Structural components, when conditions are set for its own resonant frequency, can reach extreme displacements. The Tacoma Narrows Bridge (Billah and Scanlan, 1991) is one such tragic example when winds struck the bridge at the resonant frequency of the bridge thus inducing very large displacements and rotations so much so the bridge fractured and broke into pieces. Tesla (Carlson, 2013) was the first to induce a man-made large amplitude wave that hit the resonant frequency of a building in New York causing a large earthquake. When automobiles, trucks, and motorcycles are developed, they are designed to not broach between 5 and 10 Hz, because that range induces the resonant frequencies of the human body (Brownjohn and Zheng, 2001); and nobody wants the drivers of automobiles, trucks, and motorcycles sick. Regarding the human head, most studies have focused on mechanical impacts; see Langlois et al. (2006), Madhukar and Ostoja-Starzewski (2019), and Guskiewicz and Walton (2020) for reviews but not on the resonant frequencies associated with the head, skull, and brain.

Very few experimental studies analyzing the mechanical resonant frequencies of the human head have been conducted. Probably the first researcher to analyze frequencies of the human skull was by von Bekesy (1948) but his experimental work was corrected by Franke (1956). Franke (1956) examined skulls of living subjects, a dry skull preparation, and a human cadaver. The dry skull incurred a greater resonant frequency when compared to the living subjects and the human cadaver due to the tissue in the living subjects and human cadaver dampening the result. The first fundamental mode of the dry skull was a "breathing" mode of 300 Hz; the second fundamental frequency was approximately 500 Hz in a bending mode; and the third mode was found to be 1200 Hz. Both von Bekesy (1948) and Franke (1956) had limitations on their experiments as frequencies below 200 Hz could not be garnered. Håkansson et al. (1994) experimentally found that the two first resonant frequencies were 972 (range

Multiscale Biomechanical Modeling of the Brain.
DOI: https://doi.org/10.1016/B978-0-12-818144-7.00006-2
Copyright © 2022 Elsevier Inc. All rights reserved.

240   Chapter 14

828–1164) Hz and 1230 (range 981–1417) Hz for six different skulls by themselves. The relative damping ratios were found to be between 2.6% and 8.9%, respectively. Because of the low number of skulls tested, the authors did not find a correlation with skull size and the resonant frequencies.

Now, when the skull/brain interaction were combined, resonant frequencies dropped from 300 Hz to values much lower than the skull ranges. For example, Laksari et al. (2015) numerically studied the first resonant frequency of the brain to be approximately 15 Hz, using a first-order reduced dynamic model of the skull–brain combination based on in vivo MRI data. Assuming a rigid-body, the skull–brain dynamics was approximated by an under-damped system with a low-frequency resonance at around 15 Hz. Later, Laksari et al. (2018) conducted skull/brain finite element (FE) simulations to quantify the very first fundamental frequency of the human head to be 28 Hz with a low end of 20 Hz and a high end at 40 Hz. Laksari et al. (2018) applied a dynamic mode decomposition with FE analysis to extract the dominant spatiotemporal characteristics of the brain's nodal displacements by comparing a randomly selected subset of 187 noninjury collisions and the two injury collisions where loss of consciousness occurred in American football. The scope of this Laksari et al. (2018) study was limited by the FE mesh size and large amplitude impacts.

Some frequency analyses at lower length scales have been conducted on some animals' neurons. Puil et al. (1994) experimentally found the first fundamental frequency in thalamo-cortical neurons from Guinea pigs was between 2 and 4 Hz, and a second mode exist in the range of 12–26 Hz. Hutcheon et al. (1996) determined that the resonant frequency of cortical neurons of juvenile rats ranged from 1 to 2 Hz. Hutcheon and Yarom (2000) studied the mechanical resonant frequencies of neurons to compare with the modulation of normal brain rhythms with the notion of developing medicinal drugs to address disorders like epilepsy and insomnia. They showed an olivary neuron had a resonant frequency of 4 Hz. These small neuronal frequencies are expected to be quite different from a full brain indicating the viscosity present in the tissue.

Although not directly related to the head, skull, or brain resonant frequencies, Gabler et al. (2019) developed a Diffuse Axonal Multi-Axis General Evaluation model based upon the kinematics and deformation history of the brain. Diffuse Axonal Multi-Axis General Evaluation employs the equations of motion of a three-degree-of-freedom, coupled second-order system (which is used to determine resonant frequencies and mode shapes) and predicted maximum brain strain using the directionally dependent angular acceleration time-histories from a head impact. Hence, Gabler et al. (2019) related the strain in the brain as a metric for evaluation of brain damage.

The current study is the first to conduct a FE analysis using a mesh garnered from MRI data for the whole head, skull, and brain of a human (M.F.H.). Herein we employ strain as a metric to separately evaluate the different resonant frequencies and mode shapes of the head, skull,

Resonant frequencies of a human brain, skull, and head **241**

and brain as separate entities since strain measures can be used to quantify brain damage (Gabler et al., 2019).

14.2 Problem set-up for the finite element simulations

Three different free-end modal analyses of an adult male human head (actually it was an author of this writing, M.F.H.) were performed using eigenvalue extraction in order to calculate the natural frequencies and the corresponding mode shapes. One calculation included the whole head; another calculation included just a confined brain; and the final calculation included a free boundary condition brain. Since the results from the whole head that confined the brain gave almost identical results as the confined brain results, we just describe the two conditions from hereafter. The human head mesh was created using Simpleware ScanIP software into which a MRI scan data was imported for segmentation and then converted into a multipart mesh producing an unstructured, fully tetrahedron mesh with contact surfaces and then imported into the FE preprocessor Abaqus/CAE. Table 14.1 summarizes the number of elements of associated volume for each part of the head.

The adult male brain mesh had 185,858 elements covering a volume of 1189 cm^3. Raven and Johnson (1995) garnered data claiming that the adult male human brain was approximately 1450 cm^3. Bartley et al. (1997) quantified the brain volumes of different males and females and garnered a range of 1100–1280 cm^3 in a fairly small sampling. However, Hommer (2003) found that the average brain size of a male was 1340 cm^3 with 1189 cm^3 being in the lower end of the distribution. Hence, the author's brain used in the study herein was essentially in the lower range of brain sizes. However, it is not clear if lowering of the volume affected the stiffness-to-weight ratio, which in turn, affects the resonant frequency. Finally, Laksari et al. (2015, 2018) did not mention this volume of the brain in their studies. The skull mesh herein has 279,550 elements with a volume of 861 cm^3. The cerebral spinal fluid (CSF) mesh has 124,603 elements with a volume of 260 cm^3. A typical volume of the CSF is approximately 150 cm^3 (Sakka et al., 2011), indicating that the CSF volume within the male human herein was fairly large. Adding all the parts of the adult male human head, we have a total size that has 629,372 elements with a volume of 2533 cm^3. By including the neck, we have a total size that has 1248,397 elements with a volume of 4924 cm^3.

Although FE analyses have facilitated the biomechanical modeling of the human head, they have limitations in terms of accuracy of the solution. Such limitations come from the

Table 14.1: Number of elements and volume of head components.

Material	Brain	CSF	Skull	Head	Total
Elements	185,858	124,603	279,550	629,372	1248,397
Volume (cm^3)	1188.9	259.6	861.2	2533.0	4923.62

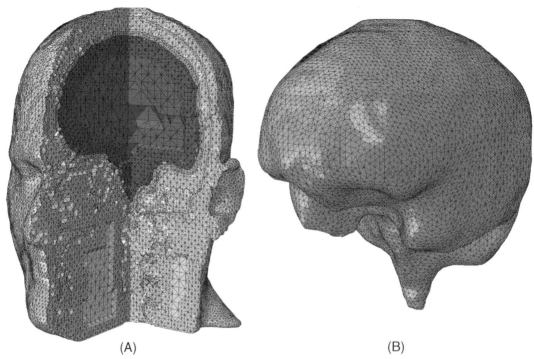

Fig. 14.1: (A) Whole head mesh showing scalp, skull, cerebral spinal fluid (CSF), and brain. (B) Brain only mesh. The brain has 185,858 elements with a volume of 1189 cm^3. The skull has 279,550 elements with a volume of 861 cm^3. The cerebral spinal fluid (CSF) has 124,603 elements with a volume of 260 cm^3. Adding all of the parts of the head, including the neck, we reach a head size that has 1248,397 elements with a total volume of 4924 cm^3.

complexities of geometries, interactions between components within the head, and material properties that are difficult to capture and characterize. In this analysis, the representative FE constitutive models that included major components of the head such as the brain, skull, and CSF as illustrated in Fig. 14.1.

The interactions and contact between the head components as well as the fluid-solid-type interactions of CSF with the brain and skull were modeled using tie constraints between their respective contact surfaces (*TIE option). Table 14.2 summarizes the different head

Table 14.2: Material properties of head components.

Material	Density (tons/mm^3)	Young's modulus (MPa)	Poisson's ratio
Brain	1.05e−09	0.012	0.45
CSF	1.04e−09	0.299233	0.496164
Head	1.2e−09	16.7	0.42
Skull	1.8e−09	15000.	0.21

material properties chosen as linear elastic where only the density, the Young's modulus and Poisson's ratio were defined.

The FE solution adopted the higher order 10-node quadratic tetrahedron, hybrid with constant pressure C3D10H. This frequency analysis is based on a stress-free model state, in which all external loads are assumed to be zero. Therefore, the body forces, such as the gravity, are zero, and there are no nonzero applied tractions on the boundary. In addition, all prescribed displacement components are also assumed to be zero (this is referred to as homogeneous essential boundary conditions). In addition, temperature changes are not considered either is this analysis. The material properties in this isothermal analysis are defined at the ambient temperature.

The main vibrations mode shapes were calculated in this analysis using the Lanczos (1950) eigensolver without using the ABAQUS high-performance linear dynamics software architecture called SIM (SIM = NO option). The eigenvectors were normalized so that the largest displacement or rotation entry in each vector is unity (Normalization = Displacement) essentially representing the associated strain levels.

14.3 Results

Tables 14.3 and 14.4 summarize the results for the first three fundamental vibration modes for the whole head and brain, respectively. Typically, large displacements signify the amplifications of the natural frequencies, but we show the associated strains (derivatives of the displacements) to illustrate the point since strain has been associated with damage within the brain (Gabler et al., 2019; Sheldon et al., 1998).

Table 14.3: Adult human male whole head modal frequencies for the first three fundamental modes.

Mode	Frequency (Hz)	Type	Location
1	22.3	Torsion in sagittal plane E23	Primary motor cortex region; occipital lobe; cerebellum
2	23.7	Simple shear in sagittal plane E23	Occipital lobe; primary motor cortex
3	24.0	Simple shear in coronal plane E13	Bilateral parietal lobes

Table 14.4: Unconfined adult human male brain modal analysis results for the first three fundamental mode shapes and frequencies.

Mode	Frequency (Hz)	Type	Location
1	13.9	Simple shear in sagittal plane E23	Midbrain region
2	14.2	Torsion in transversal plane E12	Cerebellum
3	14.9	Torsion in coronal plane E13	Bilateral temporal lobes; bilateral parietal lobes

244 Chapter 14

14.3.1 Whole head: fundamental frequency and mode shapes

The whole head frequency and mode shape analysis gave the same results for the whole head as for the case when the brain was confined to no displacements on the outer boundaries. Table 14.3 summarizes the first three modes being at frequencies of 22.3 Hz, 23.7 Hz, and 24.0 Hz, respectively, which are all fairly close to each other while being different torsional/ shearing modes.

Fig. 14.2 shows the whole head FE simulation illustrating the first fundamental mode at a frequency of 22.3 Hz in which the mode shape was torsion in the 2–3 (sagittal) plane. Maximum strain analyses although not shown herein confirm the peak values of this torsional mode in the 2–3 (sagittal) plane. The maximum shear strain (E23) (represented by red) located in the primary motor cortex of the frontal lobe region reached a level of 0.071. This shear strain of 0.071 was greater than the maximum normal strain of 0.034 indicating that shearing was the dominant mode.

Fig. 14.3 shows the second fundamental mode for the whole head at frequency of 23.7 Hz illustrating the simple shear (E23) in sagittal plane of the brain. Depicted is the maximum positive (torsion) shear strain of 0.035 (E23) (represented by red) located in the occipital lobe and primary motor cortex region. Note that this positive shear strain of 0.035 and negative shear strain of −0.073 was greater than the maximum positive normal strain of 0.032 and negative normal strain of -0.019 indicating that shearing was the dominant mode.

Fig. 14.4 shows the third fundamental mode for the whole head at a frequency of 24.0 Hz illustrating the shear strain (E13) under torsion of the coronal plane of the brain. Depicted is the maximum normal positive (torsion) shear strain of 0.043 (E13) (represented by red) located in bilateral parietal lobes.

14.3.2 Brain: fundamental frequency and mode shapes

Once that the whole head was examined through FE analysis, we separated the brain from the rest of the whole head and conducted a modal analysis on the brain with free boundary conditions. Table 14.4 summarizes the three first modes being at frequencies of 13.9 Hz, 14.2 Hz, and 14.9 Hz, all fairly close to each other and being different torsional/shearing modes similar to the whole head results.

Fig. 14.5 shows the brain-only-simulation showing simple shear (E23) of the sagittal plane of the brain exposing the mid-brain base next to the brain stem as the region for the first fundamental mode. Depicted is the maximum positive shear strain of 0.027 (E23) (represented by gold) and the maximum negative shear strain of −0.031 (represented by dark blue) located in the midbrain region. Hence, for both the head and brain the first fundamental mode was in torsion (simple shear, E23).

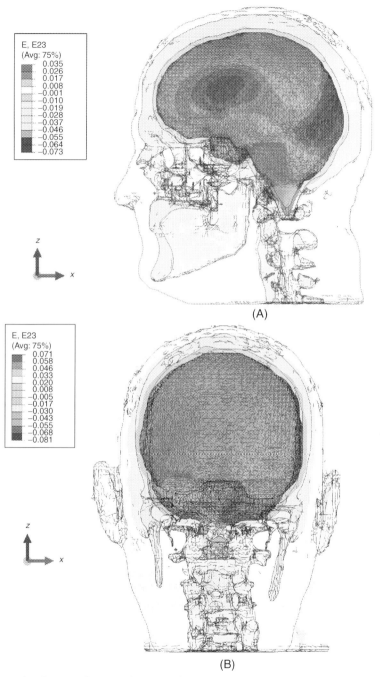

Fig. 14.2: (A) Sagittal (2–3 plane) side view of whole head showing maximum positive (torsion) shear strain (E23) under Mode 1 (22.3 Hz). The maximum positive (torsion) shear strains (E23) during Mode 1 (22.3 Hz) are found in the primary motor cortex region (represented by red) and the occipital lobe and cerebellar regions (represented by dark orange), having approximate strain levels of 0.071 and 0.058, respectively. (B) Coronal view showing the front (1–3 plane).

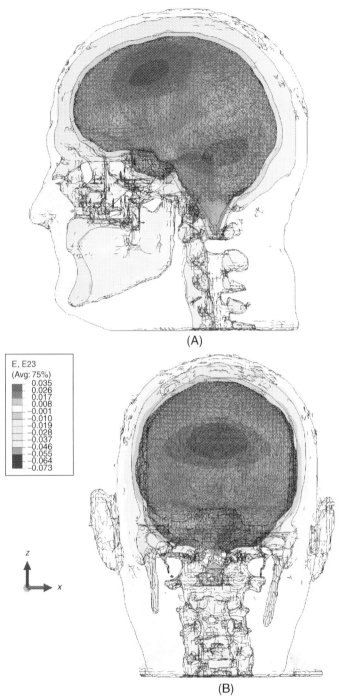

Fig. 14.3: (A) Mode 2 for the whole head at frequency of 23.7 Hz illustrating shear strain (E23) in sagittal plane of the brain. Depicted is the maximum positive (torsion) shear strain of 0.035 (E23) (represented by red) located in the occipital lobe and primary motor cortex region. (B) Coronal view of Mode 2.

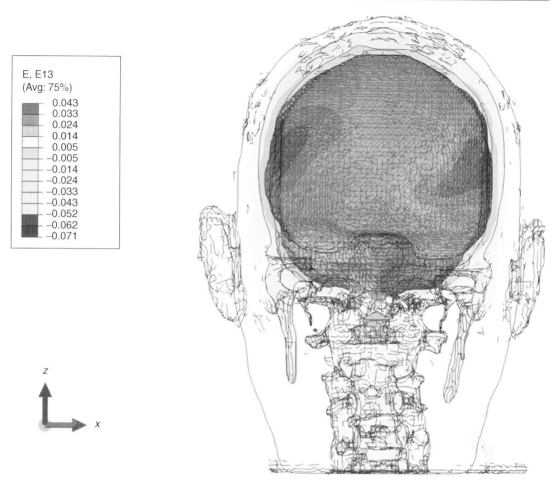

Fig. 14.4: Coronal (mid-brain) view of brain and whole head showing shear strain (E13) under mode 3 (24.0 Hz). Depicted is the maximum normal positive (torsion) shear strain of 0.043 (E13) (represented by red) located in both bilateral parietal lobes.

Fig. 14.6 shows the second fundamental mode at 14.2 Hz in the coronal plane of the brain-only-simulation. Depicted is the maximum positive (torsion) shear strain (E13) of 0.040 located in the bilateral parahippocampal gyrus and temporal lobe regions. Fig. 14.7 shows the third fundamental mode of just the brain-only-simulation at 14.9 Hz illustrating the shear strain (E13) under torsion of the coronal plane of the brain. Depicted is the maximum positive (torsion) shear strain of 0.017 (E13) (represented by red) located in the temporal and parietal lobes.

14.4 Discussion

From Figs. 14.2 and 14.5, the whole head and brain FE simulation results showed the first fundamental mode is torsion with natural frequencies of 22.3 Hz and 13.9 Hz, respectively. The sagittal (side) view of the brain in Fig. 14.2 illustrates the maximum displacements

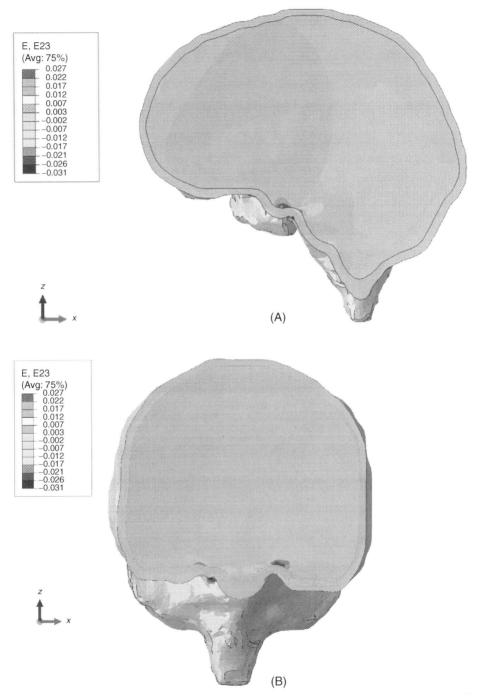

Fig. 14.5: (A) Sagittal (side 2–3 plane) view of brain showing maximum positive (torsion) shear strain (E23) under Mode 1 (13.9 Hz). (B) Coronal (front 1–3 plane) view depicting the maximum positive shear strain of 0.027 (E23) (represented by gold) and the maximum negative shear strain of −0.031 (represented by dark blue) located in the midbrain region.

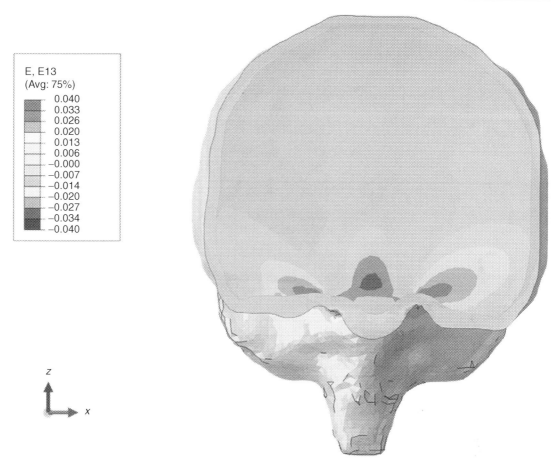

Fig. 14.6: Coronal (mid-brain) view of brain showing maximum positive (torsion) simple shear strain (E13) under mode 2 (14.2 Hz). Depicted is the maximum positive (torsion) shear strain of 0.040 (E13) (represented by red) and the maximum negative (torsion) shear strain of −0.040 (E13) located in the bilateral parahippocampal gyrus and temporal lobe regions.

located in the cerebellum and occipital lobe region, as well as in the primary motor cortex region. With damage in these areas, the functional deficits could potentially include a deficit in voluntary movements (i.e., arm and leg movements) and a deficit in "fine-tuned" movements, balance, and coordination with possible presentation of dysdiadocokinesia or dysmetria (Manto, 2009; Schmahmann, 2004). Additionally, there could be a potential decrease in visual functions, specifically with upper visual field deficit as the shear is located in Meyer's Loop (Yogarajah et al., 2009). With the dominant mode being located in the primary motor cortex, the dominant potential functional deficit would be observed in voluntary motor skills such as movement of limbs. Based on the brain motions in Fig. 14.2, the diminished motor skill is hypothesized to be in the upper limbs and face. Fig. 14.5 for the brain shows the

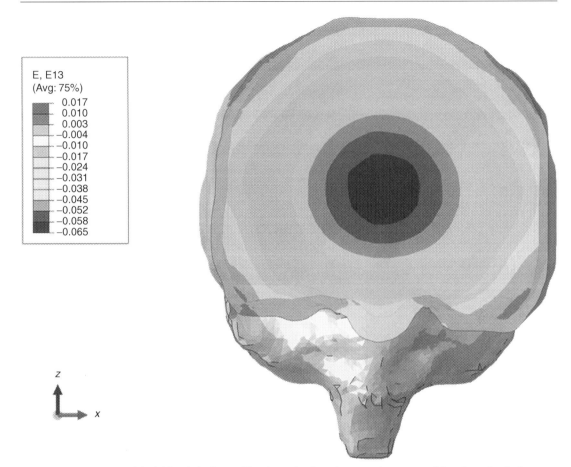

Fig. 14.7: Coronal (mid-brain) view of brain only showing maximum positive (torsion) shear strain (E13) under mode 3 (14.9 Hz). Depicted is the maximum positive (torsion) shear strain of 0.017 (E13) (represented by red) located in the temporal and parietal lobes.

damage to the midbrain, where a potential deficit in visual and auditory processing, reflexive movements, and in severe cases can lead to the loss of consciousness.

The second fundamental modes for the whole head (23.7 Hz) and brain (14.2 Hz) diverged a bit regarding their mode shapes. For the whole head, the second mode was sagittal shearing, and for the brain-only-simulation the second mode was transversal torsion in the cerebellum. Based on the shear location in Fig. 14.3, there could be a deficit in all visual fields and in voluntary motor function, most likely in the upper limb and face regions. Fig. 14.6 shows Mode 2 of just the brain-only-simulation in the bilateral parahippocampal gyrus region and temporal lobe regions. Damage to the parahippocampal gyrus and the temporal lobe regions can lead to functional deficits in memory encoding and retrieval, navigation, as well as auditory pathways (Epstein and Kanwisher, 1998).

Mode 3 for the whole head (Fig. 14.4) at a frequency of 24.0 Hz illustrated the shear strain (E13) under torsion of the coronal plane of the brain located in bilateral parietal lobes. Mode 3 of just the brain-only-simulation (Fig. 14.7) at 14.9 Hz illustrating the shear strain (E13) under torsion of the coronal plane of the brain in the temporal and parietal lobes. Damage and shearing in parietal lobes can lead to functional deficits resulting in diminished sensation and perception or diminished sensory processing/integration in the contralesional side of space (Losier and Klein, 2001) and the disengage deficit (Posner et al., 1982). With temporal lobe damage, diminished hearing, language, and memory processing are all possibilities (Yonelinas et al., 2002).

An alternative check of the fundamental vibration frequencies can be garnered by getting a very rough estimate using Euler's (Euler, 1739) equation or Franke's (Franke, 1956) equation for the first fundamental mode. These equations use the moment of inertia and the polar moment of inertias, Young's and shear moduli, the areas, lengths, and radii of the brain, CSF, skull, and head but not the density nor elastic modulus. When comparing Euler's (Euler, 1739) 5 Hz and Franke's (Franke, 1956) 2 Hz compared to 13.9 Hz for the FE result for the brain, we obtain results that were lower than the FE results of 13.9 Hz. Euler's (Euler, 1739) and Franke's (Franke, 1956) equations are for simple geometries and are expected to give large errors. Because the brain is a very complex geometry, we are comforted that the analytical solutions of Euler (Euler, 1739) and Franke (Franke, 1956) are in the "ballpark" of the more correct FE result. Furthermore, the trends from Euler (Euler, 1739) and Franke (Franke, 1956) indicated that the torsional mode was lower than the normal mode of vibration confirming the FE trends herein.

One final discussion point is warranted since the first fundamental modes were torsional. Kent et al. (2019) recently ran experiments where they quantified in-field head rotations of NFL players who experienced concussions upon head-to-ground impacts. The concussed players had an average of 8.3 m/s impact velocity and struck the ground at approximately 45 degrees to the ground surface. Their video results revealed angular velocities up to 54 degrees/s (0.15 Hz) at impact. Since these in-field high amplitude, low frequency impacts might have secondary and tertiary higher frequencies that associate with the aforementioned head and brain resonant frequencies, brain damage can be exacerbated as induced torsional strains can be large. More direct analyses of the torsional impacts related to the resonant frequencies is warranted.

14.5 Conclusions

A FE modal analysis was conducted in Abaqus/CAE on an adult human male whole head and the unconfined brain using MRI scan data of one of the authors (M.F.H.). The first three fundamental mode shapes of the whole head and brain are torsional modes oriented on different planes. The whole head first three resonant frequencies are 22.3, 23.8, and 24.0 Hz. The

252 Chapter 14

brain first three resonant frequencies are 13.9, 14.2, and 14.3 Hz. Mode 1 mostly affected the midbrain region; Mode 2 mostly affected most the cerebellum; and Mode 3 mostly affected the parietal lobes.

References

Bartley, A.J., Jones, D.W., Weinberger, D.R., 1997. Genetic variability of human brain size and cortical gyral patterns. Brain 120 (2), 257–269.

Billah, K., Scanlan, R., 1991. Resonance, Tacoma narrows bridge failure, and undergraduate physics textbooks" (PDF). Am. J. Phys. 59 (2), 118–124. doi:10.1119/1.16590 Bibcode:1991AmJPh.59.118B.

Brownjohn, J.M., Zheng, X., 2001. Discussion of human resonant frequency. SPIE Second Int. Conf. Exp. Mech. 4317, 469–474.

Carlson, W.B., 2013. Tesla: Inventor of the Electrical Age. Princeton University Press, pp. 181–185.

Epstein, R., Kanwisher, N., 1998. A cortical representation of the local visual environment. Nature 392 (6676), 598–601.

Euler, L., (1739). "Tentamen novae theoriae musicae ex certissimis harmoniae principiis dilucide expositae", Ex typographia Academiae scientiarum, Saint Petersburg.

Franke, E.K., 1956. Response of the human skull to mechanical vibrations. J. Acoust. Soc. Am. 28 (6), 1277–1284.

Gabler, L.F., Crandall, J.R., Panzer, M.B., 2019. Development of a second-order system for rapid estimation of maximum brain strain. Ann. Biomed. Eng. 9, 1–11.

Guskiewicz, K.M., Walton, S.R., 2020. The changing landscape of sport concussion. Kinesiol. Rev. 9 (1), 79–85.

Hommer, D.W., 2003. Male and female sensitivity to alcohol-induced brain damage. Alcohol Res. Health 27 (2), 181–185.

Håkansson, B., Brandt, A., Carlsson, P., Tjellström, A., 1994. Resonance frequencies of the human skull in vivo. J. Acoust. Soc. Am. 95 (3), 1474–1481.

Hutcheon, B., Miura, R.M., Puil, E., 1996. Subthreshold membrane resonance in neocortical neurons. J. Neurophysiol. 76 (2), 683–697.

Hutcheon, B., Yarom, Y., 2000. Resonance, oscillation and the intrinsic frequency preferences of neurons. Trends Neurosci. 23 (5), 216–222.

Kent, R., Forman, J., Bailey, A.M., Funk, J., Sherwood, C., Crandall, J., Arbogast, K.B., Myers, B.S., 2019. The biomechanics of concussive helmet-to-ground impacts in the National Football league. J. Biomech., 109551.

Laksari, K., Kurt, M., Babaee, H., Kleiven, S., Camarillo, D., 2018. Mechanistic insights into human brain impact dynamics through modal analysis. Phys. Rev. Lett. 120 (13), 138101.

Laksari, K., Wu, L.C., Kurt, M., Kuo, C., Camarillo, D.C., 2015. Resonance of human brain under head acceleration. J. R. Soc. Interface 12 (108), 20150331.

Lanczos, C., 1950. An iteration method for the solution of the eigenvalue problem of linear differential and integral operators. J. Res. Nat'l Bur. Std. 45, 255–282.

Langlois, J.A., Rutland-Brown, W., Wald, M.M., 2006. The epidemiology and impact of traumatic brain injury: a brief overview. J. Head Trauma Rehab. 21 (5), 375–378.

Losier, B.J.W., Klein, R.M., 2001. A review of the evidence for a disengage deficit following parietal lobe damage. Neurosci. Biobehav. Rev. 25 (1), 1–13.

Madhukar, A., Ostoja-Starzewski, M., 2019. Finite element methods in human head impact simulations: a review. Ann. Biomed. Eng. 47 (9), 1832–1854.

Manto, M., 2009. Mechanisms of human cerebellar dysmetria: experimental evidence and current conceptual bases. J. Neuroeng. Rehab. 6 (1), 10.

Posner, M.I., Cohen, Y., Rafal, R.D., 1982. Neural systems control of spatial orienting. Philos. Trans. R. Soc. Lond. B Biol. Sci. 298 (1089), 187–198.

Puil, E., Meiri, H., Yarom, Y., 1994. Resonant behavior and frequency preferences of thalamic neurons. J. Neurophysiol. 71 (2), 575–582.

Raven, P.H., Johnson, G.B., Biology, Iowa, Brown, 1995: 443.

Sakka, L., Coll, G., Chazal, J., 2011. Anatomy and physiology of cerebrospinal fluid. Eur. Ann. Otorhinol. Head Neck Dis. 128 (6), 309–316.

Schmahmann, J.D., 2004. Disorders of the cerebellum: ataxia, dysmetria of thought, and the cerebellar cognitive affective syndrome. J. Neuropsychiatry Clin. Neurosci. 16 (3), 367–378.

Sheldon, R.A., Sedik, C., Ferriero, D.M., 1998. Strain-related brain injury in neonatal mice subjected to hypoxia–ischemia. Brain Res. 810 (1-2), 114–122.

von Bekesy, G., 1948. Vibration of the head in a sound field and its role in hearing by bone conduction. J. Acoust. Soc. Am. 20 (749).

Yogarajah, M., Focke, N.K., Bonelli, S., Cercignani, M., Acheson, J., Parker, G.J.M., Alexander, D.C., et al., 2009. Defining Meyer's loop–temporal lobe resections, visual field deficits and diffusion tensor tractography. Brain 132 (6), 1656–1668.

Yonelinas, A.P., Kroll, N.E.A., Quamme, J.R., Lazzara, M.M., Sauvé, M.-J., Widaman, K.F., Knight, R.T., 2002. Effects of extensive temporal lobe damage or mild hypoxia on recollection and familiarity. Nat. Neurosci. 5 (11), 1236–1241.

CHAPTER 15

State-of-the-art of multiscale modeling of mechanical impacts to the human brain

Mark F. Horstemeyer

Liberty University, Lynchburg, VA, United States

15.1 Introduction

As mentioned earlier on in this book, over 1.7 million traumatic brain injury (TBI) cases occurred in the United States every year (Faul et al., 2010) arising from mechanical impacts. As one who personally experienced 12 concussions, I am thrilled with the progress that researchers have made toward quantifying the effects of TBI. Progress in knowledge and understanding of TBI using computational tools has been made as illustrated in this book at the atomistic level (Chapter 3), molecular level (Chapter 4), cellular level (Chapter 5), sulci/ gyri level (Chapter 6), and the full brains (Chapters 7–14). Now that we understand the current state of multiscale modeling of the human brain undergoing mechanical impacts in the previous chapters, let us consider the work that is still needed.

15.2 Work to be completed

15.2.1 Multiphysics aspects of the brain

Although the multiphysics internal state variable was presented in Chapter 8, which comprised a first-order coupled thermal–mechanical–chemical constitutive model, there is still needed electricity to the model. Experimentally, brain measurements can arise from brain signals that are electrical in nature. Hence, the thermal–mechanical–chemical behavior induces electrical signals that can and need to be quantified. Then, multiphysics experiments need to be set up to measure the stress–strain behavior of the brain based upon temperature, strain rate, chemistry, and electrical differences used as boundary conditions. Different multiphysics experiments can be set up to develop, calibrate, and validate different formulations of the next-generation multiphysics internal state variable constitutive model.

15.2.2 Multiscale structure–property relationships of the brain

In this book, we included the multiscale modeling structure–property relationships almost solely from a mechanical perspective and only related to one particular damage source,

Multiscale Biomechanical Modeling of the Brain.
DOI: https://doi.org/10.1016/B978-0-12-818144-7.00014-1
Copyright © 2022 Elsevier Inc. All rights reserved.

256   Chapter 15

like the neuronal membrane in Chapter 4. This is a good start. However, to model the brain in a more accurate manner, the multiscale heterogeneous structure–property relationships need examination for the different structures, like other parts of the neuron beyond the neuronal membrane and in different multiphysics environments. This requires experiments to quantify the different structure–property relationships at each length scale to help calibrate and validate the different constitutive models at each different length scales. Furthermore, the bridges of information that passes from length scale to length scale requires the additional concept of uncertainty analysis in order to consider the distribution and/or variations in the particular structures at that particular length scale. This in turn will lead to a distribution of properties.

15.2.3 Different biological effects on the brain

Somebody once asked me how come Ken Stabler got chronic traumatic encephalopathy (CTE) where Roger Staubach did not. They both played quarterback in the National Football League about the same time and played about the same number years, yet the former got CTE, and the latter did not. *The New York Times* (Branch, 2016) reported that researchers at Boston University discovered CTE in Stabler's brain after his death. Well, first of all, Roger Staubach is not dead, so nobody has checked whether he got CTE or not. Furthermore, Roger Staubach played 11 years while Ken Stabler played 15 years and as mentioned, the history matters. Although each individual human brain is similar, there are differences in individual size and morphology arising from the deoxyribonucleic acid (DNA), the main constituent of one's chromosomes that carries a person's genetic information. How one's brain responds to a mechanical can be very different based upon their DNA and, as such, the history of one's deteriorated Tau proteins can be strongly dependent upon one's DNA.

Biology not only can affect the structure–property relationships at different length scales within the brain, but it is also involved in creating the size and geometry of one's brain. Rushton (1992) and Rushton and Ankney (1996) showed that there were distinct brain differences between races and sex. Furthermore, just in white males above 22 years old there are differences in brain size based purely upon the genetic make-up of a person (Bond and Woods, 2006; Tang, 2006).

Biological differences also are important when considering the mechanical properties of brain tissue. The experiments that quantify the stress–strain behavior on brain tissue either comes from animals such as pigs (Prabhu et al., 2011; Pervin and Chen, 2011), rats (Karimi and Navidbakhsh, 2014), etc. or from human candavers (Donnelly and Medige, 1997). One question that arises is which material represents the closest behavior to live human brain tissue, fresh animal brain tissue, or human cadaver brain tissue? This is a difficult question to answer

State-of-the-art of multiscale modeling of mechanical impacts to the human brain **257**

considering that we will not (and ethically should not) test live human brain tissue with mechanical impacts to quantify its constitutive behavior.

15.2.4 The liquid–solid aspects of the brain

Most the chapters herein focused on solid modeling of the brain, although Chapter 13 addressed the fluid aspects of the brain. We need not only to analyze the multiphysics, multiscale, and constitutive behavior of the brain, we also have to couple this information with the liquid–solid interactions within the brain. Since the brain is made of approximately 70% by volume liquid in both the free-flowing and cellular form, the fully coupled thermal–mechanical–chemical–electrical aspects of the brain becomes more complicated. In addition, the multiscale, multiphysics material model would need to be implemented in an Arbitrary Lagrangian–Eulerian (ALE) code. Typically, the numerical modeling of fluids occurs in an Eulerian framework as space watches material through. Alternately, the numerical modeling of solid occurs in a Lagrangian framework as material moves through space. Hence, both numerical formulations would require the sharing of information at Gauss points, nodal points, or some fashion. Now, they would have to admit all of the multiscale structures, their structure–property relations, and the associated multiphysics effects on the multiscale structure–property relations.

15.2.5 Different human ages

Although Chapter 11 focused on infants, it included only mechanical boundary conditions and only one age. The experimental data missing requires a parameter study for different age groups regarding the multiphysics, multiscale, and liquid–solid interface effects. Because brain development is highly nonlinear over time (Rushton and Ankney, 1996), different mechanical impacts onto a head would produce different levels of damage. A baby's brain will double in size within just 1 year after birth and keeps growing to approximately 80% of an adult's brain by 3 years of age. It is almost 90% by 5 years of age, but continuous grows slowly up to 22 years old, where it reaches its peak power (Guberti, 2012). One probably can assume that the age-related TBI incidents would give a similar response after 22 years of age, unless there was a history of accumulated TBI incidences for the individual (Horstemeyer et al., 2019).

As per Horstemeyer et al. (2019), the damage level in a brain is cumulative over time. For example, I have had about 12 concussions in my life time from sports and total damage state has nonlinearly accumulated after each impact. The next impact on my head would cause more damage than another person at my age who has never experienced a concussion. In fact, based upon the damage model in Horstemeyer et al. (2019), another person might receive a much larger mechanical impact to the brain to reach the current level of damage in my own brain. As such, the history matters when considering the mechanical damage to the human brain.

258 Chapter 15

15.3 Conclusions

As denoted in this book, we have matured in our understanding of the structure–property relationships in the brain at different length scales given certain mechanical loading environments. However, we still have a ways to go as expressed in this chapter. In particular researchers should focus on the following:

1. Develop a multiphysics model based on mechanical–thermal–chemical–electrical properties at each length scale.
2. Quantify the different structures at each length scale and model their behavior to garner the resulting multiphysics properties.
3. Tie into the multiphysics model the biological feature of DNA.
4. Update numerical methods that capture the liquid–solid interactions in the model.
5. Conduct experiments to calibrate and validate the model at different length scales in different age groups.

References

Bond, J., Woods, C.G., 2006. Cytoskeletal genes regulating brain size. Curr. Opin. Cell Biol. 18 (1), 95–101.

Branch, J., 2016. Ken Stabler, A magnetic N.F.L. star was sapped of spirit by C.T.E. The New York Times.

Donnelly, B.R., Medige, L., 1997. Shear properties of human brain tissue. Transactions of the ASME Journal of Biomechanical Engineering 119 (4), 423–432.

Faul, M., Wald, M.M., Xu, L. and Coronado, V.G., 2010. Traumatic brain injury in the United States; emergency department visits, hospitalizations, and deaths, 2002-2006.

Guberti, N., 2012, https://nancyguberti.com/5-stages-of-human-brain-development/.

Horstemeyer, M.F., Berthelson, P.R., Moore, J., Persons, A.K., Dobbins, A., Prabhu, R.K., 2019. A mechanical brain damage framework used to model abnormal brain tau protein accumulations of National Football League players. Ann. Biomed. Eng. 47 (9), 1873–1888.

Karimi, A., Navidbakhsh, M., 2014. An experimental study on the mechanical properties of rat brain tissue using different stress–strain definitions. J. Mater. Sci.: Mater. Med. 25 (7), 1623–1630.

Pervin, F., Chen, W.W., 2011. Effect of inter-species, gender, and breeding on the mechanical behavior of brain tissue. NeuroImage 54, S98–S102.

Prabhu, R., Horstemeyer, M.F., Tucker, M.T., Marin, E.B., Bouvard, J.L., Sherburn, J.A., Liao, J., Williams, L.N., 2011. Coupled experiment/finite element analysis on the mechanical response of porcine brain under high strain rates. J. Mech. Behav. Biomed. Mater. 4 (7), 1067–1080.

Rushton, J.P., 1992. Cranial capacity related to sex, rank, and race in a stratified random sample of 6,325 U.S. military personnel. Intelligence 16 (3–4), 401–413. doi:10.1016/0160-2896(92)90017-l.

Rushton, J.P., Ankney, C.D., 1996. Brain size and cognitive ability: Correlations with age, sex, social class, and race. Psychonom. Bull. Rev. 3 (1), 21–36.

Tang, B.L., 2006. Molecular genetic determinants of human brain size. Biochem. Biophys. Res. Commun. 345 (3), 911–916.

Index

Page numbers followed by "*f*" and "*t*" indicate, figures and tables respectively.

A

Acceleration, 188
Arachnoid granulations, 232
Arachnoid mater, 3–4
Arachnoid trabeculae, 232
Arbitrary Lagrangian-Eulerian
(ALE) code, 256–257
Archicortex (allocortex), 5–6
Astrocytes, 8–9, 213–214
  fibrous, 8–9
  protoplasmic, 8–9
Atomic force fields, 57
  classical, 57
  coarse-grained force field, 57
  reactive, 57
Atoms, simulation ensembles of, 59
Axolemma, 12–13
Axon initial segment (AIS), 12–13
Axoplasm, 12–13

B

Beta amyloid, 214–215
Biomechanical systems, physical
chemistry of, 41
Blast injuries
  primary, 16
  quaternary, 16
  secondary, 16
Brain
  biological effects, 256
  constitutive modeling, 27
  deformation experiments,
    199–200
  deformation measurements, 196
  flow abnormalities, 213–214
  fundamental frequency, 244
  gross anatomy, 3

brainstem, 7
cerebellum, 6
cerebrum, 4
diencephalon, 7
liquid-solid aspects of, 256–257
lobes of, 5*f*
microanatomy, 8
neuroglia, 8
neurons, 9
mode shapes, 244
multiphysics aspects, 255
multiscale modeling, 54*f*
multiscale structure, 2, 3*f*, 255
pseudo-elastic viscoelastic model
  for, 29–30
schematics of, 4*f*
skull, and head, 240
tissue experiments, 32
Brainstem, 3–4, 7
Brownian motion, 131
Building surrogate models, 162

C

Cauchy-Green strain tensor,
  30–31, 108
Cauchy stress, 89
Cerebellar cortex, 6–7
Cerebellum, 3–4, 6
Cerebral cortex, 5–6
Cerebral spinal fluid, 3–4, 87,
  184, 241
Cerebrum, 3, 4
Chiari malformation, 216, 222–223
Cholesterols, 11–12
Chronic traumatic encephalopathy
  (CTE), 86, 256
Circulatory flow, 217

Classical force field atomistic
  models, 43
  upscaling properties, 44
Closed-head injury, 16
Computational fluid dynamics
  (CFD), 223–228
Constitutive models, 27–28
Corpus callosum, 4–5
Cytoplasm, 11

D

Damage evolution equation, 122
Dendrites, 12–13
Dendritic spine, 12–13
Density functional theory (DFT),
  39, 42
Deoxyribonucleic acid (DNA), 256
Diencephalon, 3–4, 7
  epithalamus, 7
  hypothalamus, 7
  subthalamus, 7
  thalamus, 7
Diffuse axonal injury (DAI),
  15, 107
Diffuse Axonal Multi-Axis General
  Evaluation model, 240–241
Diffuse tensor imaging (DTI), 1–2,
  30–31
Diffusion-weighted MRI (DW-MRI),
  105–106
Dura mater, 3–4
Dynamic ultrasound imaging,
  197–198

E

Endoplasmic reticulum (ER), 10–11
Epithalamus, 7

Index

Euler's equation, 251
Exchange-correlation (XC) energy, 42

F

Falx cerebri, 4–5
Fast axonal transport tracks, 214f
Fibrous astrocytes, 8–9
Finite element
  analysis, 20, 27, 85–86, 89t
  high-resolution magnetic
    resonance elastography, 139
  method, 139
  model, 160
Fissure, 4–5
Flocculonodular lobe, 6–7
Fluid-filled spaces, 210–211
Folia, 6–7
Fractional anisotropy (FA), 108
Franke's equation, 251
Frontal lobe, 4–5

G

Gadd severity index, 195
Generalized gradient approximation
  (GGA), 42
Glasgow coma scale (GCS), 15
Glymphatic hypothesis, 218–219
Gray matter, 5–6
Gyrus, 4–5

H

Hardy's displacement data,
  145–148
Head
  components, 242t
  deformation, 188–190
  frequency, 244
High speed digital image correlation
  (HS-DIC), 180–181
Holzapfel-Gasser-Ogden
  model, 108
  SEF, 30–31
Homogenous electron gas, 42
Hopkinson pressure bar, 19
Human brain, 1
Hydrocarbon, interatomic potential
  for, 46
  interatomic force fields, 49
  interatomic potential, 48
  MEAMBO, 46, 47
Hydrocephalus, 215–216

Hypercube sampling procedure,
  158–159
Hyperviscoelastic material, 143t
Hypothalamus, 7

I

Insular lobe, 4–5
Internal state variables (ISV), 53
Intracranial pressure (ICP), 219

K

Kelvin-Voigt model, 28–29

L

Lennard-Jones potential, 45
Limbic lobe, 4–5
Local density approximation (LDA), 42
  limitations of, 42

M

Macroglia, 8–9
Macroscale
  brain finite element model, 103
  mechano-physiological internal
    state variable model, 119
Magnetic resonance elastography
  (MRE), 194–195, 202
Magnetic resonance imaging
  (MRI), 1–2, 194–195, 219
Martini force, 59
Maxwell-Kelvin model, 28
Mechano-physiological internal
  state variable (MPISV), 34,
  53, 127
Membrane disruption, 121
Membrane failure limit diagram
  (MFLD), 69, 70
Meninges, 3–4
Mesocortex (juxtallocortex), 5–6
Mesoscale, 14
  anatomical structures and
    imaging techniques, 105
  future prospects, 110
  material-based method, 108
  structure-based method, 109
  finite element modeling, of brain
    structural heterogeneities, 93
  computational methods for
    properties, 87
  geometrical complexities, 94
  methods, 87

modeling length scale, 86
  model validation and boundary
    conditions, 92
Meta-GGA functionals, 42
Microglia, 8–9
Misfolded proteins, 214–215
Modeling mechanical behavior of
  axons, 81
Modeling microscale neurons,
  78, 79
Model validation, 205
Modified Embedded Atom
  Method with Bond Order
  (MEAMBO), 46, 47
Molecular dynamics (MD), 39,
  56–57
  methods, 57
  atomic force fields, 57
  boundary conditions, 60
  molecular dynamics analysis
    methods, 64
  molecular dynamics simulation
    method, 57
  simulation details, 63
  simulation ensembles of
    atoms, 59
  phospholipid bilayer, 69
  membrane failure limit
    diagram, 70
  stress-strain and damage
    response, 70
  simulations, 120–121
Molecular dynamics (MD)
  simulations, 119
Møller-Plesset perturbation (MP2)
  method, 47–48
Monte Carlo simulations, 42
Mooney-Rivlin model, 28–29, 31
Moving deformable barrier (MDB),
  155–156
Mullins effect, 27–28

N

Nahum's experiments, 144–145
Nanoscale cellular structures,
  modeling
  methods, 57
  atomic force fields, 57
  boundary conditions, 60
  molecular dynamics analysis
    methods, 64

Index

molecular dynamics simulation method, 57
simulation details, 63
simulation ensembles of atoms, 59
phospholipid bilayer, 69
membrane failure limit diagram, 70
stress-strain and damage response, 70
Neck injury metric analysis, 161–162
Neocortex (isocortex), 5–6
Neo-Hookean model, 28–29
Nernst-Planck diffusion equation, 131
Nernst-Planck model, 134
Neurobehavioral sequelae, 17
neuropsychiatric, 17
somatic, 17
Neuroglia, 8, 9*f*
macroglia, 8–9
microglia, 8–9
Neuronal arbors, 13
Neurons, 9, 9*f*, 10*f*
membrane damage, stereological quantification of, 67
Neuropsychiatric sequelae, 17–18
Neutral density targets (NDT), 144
Newtonian fluid, 216–217
Nissl substance, 10–11
Nonlinear inversion (NLI)
algorithm, 141
technique, 140–141
Nonpenetrating injury, 16
Nucleation of pores, 123

O

Occipital lobe, 4–5
Ogden model, 31
Oligodendrocytes, 8–9
Open-head brain injury, 16

P

Parietal lobe, 4–5
Pauli exclusion principle, 41
Péclet numbers, 212–213, 219
Penetrating brain injury, 16
Phase-contrast magnetic resonance imaging (MRI), 140
Phospholipid, 56
bilayer
cell membrane of, 55*f*

molecular dynamics analysis methods, 64
hydrophilic, 56
hydrophobic, 56
semipermeable, 56
Pia mater, 3–4
Piola-Kirchhoff stress, 89
Plastic equivalent strain, 171–172
Pore growth rate, 125
Pore number density equation, 123
Pore resealing, 126
Postmortem human surrogate (PMHS), 177–178
Pressure monitoring, 219
Protoplasmic astrocytes, 8–9
Pseudo-elastic viscoelastic model, for brain, 29–30
Python scripts, 160

Q

Quantum mechanics, 39

R

Respiratory flow, 217
Reynolds number, 210, 211–212

S

Soma, 5–6
Somatic sequelae, 17
Sonomicrometry, 198
dataset, 200
Statistical analysis, 185–186
Strain energy function (SEF), 28–29
Strain rate sensitivity, 29–30
Stress-strain behavior, of neuron membrane, 65
Subthalamus, 7
Sulci, 6–7
effects of, 96
Sulcus, 4–5
length, 98
orientation, 97
Superimposed mesh method, 109
Synapse, 13
axoaxonic, 13
axodendritic, 13
axosomatic, 13
Syringobulbia, 216
Syringomyelia, 216
System constraints, 158–159

T

Tagged magnetic resonance imaging, 200
Tangential friction coefficient, 91
Tau protein, 214–215
Temporal lobe, 4–5
Thalamus, 7
Thermal-mechanical-chemical constitutive model, 255
Traumatic brain injuries (TBIs), 1, 27, 53, 85–86, 103, 119, 193, 214–215, 255
adverse effects of, 1
causes of, 1
defined, 1
multiscale nature of, 13
computational models, 20
examples of injuries, 16
experiments, 18
multiscale injury mechanisms, 14
neurobehavioral sequelae, 17
research methods, 18
types of injury, 15
primary injury, 15
secondary injury, 15
structural anisotropy in macroscale models of, 107

V

Ventricles, subarachnoid space, 209
Ventricular system, 209
Vermis, 6–7
Virchow Robin spaces, 210–211
Volumetric plasticity, 29–30
von Mises stress, 96
Voxel-based finite element, 142*f*

W

Womersley number, 212
World Health Organization, 193

X

X-ray
imaging of radio-opaque markers, 196
imaging technology, 196
method, 197

Z

Zero-point energy, 44

Printed in the United States
by Baker & Taylor Publisher Services